Welding technology

2nd edition

J. W. Giachino

Professor Emeritus and Former Head
Department of Engineering and Technology
Western Michigan University
Kalamazoo, Michigan

W. Weeks

Associate Professor of Welding
Department of Engineering and Technology
Western Michigan University
Kalamazoo, Michigan

G. S. Johnson

Associate Professor of Mechanical Engineering
Department of Engineering and Technology
Western Michigan University
Kalamazoo, Michigan

American Technical Publishers, Inc.
Alsip, Illinois 60658

Preface
to the Second Edition

Welding is universally recognized as the principal joining process in the fabrication of numerous metal products ranging in size from small electronic components to complex building structures, missiles, and transportation equipment. A variety of welding processes have been developed over the years to meet the increasing demands for greater production economy and more effective metal bonding. Today, these processes include shielded metal-arc, gas shielded-arc, resistance, and a host of other special techniques intended specifically for welding sophisticated metals used in aerospace and nuclear industries.

Welding Technology has been designed to provide a comprehensive coverage of current welding practices found in many industries. The text is planned particularly for people who may eventually supervise production welding.

In addition to descriptions of modern joining processes, other related welding data are included so students will have a better technical knowledge of various welding phenomena. This information deals with such topics as metallurgy of welding, testing of weldments, strength of materials, joint design, welding symbols, safety, and cost estimating.

In the new Second Edition of *Welding Technology*, changes have been made throughout to conform to current industrial practices. A new chapter, dealing with automated welding, has been added. This area of welding is becoming increasingly important in industry, and students of welding should have a basic understanding of its nature and significance. A new appendix, Appendix A, on metric conversion has been added. Welding Tables are given in Appendix B.

The Publishers

Contents

Chapters   1   Gas Welding 1

2   Shielded Metal-Arc Welding 15

3   Gas Tungsten-Arc Welding 45

4   Gas Metal-Arc Welding 70

5   Resistance Welding 109

6   Special Welding Processes 132

7   Metallurgy of Welding 153

8   Weldability of Metals 196

9   Brazing and Soldering 236

10   Surfacing 261

11   Flame and Arc Cutting 283

12   Strength of Materials 306

13   Design of Weldments 344

14   Testing Welds 369

15   Production Economy and Cost Estimating 396

16   Safety in Welding 424

17   Welding Symbols 437

18   Automated Welding 454

Appendix A   Metric Conversion 466

Appendix B   Welding Tables 468

Index 481

Singer Safety Products, Chicago, Ill.

SAFETY IS A FULL TIME RESPONSIBILITY!

The Occupational Safety and Health Act of 1970, has made it mandatory that employers comply with job safety and health standards. A concern for safety in welding is perhaps more important than in other industrial fields. The welders above are protected by some of the most modern safety equipment available, including specially treated fire retardant leather tunics. The area is largely shielded off with canvas curtains that are fire resistant to protect other employees from the effects of the welding.

Chapter 1

Gas Welding

Gas welding is a process in which coalescence is achieved by directing a gas flame over metal where a filler rod may or may not be used to intermix with the molten puddle. The energy required for welding develops from a combustion of a fuel with oxygen or air. The most commonly used fuels are acetylene, Mapp gas (stabilized methylacetylene-propadiene), and hydrogen.

Gas welding has limited application for industrial production purposes because it is much slower than other welding processes. It is used considerably more for general maintenance work, welding metals of low melting points, and performing such operations as brazing, soldering, and metallizing.

OXYACETYLENE WELDING

The oxyacetylene flame is probably the most widely used for welding because of its higher flame temperature as compared with other oxygen-fuel flames. The flame is generated by burning a mixture of acetylene (C_2H_2) and oxygen (O_2). Acetylene originates from a chemical reaction of calcium carbide (CaC_2) and water (H_2O) or:

$$CaC_2 + 2H_2O = C_2H_2 + Ca\,(OH)_2$$

Complete combustion of acetylene is represented by the chemical equation:

$$C_2H_2 + 2.5\,O_2 = 2CO_2 + H_2O$$

where one volume of acetylene and two and a half volumes of oxygen

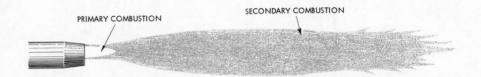

Fig. 1-1. In an oxyacetylene flame primary combustion provides the welding heat and secondary combustion the protective shield.

produce two volumes of carbon dioxide and one volume of water vapor. This ratio of 2½ oxygen to 1 acetylene produces a high temperature flame of approximately 5800° F.

Combustion occurs in two stages as shown in Fig. 1–1. Primary combustion, which is the actual flame for welding, is the inner cone. Secondary combustion is at the outer portion of the flame and serves as a protective shield to prevent the air from contaminating the molten puddle as well as to provide preheating effects to the welding operation.

The primary flame results when oxygen and acetylene are mixed in the welding torch in a one to one ratio. This mixture flowing from the torch tip burns to form the inner-cone because of the following reaction:

$$C_2H_2 + O_2 = 2CO + H_2$$

where one volume of acetylene combines with one volume of oxygen to produce two volumes of carbon monoxide and one volume of hydrogen.

In the secondary flame the carbon monoxide and hydrogen produced by the first reaction combine with oxygen in the atmosphere as shown below:

$$2CO \text{ (carbon monoxide)} + H_2 \text{ (hydrogen)} + 1\tfrac{1}{2} O_2 \text{ (oxygen)} =$$
$$2CO_2 \text{ (carbon dioxide)} + H_2O \text{ (water vapor)}$$

Thus complete combustion of acetylene results from one volume of oxygen provided by the torch and 1½ volumes provided from the surrounding atmosphere.

Equipment

Equipment for oxyacetylene welding consists of oxygen and acetylene cylinders, pressure regulators which reduce the high cylinder gas

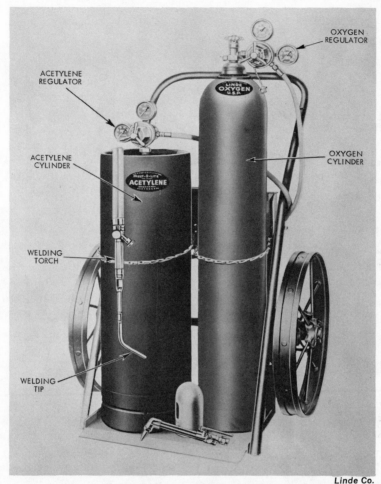

Fig. 1-2. Basic units of an oxyacetylene welding outfit.

pressure to the required working pressure, a torch where the two gases are mixed, and hoses which connect the regulators to the torch. See Fig. 1–2.

Cylinders. Oxygen and acetylene cylinders are made from seamless drawn steel and tested very carefully. Sizes of cylinders are based on the cubic feet of gas they contain. The three common sizes of oxygen

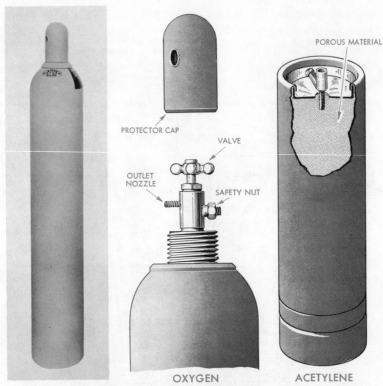

POROUS MATERIAL

PROTECTOR CAP

VALVE

OUTLET
NOZZLE

SAFETY NUT

OXYGEN   ACETYLENE

Fig. 1-3. Oxygen and acetylene are stored in special cylinders.

cylinders are 244, 122, and 80 cu ft. Acetylene cylinders are made to
hold 300, 100, and 60 cu ft. See Fig. 1–3.

Cylinders are equipped with a safety device to permit the gases to
drain slowly in the event the temperature increases the pressure beyond
the safety load of the cylinder. Since gases expand when heated and contract when cooled the incorporation of a safety device is imperative.

The flow of oxygen from the cylinder is controlled by a high pressure
valve which is opened and closed by turning a hand wheel. The hand
wheel must be turned slowly to permit a gradual pressure load on the
regulator and opened as far as the valve will turn to full gas pressure.

Acetylene cylinders are packed with a porous material saturated with
acetone. The acetone absorbs large quantities of acetylene under pres-

sure without changing the nature of the gas itself. This is necessary because free acetylene cannot be stored in cylinders as other gases since it becomes dangerously unstable when under pressure in a free state. Actually acetylene should never be used beyond a 15 pound pressure.

The cylinder valve on acetylene cylinders is opened with a T-wrench. The valve should be opened to not more than one and one-half turns. A slight opening is advisable since it permits closing the valve in a hurry in case of an emergency.

When the volume demand of gas is high, especially for continuous operation over a long period of time, the oxygen and acetylene cylinders are attached to separate manifold systems. See Fig. 1–4. The gas in

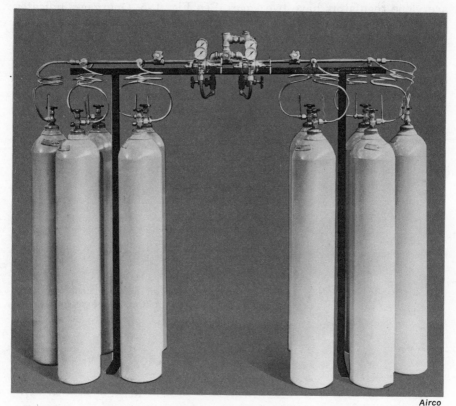

Airco

Fig. 1-4. For high volume consumption cylinders are often attached to a manifold.

each system flows into the main stream and is controlled by a single master regulator.

Oxygen and acetylene cylinders are usually placed on a two-wheel cart to facilitate moving wherever welding is to be done. If cylinders are to be used at a permanent work station precautions must be taken to chain them so they cannot be overturned accidentally.

In setting up cylinders for welding, the cylinder valves should first be "cracked" (opened slightly) to blow any dirt out of the nozzles.

Regulators. Oxygen and acetylene regulators are fastened to cylinders and their function is to reduce cylinder gas pressure to suitable values for welding purposes. For example, if an oxygen cylinder has a pressure of 1800 lbs per sq in. and a pressure of 10 lbs is needed at the torch, the regulator must maintain a constant pressure of 10 lbs even if the cylinder pressure drops down to 500 lbs. Both oxygen and acetylene regulators have two gages. See Fig. 1–5. One gage indicates the actual pressure in the cylinder and the other shows the working or line pressure used at the torch. Gas flow from a cylinder is regulated by an adjusting screw. This adjusting screw must always be released before a cylinder valve is opened otherwise the tremendous pressure of the gas in the cylinder is forced on to the working pressure gage which may damage the regulator.

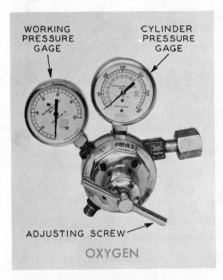

WORKING
PRESSURE
GAGE

CYLINDER
PRESSURE
GAGE

ADJUSTING SCREW

OXYGEN

Fig. 1-5. Oxygen and acetylene regulators control the flow of gas for welding.

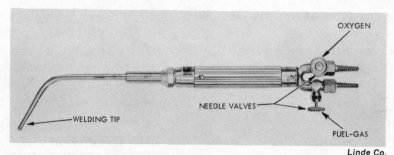

OXYGEN

NEEDLE VALVES

WELDING TIP

FUEL-GAS

Linde Co.

Fig. 1-6. A typical oxyacetylene torch and welding tip.

Table I.   Torch Operating Pressures.

TIP NUMBER	THICKNESS OF METAL (INCHES)	OXYGEN PRESSURE (POUNDS)	ACETYLENE PRESSURE (POUNDS)
00	1/64	1	1
0	1/32	1	1
1	1/16	1	1
2	3/32	2	2
3	1/8	3	3
4	3/16	4	4
5	1/4	5	5
6	5/16	6	6
7	3/8	7	7
8	1/2	7	7
9	5/8	7 1/2	7 1/2
10	3/4 & up	9	9

(Airco)

Welding Torch. The essential elements of a welding torch are a mixing chamber where the two gases are brought together and mixed, two needle control valves which control the flow of oxygen and acetylene, hose connections, and a welding tip. See Fig. 1–6. Different sizes of tips can be inserted in the mixing chamber to permit welding a range of metal thicknesses. The size of a tip is indicated by the diameter of its orifice which is usually designated by a number. Manufacturers of welding equipment provide charts to show the normal operating pressures for various tip sizes. See Table 1.

Hose. Special designed non-porous hose is used to convey oxygen and acetylene from the cylinders to the torch. To prevent any interchange of hoses the oxygen hose is either green or black in color and the acetylene red. Acetylene hose connections have left-hand threads and oxygen hose connections right-hand threads.

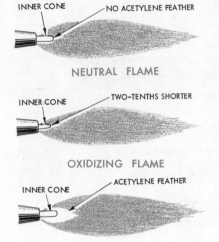

INNER CONE   NO ACETYLENE FEATHER

NEUTRAL FLAME

INNER CONE   TWO-TENTHS SHORTER

OXIDIZING FLAME

INNER CONE   ACETYLENE FEATHER

REDUCING OR CARBONIZING FLAME

Fig. 1-7. Most oxyacetylene welding is done with a neutral flame.

Welding Technique

Actual fusion of metal is achieved by plying the torch along the seam with the inner cone of the flame over the metal. For most welding operations a neutral flame is used. A neutral flame has approximately a 1 to 1 oxygen-acetylene ratio. When the amount of oxygen is increased the flame is oxidizing. A reducing flame develops when there is an excessive amount of acetylene. See Fig. 1–7.

A flame can also be made soft or harsh by varying the gas flow. If the flow of gas is too low for a given tip size the flame will be ineffective in bringing about proper fusion and will be sensitive to burnback (popping). Too high a gas flow produces a high-velocity flame that blows the molten metal from the puddle.

Welding is done by using either a *forehand* or *backhand* technique. Forehand welding is performed with the torch tip pointed forward in the direction in which the weld progresses and the rod precedes the flame. In backhand welding the tip is pointed back toward the weld which has been deposited and the filler rod is interposed between the flame and the weld. See Fig. 1–8. Generally, the forehand method is best for welding thin materials of ⅛″ and less in thickness since there

Fig. 1-8. Oxyacetylene welding is done by using a forehand (top) or backhand (bottom) technique.

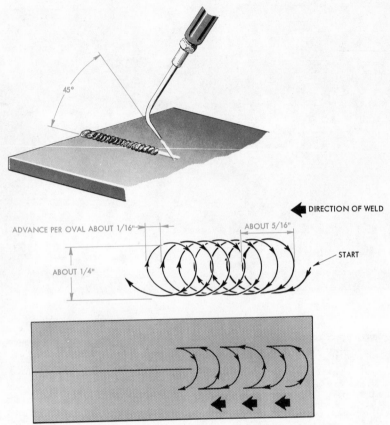

Fig. 1-9. Position and motion of the torch for welding.

is better control of the weld puddle and smoother welds are possible. The backhand technique is usually recommended for metals that are heavier than ⅛″ because sound welds can be made at a greater speed.

The torch is held at an angle of approximately 45° with the completed part of the weld. As the torch is moved over the weld seam it is often rotated in a circular or semi-circular motion as shown in Fig. 1–9. A filler rod of the same composition as the base metal is used in welding most types of joints. The rod is held at about the same angle as the torch but slanted away from the torch. See Fig. 1–10. The diameter of the rod should be equal to the thickness of the base metal. If the rod

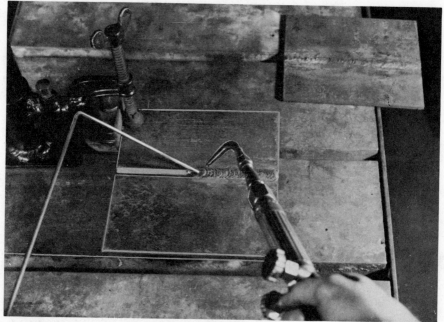

Fig. 1-10. Position of the welding rod.

is too large the heat of the molten pool will be insufficient to melt the rod, and if the rod is too small the heat cannot be absorbed by the rod with the result that a hole is burned in the plate.

As the molten puddle is carried forward along the seam the rod is dipped in and out of the molten metal. When the rod is not in the puddle the tip is kept just inside the outer envelope of the flame.

OXY-MAPP WELDING

Mapp gas is a Dow Chemical Company product developed as a fuel for welding, brazing, cutting, flame hardening, and metallizing operations. This fuel gas is a stabilized methylacetylene-propadiene product which has many of the physical properties of acetylene but lacks the shock sensitivity of acetylene. The gas is the result of rearranging the molecular structure of acetylene and propane.

Table II. Flame and Temperature Comparison of Fuel Gases.

FUEL	MAXIMUM FLAME TEMPERATURE IN O_2°F	MAXIMUM BURNING VELOCITY IN O_2
Acetylene	5589	9.8 ft/sec
Mapp Gas	5301	7.9 ft/sec
Propane	4579	5.5 ft/sec

(Dow Chemical Company)

Acetylene is thermodynamically unstable because of the strained chemical linkage connecting its carbon atoms. But as an unstable fuel, it produces a very high flame temperature which is one reason for its extensive use as a welding gas. Although propane itself has the desired stability characteristic its limiting factor for welding is its low flame temperature. Table II shows the flame temperatures and burning velocities of several gases. It will be noticed that while acetylene produces a hotter flame it has a higher burning rate in oxygen which often reflects greater gas consumption.

Since Mapp gas is not sensitive to shock it can be stored and shipped in lighter cylinders. Acetylene to be kept safe must be stored in cylinders filled with a porous filler and acetone. Where empty acetylene cylinders weigh around 220 pounds Mapp cylinders weigh only 50 pounds. Normally a filled cylinder of acetylene weighs 240 pounds while a filled cylinder of Mapp gas weighs 120 pounds.

Mapp gas has a very distinct odor so any leakage can readily be detected. Insofar as welding is concerned the same equipment and operations can be used. However, a slightly larger tip is required because of the greater gas density and slower flame propagation rate. The only

Table III. Explosive Limits of Industrial Gas.

FUEL	LIMITS IN AIR %	LIMITS IN OXYGEN %
Mapp Gas	3.4 to 10.8	2.5 to 60
Acetylene	2.5 to 80.0	3.0 to 93
Propane	2.3 to 9.5	2.4 to 57
Natural Gas	5.3 to 14.0	5.0 to 59

(Dow Chemical Company)

significant difference is in the flame appearance. A neutral flame will produce a long inner cone and needs fuel-to-oxygen ratio of 1:2:3.

The explosive limits of the vapor of Mapp gas in air and oxygen are much narrower than acetylene and about the same as propane and natural gas. See Table III. It can be used at full pressure safely (up to 375 psi) for jobs that are dangerous with acetylene. Table III lists lower and upper limits of some gases. The range indicates the safety of a gas.

OXY-HYDROGEN WELDING

The oxy-hydrogen flame is produced by burning two volumes of hydrogen (H_2) with one volume of oxygen (O_2). The resulting combustion generates a low temperature flame which is used primarily for welding thin sections of aluminum, aluminum alloys, and lead, and for brazing operations where lower temperatures are required.

The equipment for oxy-hydrogen welding is essentially the same as used in oxyacetylene welding except that a special hydrogen pressure regulator is used.

One of the unusual characteristics of oxy-hydrogen is that the flame is practically non-luminous. Consequently, difficulty is often experienced in adjusting for a neutral flame. To avoid welding with what might be an oxidizing flame the practice is to adjust the regulator so it provides a greater flow of hydrogen. The excess hydrogen is harmless since the reducing atmosphere does not leave carbon deposits.

AIR-ACETYLENE WELDING

The air-acetylene flame is generated by burning a mixture of acetylene with air. The torch operates on the same principle as the Bunsen burner. As the acetylene flows to the torch under pressure it draws in the right amount of air for proper combustion. The temperature of the air-acetylene flame is even lower than that of the oxy-hydrogen and consequently has restricted usage for welding purposes. Generally, the air-acetylene flame is used for soldering and brazing very light metal sections. It has wider applications in the plumbing industry where it is used extensively for soldering and brazing copper tubing.

REVIEW QUESTIONS

1. Why does gas welding have limited applications for production work?
2. Acetylene is a product of what chemical reaction?
3. How does the flame temperature of acetylene compare with other gases?
4. In a welding flame, what is the function of the inner and outer combustion stages?
5. Why cannot free acetylene be compressed and stored in cylinders as other gases?
6. Why are gas cylinders equipped with safety devices?
7. Why is most gas welding done with a neutral flame?
8. Why should cylinder valves be "cracked" slightly before attaching the pressure regulators?
9. Why should cylinder valves be opened slowly?
10. What determines the size of tip to be used for welding?
11. What are the effects of an oxidizing flame in a welding process?
12. How does the forehand differ from the backhand welding technique?
13. What is the limitation of the air-acetylene welding process?
14. For what is the oxy-hydrogen welding process used?
15. Why is the oxy-hydrogen flame adjusted to provide an excessive flow of hydrogen?
16. What is Mapp gas and how does it compare with acetylene?

Shielded Metal-Arc Welding

Chapter 2

Shielded metal-arc welding, sometimes referred to as stick or just plain arc welding, is a process where coalescence is achieved by generating an electric arc between a coated metallic electrode and the workpiece. The heat produced by the electric arc melts the metal which mixes with the molten deposits of the coated electrode. The arc energy is provided by a power supply unit that furnishes direct or alternating current. The metallic electrode carries the current to form the arc, produces a gas which shields the arc from the atmosphere, and adds metal to control the bead shape. See Fig. 2–1. When an arc is struck with the electrode the intense heat melts the tip of the electrode. The tiny drops of metal from the electrode enter the arc stream and are deposited on the base metal. As the molten metal is deposited a slag forms over the bead which serves as an insulation against the contaminants of the air while cooling takes place.

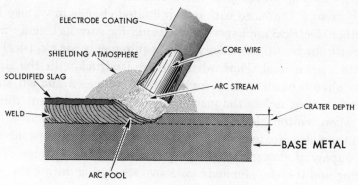

ELECTRODE COATING
SHIELDING ATMOSPHERE
SOLIDIFIED SLAG
WELD
ARC POOL
CORE WIRE
ARC STREAM
CRATER DEPTH
BASE METAL

Fig. 2-1. Cross-section of a coated electrode in the process of welding.

Application

The shielded metal-arc welding process is used extensively in welding both ferrous (iron or steel) and nonferrous (other) metals. It has many applications in producing a vast assortment of metal products. A considerable amount of arc welding will be found in the ship construction industry, and in welding girders, beams, and columns for buildings and bridges. Shielded metal-arc welding is universal in repairing and servicing equipment such as locomotives, farm implements, automobiles, machinery, and a host of other items.

Most shielded metal-arc welding is done manually where the operator establishes his welding conditions and then manipulates an electrode over the workpiece. A limited amount of shielded metal-arc welding is done with automatic equipment especially in situations where considerable production is involved. In these circumstances the electrode is mounted on a carriage which moves over the workpiece or else the carriage head is stationary and the workpiece travels under the electrode. With automatic equipment current, speed, and length of travel are completely preset and function automatically.

Equipment

The basic units of any shielded metal-arc welding system consist of a power supply, an electrode holder, a ground clamp, and a protective shield.

Power Supply. For effective shielded metal-arc welding a constant-current type of machine is required. This welding machine is characterized by a drooping volt-ampere curve in which a relatively constant supply of current is produced with only a limited change in voltage load.

In other electrical appliances the demand for current usually remains fairly constant but in arc welding the requirements for electrical power fluctuate a great deal. Thus, when the arc is struck with the electrode a short circuit results which immediately induces a sudden surge of electrical current unless the machine is designed to prevent this. Similarly, when molten globules of weld metal are carried across the arc stream they also create a short circuiting condition. A constant-current power supply is designed to minimize these sudden surges of short-circuiting and thereby eliminate excessive spattering during the welding process.

In shielded metal-arc welding the open-circuit voltage (voltage when the machine is running and no welding is being done) is considerably higher than the arc voltage (voltage after the arc is struck). The open voltage may vary from 50 to 100 and the arc voltage 18 to 36. During the welding process the arc voltage will change with differences in arc length. Since it is difficult for even the skilled operator to maintain a uniform arc length at all times, a machine with a steep volt-ampere curve will produce a more stable arc, because there will be almost no change in welding current even with changes in arc voltage. A volt-ampere curve shows the output voltage available at any given output current within the limits of the minimum and maximum current control setting on each range. For example the curve in Fig. 2–2 indicates that a high open circuit voltage is available at O which facilitates starting the arc. As welding proceeds the voltage drops to the arc voltage at A and at this point fluctuation in the arc length will not appreciably affect the current. If the electrode is short circuited on the work the current will not become excessive as indicated at B.

The magnitude of the current used directly affects the melting rate. As the rate of current is increased the current density at the electrode tip also is increased. The amount of current required for any welding operation is governed by the thickness of the parent metal and the position of the weld seam. Control of the current is by a wheel or lever arrangement. One control sets the machine for an approximate current setting and another control provides a more accurate current adjustment.

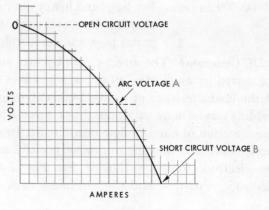

Fig. 2-2. A volt-amp curve shows the relationship between the output voltage and output current.

The three basic welding machines are referred to as generators, transformers, and rectifiers. Sizes of welding machines are designated according to their output rating which may range from 150 to 600 amperes. The output rating is based on a 60 percent duty cycle. This means that a power supply can deliver its rated load output for six minutes out of every ten minutes. In any manual welding operation a power source is not required to provide the current continuously as other electrical machines. For some electrical devices once the power is turned on it must deliver its rated output until it is shut off. With a welding power supply the machine often is idle part of the time while the operator changes electrodes, adjusts the work, or shifts welding positions. Thus the standard method of rating a machine is to show the percentage of time power actually must be delivered. Although manual and semi-automatic operated welding machines are rated at a 60 percent duty cycle, fully automatic power supply units are usually rated at 100 percent duty cycle.

The size of welding machine to be used depends on the kind and amount of welding to be done. The following is a general guide for selecting a welding machine:

150-200 ampere—For light to medium duty welding. Excellent for all fabrication purposes and rugged enough for continuous operation on light or medium production work.

250-300 ampere—For average welding requirements. Used in plants for production, maintenance, repair, tool-room work, and all general shop welding.

400-600 ampere—For large and heavy duty welding. Especially adaptable for structural work, fabricating heavy machine parts, pipe and tank welding.

DC Generator. The direct-current power supply consists of a generator driven by an electric motor or a gasoline engine. See Fig. 2–3. One of the characteristics of a direct-current welding generator is that the welding can be done with straight or reverse polarity. *Polarity* indicates the direction of current flow in a circuit. In straight polarity the electrode is negative and the workpiece positive and the electrons flow from the electrode to the workpiece. See Fig. 2–4. In reverse polarity the electrode is positive and the workpiece negative and the electrons flow

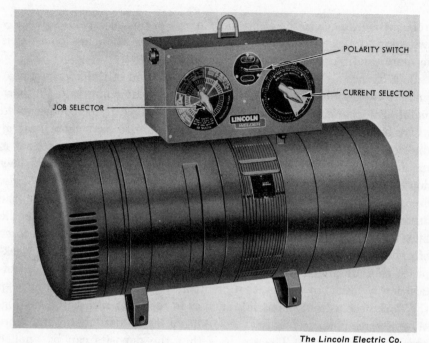

POLARITY SWITCH

CURRENT SELECTOR

JOB SELECTOR

The Lincoln Electric Co.

Fig. 2-3. A common power supply of direct current for welding is a motor generator.

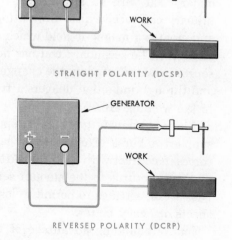

GENERATOR

WORK

STRAIGHT POLARITY (DCSP)

GENERATOR

WORK

REVERSED POLARITY (DCRP)

Fig. 2-4. With a DC power supply welding can be done with straight or reverse polarity.

from the workpiece to the electrode. Change of polarity may be done by switching cables although on modern machines changing polarity is accomplished by simply turning a switch.

Polarity has a direct relationship to the location of the liberated heat since it is possible to control the amount of heat going into the base metal. By changing polarity the greatest heat can be concentrated where it is most needed. Usually it is preferable to have more heat at the workpiece because the area of the work is greater and more heat is required to melt the metal than the electrode. Thus, if large heavy deposits are to be made the work should be hotter than the electrode. For this purpose, straight polarity would be more effective. On the other hand, in overhead welding it is necessary to quickly freeze the filler metal to help hold the molten metal in position against the force of gravity. By using reverse polarity less heat is generated at the workpiece, thereby giving the filler metal greater holding power for out-of-position welding. In other situations it may be more expedient to keep the workpiece as cool as possible such as in repairing a cast iron casting. With reverse polarity less heat is produced in the base metal and more heat at the electrode. The result is that the deposits can be applied rapidly while the base metal is prevented from overheating.

Transformer. The transformer type of welding machine produces alternating current where power is taken directly from a power supply line and transformed into a voltage required for welding. See Fig. 2–5. In its simplest form an AC transformer consists of a primary and a secondary coil with an adjustment to regulate the current output. The primary coil receives the alternating current from the source of supply and creates a magnetic field which constantly changes in direction and strength. The secondary coil has no electrical connection to the power source but is affected by the changing lines of force in the magnetic field and through induction delivers a transformed current at a higher value to the welding arc.

Some AC transformers are equipped with an arc booster switch which supplies a burst of current for easy arc starting when the electrode comes in contact with the work. After the arc is struck the current automatically returns to the amount set for the job. The arc booster switch has several settings to permit quick arc starting for welding either thin sheets or heavy plates.

One outstanding advantage of the AC welder is freedom from *arc*

CURRENT ADJUSTER
ARC BOOSTER
SWITCH

Fig. 2-5. AC transformer type of welding unit.

Hobart Brothers Co.

blow which often occurs when welding with DC machines. Arc blow is a phenomenon that causes the arc to wander while welding in corners on heavy metal or when using large size coated electrodes. Direct current when flowing through the electrode, workpiece, and ground clamp generates a magnetic field around each of these units which may cause the arc to deviate from its intended path. The arc usually is deflected either forward or backward along the line of travel and may be severe enough to cause excessive spatter and incomplete fusion. It also tends to pull atmospheric gases into the arc resulting in porosity. The bending of the arc is due to the effects of an unbalanced magnetic field. Thus, when a greater concentration of magnetic flux develops on one side of the arc it tends to blow away from the source of greatest concentration. Arc blow can often be corrected by changing the position of the ground clamp, welding away from the ground clamp, or changing the position of the weldpiece on the workbench.

Rectifiers. Rectifiers are essentially transformers containing an electrical device which changes alternating current into direct current. Some

Fig. 2-6. A combination AC-DC rectifier is often used as a power supply unit.

types are designed to provide a choice of voltages for shielded metal-arc and submerged arc welding, or a voltage characteristic for gas tungsten-arc and gas metal-arc welding.

The rectifiers for shielded metal-arc welding are usually of the constant-current type where the welding current remains reasonably constant for small variations in arc length. The other type is classified as a constant-potential welder and is used for semiautomatic and automatic gas metal-arc welding where a continuous wire electrode is fed at a constant speed into the weld seam. A more detail description of constant-potential power sources will be found in Chapter IV.

Rectifiers are made to provide DC current only or both DC and AC welding current. See Fig. 2–6. By means of a switch the output terminals can be changed to the transformer or to the rectifier and produce either AC current or DC straight or DC reverse polarity current.

At present the two rectifier materials used for welding machines are selenium and silicon. Both provide excellent performance although silicon will often permit operation at higher current densities.

Electrode Holder. The electrode holder is used to hold the electrode and guide it over the seam to be welded. See Fig. 2–7. A properly designed holder should be light to reduce excessive fatigue while welding, receive and eject electrodes easily, and be properly insulated. Some holders are fully insulated while in others only the handle is insulated.

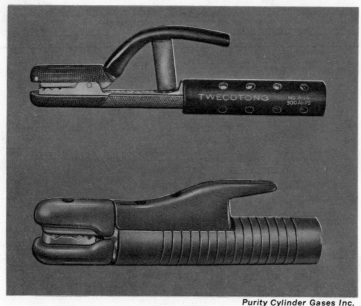

Purity Cylinder Gases Inc.

Fig. 2-7. Two common insulated electrode holders.

When using a holder with uninsulated jaws, care must be taken never to lay it on the bench plate while the machine is running because it will cause a flash.

Holders should be firmly connected to the cable. A loose connection where the cable joins the holder may result in overheating the holder. The diameter of the cable is fixed by the manufacturer of the welding equipment. The use of ample current carrying cable is necessary for proper welding. Thus a 30 ft lead cable of a given size may be satisfactory to carry the required current but if another 30 ft of cable were added the combined resistance of the two leads would reduce the current output of the machine. If the machine is then readjusted for a higher

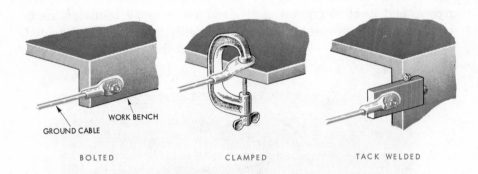

GROUND CABLE   WORK BENCH

BOLTED   CLAMPED   TACK WELDED

GROUND CLAMP

Fig. 2-8. A good ground is essential for proper welding.

output under this condition the additional load may cause the power supply to overheat as well as to increase its power consumption.

The primary cable which connects the welding machine to the source of electricity is also significant. The length of this cable has been determined by the manufacturer of the power supply unit and represents a length that permits the machine to operate efficiently without an appreciable drop in voltage. If a longer cable is used more voltage will be required for the work to be done, and if no more voltage is available the resulting voltage drop will seriously affect the welding process.

Ground Clamp. A very important unit of any electrical welding equipment is the ground clamp. Without proper grounding the full potential of the circuit will fail to provide the required heat for welding.

A good ground connection can be made in several ways. The ground cable may be fastened to the work bench by a C clamp, a special ground clamp, or by bolting or welding a lug on the end of the cable to the bench. See Fig. 2–8.

Protective Shield. A suitable helmet or handshield is required for all arc welding. See Fig. 2–9. An electric arc produces a brilliant light and

Jackson Products

Fig. 2-9. Shields should always be used when arc welding.

also gives off invisible ultraviolet and infrared rays which are extremely dangerous to the eyes and skin. The arc should never be looked at with the naked eye within a distance of 50 feet.

Both the helmet and handshield are equipped with special colored lenses that reduce the brilliancy of the light and screen out the infrared and ultraviolet rays. Lenses come in different shades and the types used depends on the kind of welding done. In general, the recommended practice is as follows:

Shade 5 for light spot welding
Shade 6 and 7 for welding up to 30 amperes
Shade 8 for welding 30 to 75 amperes
Shade 10 for welding 75 to 200 amperes
Shade 12 for welding 200 to 400 amperes
Shade 14 for welding over 400 amperes.

Welding Technique

Shielded metal-arc welding can be performed in various positions such as horizontal, flat, vertical, and overhead. See Fig. 2–10. Generally speaking, greater economies are realized if the welding can be done in a flat position because of less operator fatigue, greater welding speed, and better penetration.

Types of Joints. The basic types of joints used in joining metal products are illustrated in Fig. 2–11. Joint design is discussed more fully in Chapter XIII. Notice in Fig. 2–11 that butt joints are either closed or opened. A *closed butt joint* has the edges of the two plates in direct contact with each other. This joint is suitable for welding steel plates that do not exceed $\frac{1}{8}''$ to $\frac{3}{16}''$ in thickness. Heavier metal can be welded but only if the machine has sufficient amperage capacity and if heavier electrodes are used. The *open butt joint* has the edges slightly apart to provide for better penetration. Very often a backup bar of steel, copper, or brick is placed under an open joint to prevent the bottom edges from burning through. When the thickness of metal exceeds $\frac{1}{8}''$ or $\frac{3}{16}''$, the edges have to be beveled for more adequate penetration. Beveling may be confined to one plate or the edges of both plates are beveled, depending on the thickness of the metal. Angle of bevel is usually 30°.

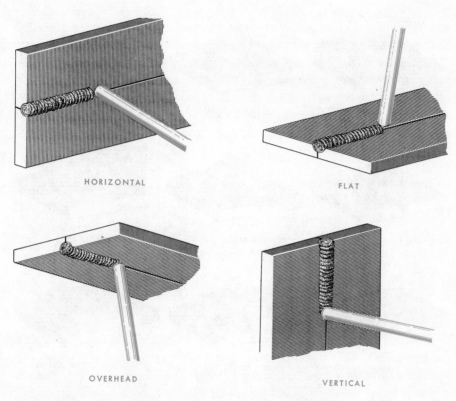

HORIZONTAL

FLAT

OVERHEAD

VERTICAL

Fig. 2-10. Four basic welding positions.

Starting the Arc. Generation of the arc is accomplished by tapping or scratching the end of the electrode on the workpiece. See Fig. 2–12. As soon as the arc is started the electrode is immediately raised a distance equal to the diameter of the electrode. Failure to raise the electrode causes it to stick to the metal and if it is allowed to remain in this position with the current flowing the electrode will become red hot. When an electrode does stick it can be broken loose by quickly twisting or bending it. If it does not dislodge the electrode should be released from the holder.

Current Setting. The amount of current to be used depends upon the thickness of the metal to be welded and the actual position of welding. As a rule, higher currents and larger diameter electrodes can be used for welding in flat positions than in vertical or overhead welding. Electrode

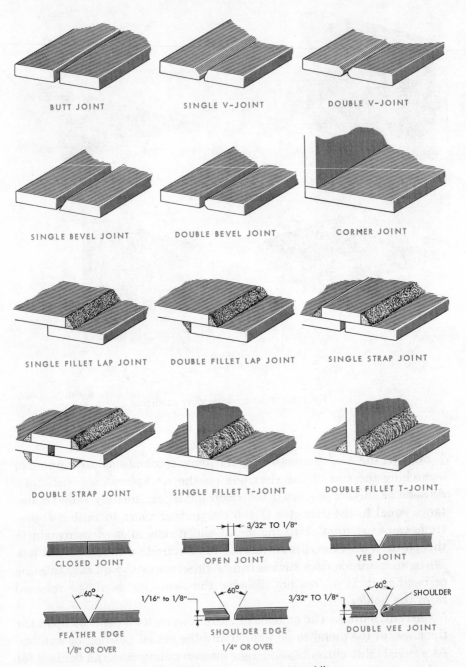

Fig. 2-11. Types of joints used in welding.

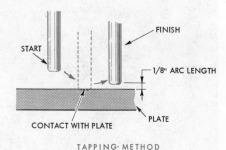

TAPPING METHOD

Fig. 2-12. Two methods of starting the arc.

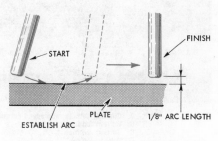

SCRATCHING METHOD

diameter is governed by the thickness of the plate and welding position. For most flat position welding the maximum diameter electrodes should be ⁵⁄₁₆″ or ⅜″, while ³⁄₁₆″ should be the maximum for vertical and overhead welding.

Manufacturers of electrodes generally specify a range of current values for various diameter electrodes. See Appendix. However, since the recommended current setting is only approximate final current adjustment is made as the welding operation proceeds. For example, if the current range for a particular size electrode is 90–100 amperes, the usual practice is to set the control midway between the two limits. After the welding is started a final adjustment is made by either increasing or decreasing the current.

Whenever the current is too high the electrode will melt too fast and the molten puddle will be too large and irregular. Where the current is too low there will not be enough heat to melt the base metal and the molten pool will be too small. The result is not only poor fusion but the beads will pile up and be irregular in shape. See Fig. 2–13. Too high a

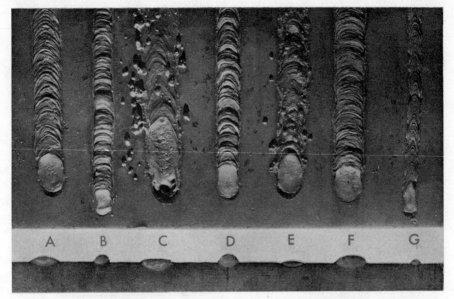

A. Current, voltage, and speed normal.
B. Current too low.
C. Current too high.
D. Voltage too low.

E. Voltage too high.
F. Speed too slow.
G. Speed too fast.

Fig. 2-13. Examples of properly and improperly formed beads.

current may also produce *undercutting* which leaves a groove in the base metal along both sides of the bead. A current that is set too low will cause *overlaps* to form where the molten metal from the electrode falls on the workpiece without sufficiently penetrating the base metal. Both undercutting and overlapping result in weak welds. See Fig. 2–14.

Arc Length. If the arc is too long the metal melts off the electrode in large globules which wobble from side to side as the arc wavers. This produces a wide, spattered, and irregular bead without sufficient fusion between the original metal and the deposited metal. An arc that is too short fails to generate sufficient heat to melt the base metal properly. Furthermore, the electrode will stick frequently and produce uneven beads with irregular ripples.

The length of the arc depends on the type of electrode used and the kind of welding done. Thus, for small diameter electrodes a shorter arc is necessary than for larger electrodes. Generally, the length of the arc should be approximately equal to the diameter of the electrode. A

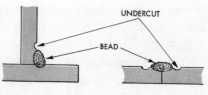

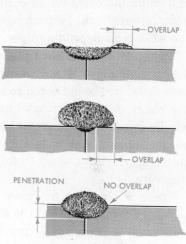

Fig. 2-14. Undercuts and overlaps should be avoided in any welding process.

shorter arc is usually better for vertical and overhead welding than for most flat position welding because better control of the molten puddle is achieved with the shorter arc.

The use of a short arc also prevents impurities from the atmosphere entering the weld. A long arc allows the atmosphere to flow into the arc stream permitting nitrides and oxides to form. Moreover, when the arc is too long heat from the arc stream is dissipated too rapidly causing considerable metal spatter. When the electrode, current, and polarity are correct, a short arc will produce a sharp, crackling sound. A long arc can be recognized by a steady hiss very much like escaping steam.

Travel Speed. Travel speed is the rate the electrode travels along the weld seam. The maximum speed of welding is influenced by the skill of the operator, the position of the weld, the type of electrode, and the required joint penetration. Normally if the speed is too fast, the molten pool does not last long enough and impurities become locked in the weld. The weld bead usually is narrow with pointed ripples. On the other hand if the travel is too slow the metal piles up excessively and the weld bead is high and wide. In most instances the limiting speed is the highest speed that produces a satisfactory surface appearance. See Fig. 2–13.

Crater Formation. As the arc comes in contact with the base metal a pool or pocket is formed which is called a crater. The size and depth of a crater indicates the amount of penetration. In general, the depth of penetration should be from one-third to one-half the total thickness of the weld bead depending upon the size of the electrode. See Fig. 2–15.

To secure a sound weld the metal deposited from the electrode must fuse completely with the base metal. Fusion will result only when the base metal has been heated to a liquid state and the molten metal from the electrode readily flows into it. Thus if the arc is too short there will be insufficient spread of heat, or if the arc is too long the heat is not centralized enough to form the desired crater. An improperly filled crater may cause a weld to fail when a load is applied on the welded structure.

There is always a tendency when starting an electrode for a large globule of metal to fall on the surface of the plate with little or no penetration. This is especially true when beginning a new electrode at the crater left from a previously deposited weld. To ensure that the crater is filled the arc should be struck approximately ½ inch in front of the crater as in A, Fig. 2–16, brought through the crater to a point B beyond

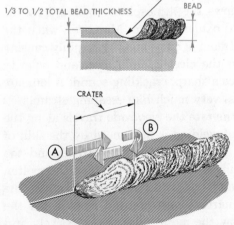

Fig. 2-15. Correct penetration and fusion are important for sound welds.

Fig. 2-16. Method for properly filling a crater.

the crater and then the weld carried back through the crater. When the electrode reaches the end of a seam care must be taken to make certain that the crater is filled. This means breaking the arc at the right moment. Two procedures are used to break the arc to secure a full crater. With one procedure the arc is shortened and the electrode quickly moved sideways and out of the crater. In the other method the electrode is held stationary just long enough to fill the crater and then gradually withdrawn from it.

Occasionally the crater may become too hot and the molten metal has a tendency to run. When this happens the electrode should be lifted and shifted quickly to the side or ahead of the crater. Such a movement reduces the heat, allows the crater to solidify momentarily, and stops the deposit of the metal from the electrode.

Electrode Position. The angular position of the electrode has a direct influence in the quality of weld made. Very often the position of the electrode is a factor that determines the ease with which filler metal is deposited and contributes to the elimination of undercutting, of slag inclusions, and helps weld contour uniformity. The two guiding considerations of electrode position are lead angle and work angle. Lead angle is the angle between the joint and the electrode when viewed in a longitudinal plane. Work angle is the angle between the electrode and the work when viewed from an end plane. See Fig. 2–17. Some of the basic electrode positions for flat, horizontal, vertical, and overhead welding are illustrated in Fig. 2–18.

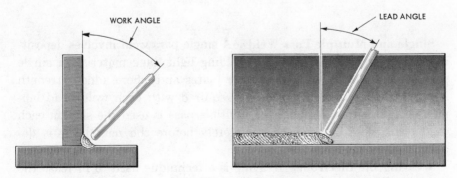

Fig. 2-17. An electrode must be held at the proper work and lead angle.

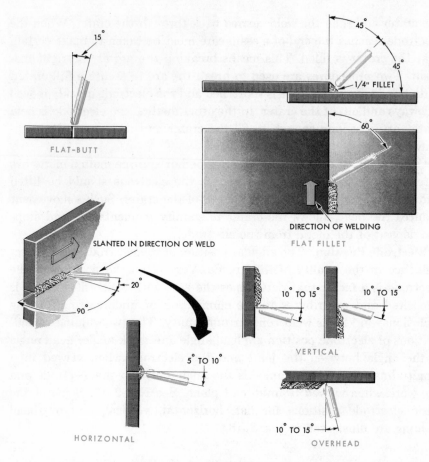

15°

FLAT-BUTT

45°

45°

1/4" FILLET

60°

DIRECTION OF WELDING

FLAT FILLET

SLANTED IN DIRECTION OF WELD

20°

90°

10° TO 15°   10° TO 15°

VERTICAL

5° TO 10°

10° TO 15°

HORIZONTAL

OVERHEAD

Fig. 2-18. Electrode positions for flat, horizontal, vertical, and overhead welding.

Single and Multiple Pass Welds. A single pass weld involves depositing one layer of weld metal. In welding light gage materials a single pass is usually sufficient. On heavier plates and where added strength is required two or more layers are required with each weld bead lapping over the other. Whenever a multiple pass is used the slag on each weld bead must be removed completely before the next layer is deposited. See Fig. 2–19.

Weaving the Electrode. Weaving is a technique used to increase the width and volume of the weld bead. Enlarging the size of the weld bead

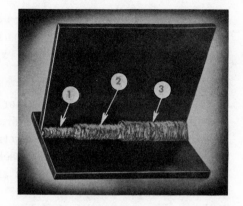

Fig. 2-19. A typical multipass weld. This has three beads.

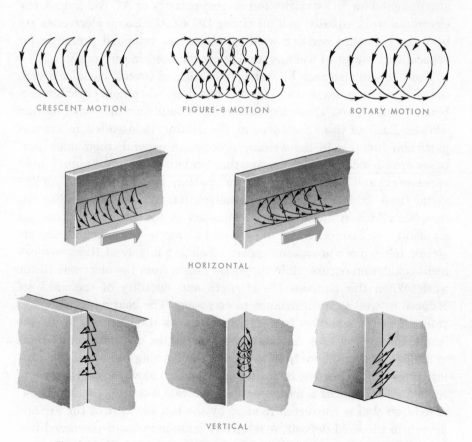

CRESCENT MOTION FIGURE-8 MOTION ROTARY MOTION

HORIZONTAL

VERTICAL

Fig. 2-20. Typical weaving patterns used for shielded metal-arc welding.

is often necessary on deep grooves or fillet welds where a number of passes must be made. Fig. 2–20 illustrates several weaving patterns. The patterns used depend to a large extent on the position of the weld.

Selecting the Electrode

The type of electrode selected for shielded metal-arc welding is governed by the quality of weld required, weld position, joint design, and welding speed. In general, all electrodes are classified into five main groups: mild steel, high carbon steel, special alloy steel, cast iron, and nonferrous. The greatest range of arc welding is done with electrodes in the mild steel group.

Electrodes are manufactured to weld different metals and they are also designed for DC straight and reverse polarity or AC welding. A few electrodes work equally well on either DC or AC. Some electrodes are best suited for flat position welding, others are intended primarily for vertical and overhead welding, and some are used in any position.

The shielded electrode has a heavy coating of several chemicals such as cellulose, titanium dioxide, ferro-manganese, silica flour, calcium carbonate, and others. These ingredients are bound together with sodium silicate. Each of the substances in the coating is intended to serve a particular function in the welding process. In general, their main purposes are to induce easier arc starting, stabilize the arc, improve weld appearance and penetration, reduce spatter, and protect the molten metal from oxidation, that is, contamination by the surrounding atmosphere. Molten metal as it is deposited in the welding process has an affinity, or attraction, for oxygen and nitrogen. Inasmuch as the arc stream takes place in an atmosphere consisting largely of these two elements oxidation occurs while the metal passes from the electrode to the work. When this happens the strength and ductility of the weld are reduced as well as its resistance to corrosion. The coating on the electrode prevents oxidation from taking place. As the electrode melts the heavy coating releases an inert gas around the molten metal which excludes the atmosphere from the weld. The burning residue of the coating forms a slag over the deposited metal which slows down the rate of cooling and produces a more ductile weld. Some coatings include powdered iron that is converted to steel by the intense heat of the arc and flows into the weld deposit. A relatively large amount of powdered iron helps increase the rate of electrode deposition.

Identifying Electrodes. Electrodes are often referred to by a manufacturer's trade name. To ensure some degree of uniformity in manufacturing electrodes the American Welding Society (AWS) and the American Society for Testing and Materials (ASTM) have set up certain requirements for electrodes. Thus, different manufacturer's electrodes which are within the classification established by the AWS and ASTM may be expected to have the same welding characteristics. In this classification each type of electrode has been assigned specific symbols such as E-6010, E-7010, E-8010, etc. The prefix E identifies the electrode for electric arc welding. The first two digits in the symbol designate the minimum allowable tensile strength of the deposited weld metal in thousands of pounds per square inch. For example, the 60 series electrodes have a minimum pull strength of 60,000 pounds per square inch; the 70 series, a strength of 70,000 pounds per square inch.

The third digit of the symbol indicates possible welding positions. Three numbers are used for this purpose: 1, 2, and 3. Number 1 is for an electrode which can be used for welding in any position. Number 2 represents an electrode restricted for welding in horizontal and flat positions. Number 3 represents an electrode to be used in the flat position only. The fourth digit of the symbol simply shows some special characteristic of the electrode such as weld quality, type of current, and amount of penetration. See Table I.

Most manufacturers currently make it a practice to stamp on each electrode the regular AWS classification. In addition to this classifica-

Table I.   Interpretation of Last Digit in AWS Electrode Classification.

LAST DIGIT	0	1	2	3	4	5	6	7	8
Power supply	(a)	AC or DC rev polarity	AC or DC	AC or DC	AC or DC	DC rev polarity	AC or DC rev. polarity	AC or DC	AC or DC rev. polarity
Type of slag	(b)	Organic	Rutile*	Rutile*	Rutile*	Low Hydrogen	Low Hydrogen	Mineral	Low Hydrogen
Type of arc	Digging	Digging	Medium	Soft	Soft	Medium	Medium	Soft	Medium
Penetration	(c)	Deep	Medium	Light	Light	Medium	Medium	Medium	Medium
Iron Powder in Coating	0–10%	None	0–10%	0–10%	30–50%	None	None	50%	30–50%

Notes: (a) E–6010 is DC reverse polarity; E–6020 is AC or DC
       (b) E–6010 is organic; E–6020 is mineral
       (c) E–6010 is deep penetration; E–6020 is medium penetration

*A hard titanium dioxide slag.

tion a color marking is often used for some electrodes. The color code is one established by the National Electrical Manufacturers Association (NEMA). See Appendix.

Selecting the Electrode. Several factors must be taken into consideration in choosing an electrode for welding purpose. The welding position is particularly significant. Table II lists the general recommendations for types of current and welding positions of some of the more common electrodes.

As a rule, an electrode should never be used which has a diameter larger than the thickness of the metal to be welded. Some operators prefer larger electrodes because they permit faster travel along the joint and thus speed up the welding operation; but this takes considerable skill.

Position and the type of joint are also factors in determining the size of the electrode. For example, in a thick metal section with a narrow vee, a small diameter electrode is always used to run the first weld bead or root pass. This is done to ensure thorough penetration at the root of the weld. Successive passes are then made with larger diameter electrodes.

For vertical and overhead welding, $\frac{3}{16}''$ is the largest diameter electrode that should be used regardless of plate thickness. Larger electrodes make it too difficult to control the deposited metal. From the standpoint of economy it is always a good practice to use the largest size electrode that is practical for the work to be done. It takes approximately half the time to deposit a quantity of weld metal from $\frac{1}{4}''$ coated mild steel electrodes than from $\frac{3}{16}''$ electrodes of the same type. The larger sizes not only make possible the use of higher currents but require fewer stops to change the electrode.

Deposition rate and joint preparation are also important features which influence the selection of electrode. Electrodes for welding mild steel are sometimes classified as fast-freeze, fill-freeze, and fast-fill[1] (See Table II) The *fast-freeze* electrodes are those which produce a snappy, deep penetrating arc and fast-freezing deposits. They are commonly called reverse polarity electrodes even though some can be used on AC. These electrodes have little slag and produce flat beads. They are widely used for all types of all-position welding for both fabrication and

---

[1]The Lincoln Electric Company.

Table II. Electrode Characteristics.

TYPE	AWS CLASS	CURRENT TYPE	WELDING POSITION	WELD RESULTS
Mild Steel	E6010 E6011	DCR DCR, AC	F, V, OH, H F, V, OH, H	Fast freeze, deep penetrating, flat beads, all-purpose welding
	E6012 E6013 E6014	DCS, AC DCR, DCS, AC DCS, AC	F, V, OH, H F, V, OH, H F, V, OH, H	Fill-freeze, low penetration, for poor fit-up, good bead contour, minimum spatter
	E6020 E6024	DCR, DCS, AC DCR, DCS, AC	F, H F, H	Fast-fill, high deposition, deep groove welds, single pass
	E6027	DCR, DCS, AC	F, H	Iron powder, high deposition, deep penetration
	E7014	DCR, DCS, AC	F, V, OH, H	Iron powder, low penetration, high speed
	E7024	DCR, DCS, AC	F, H	Iron powder, high deposition, single and multiple pass
Low Hydrogen	E6015 E6016 E6018 E7016 E7018 E7028	DCR DCR, AC DCR, AC DCR, AC DCR, AC DCR, AC	F, V, OH, H F, V, OH, H F, V, OH, H F, V, OH, H F, V, OH, H F, H	Welding of high-sulphur and high carbon steels that tend to develop porosity and crack under weld bead
Stainless Steel	E308-15, 16	DC, AC	F, V, OH, H	Welding stainless steel 301, 302, 303, 304, 308
	E309-15, 16	DC, AC	F, V, OH, H	Welding 309 alloy at elevated temperature application and dissimilar metals
	E310-15, 16	DC, AC	F, V, OH, H	Welding type 310 and 314 stainless steel where high corrosion and elevated temperatures are required
	E316-15, 16	DC, AC	F, V, OH, H	Welding type 316 stainless steel and welds of highest quality. Contains less carbon to minimize carbon transfer in the weld. Type 316 reduces pitting corrosion
	E347-15, 16	DC, AC	F, V, OH, H	For welding all grades of stainless steels
Low Alloy	E7011-A1 E7020-A1	DCR, AC DCR, DCS, AC	F, V, OH, H F	For welding carbon moly steels
	E8018-C3	DCR, AC	F, V, OH, H	For low alloy, high tensile strength
	E10013-G	DCS, AC	F, V, OH, H	For low alloy, high tensile steels

DCR — Direct Current Reverse Polarity
DCS — Direct Current Straight Polarity
AC — Alternating Current
F   — flat, V — vertical, OH — overhead, H — horizontal

repair work. *Fill-freeze* electrodes have a moderately forceful arc and deposit rate between those of the fast-freeze and fast-fill electrodes. They are commonly called the "straight-polarity" electrodes even though they may be used on AC. These electrodes have complete slag coverage and weld beads with distinct, even ripples. They are the general-purpose electrode for production shop and are also widely used for repair work. They can be used in all positions, though the fast-freeze electrodes are preferred for vertical and overhead welding. The *fast-fill* group includes the heavy coated, iron powder electrodes with soft arc and fast deposit rate. These electrodes have a heavy slag and produce exceptionally smooth weld beads. They are generally used for production welding where all work can be positioned for downhand (flat) welding.

Another group of electrodes are the low-hydrogen type which contain little hydrogen in either moisture or chemical form. These electrodes have outstanding crack resistance, little or no porosity, and quality x-ray deposits.

Welding stainless steel requires an electrode containing chromium and nickel. All stainless steels have low thermal conductivity. In electrodes this causes overheating and improper arc action when high currents are used. In base metal it causes large temperature differences between weld and the rest of the work which warps the plate. A basic rule in welding stainless steel is to avoid high currents and high heat in the weld. Another reason for keeping the weld cool is to avoid carbon precipitation which, under some conditions, causes intergranular corrosion.

There are also many special-purpose electrodes for hardfacing, welding copper and copper alloys, aluminum, cast iron, manganese, nickel alloys, and nickel-manganese steels. Composition of these electrodes are usually designed to match the base metal to be welded.

STUD WELDING

Stud welding is basically a shielded metal-arc welding process except that the stud serves as the electrode to produce the arc. The heat required to fuse the stud to the workpiece is developed by the passage of current from the stud to the plate. The principal difference from shielded metal-arc is the establishment of the arc, welding time, and plunging of the stud to the plate all of which are automatically controlled.

Any direct-current welding generator or rectifier with a high open-circuit drooping voltage characteristic is suitable for stud welding. Best performance is obtained with a high amperage capacity welder (300-600). The capacity of the welding machine depends on the size of the studs. For example, a 400 amp DC welder will weld studs up to $\frac{7}{16}''$ in diameter. Stud welding can be done with portable guns or automatic machines.

Welding Process

Two methods have been developed for stud welding. With one method, the welding ends of the studs are recessed and contain flux. See Fig. 2–21A. The flux acts as an arc stabilizer as well as a deoxidizing agent. The equipment consists of a stud gun and a timing device. An individual

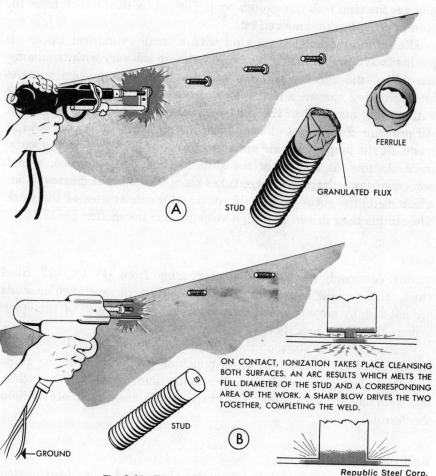

FERRULE

GRANULATED FLUX

STUD

(A)

STUD

(B)

—GROUND

ON CONTACT, IONIZATION TAKES PLACE CLEANSING BOTH SURFACES. AN ARC RESULTS WHICH MELTS THE FULL DIAMETER OF THE STUD AND A CORRESPONDING AREA OF THE WORK. A SHARP BLOW DRIVES THE TWO TOGETHER, COMPLETING THE WELD.

Republic Steel Corp.

Fig. 2-21. Two methods used in stud welding.

porcelain ferrule is used with each stud. The function of the ferrule is to concentrate the heat, act with the flux to restrict air from the molten metal, and confine the molten metal to the weld zone.

In operation the stud is loaded in the chuck of the gun and a ferrule fastened over the stud. The stud is then positioned on the workpiece and a trigger-switch causes the current to energize a solenoid coil which lifts the stud away from the plate. As the stud pulls away from the plate an arc is generated. The arc melts the end of the stud and a corresponding area on the plate. At the proper time, the timing device shuts off the current, the solenoid releases the stud, and spring action plunges the stud into the molten pool. The gun is then pulled from the stud and the ferrule knocked off.

The second method uses a stud with a small cylindrical tip on its joining face. See Fig. 2–21B. The size of the tip will vary with the diameter of the stud and the material being welded. This process operates on alternating current and employs an air gun. The gun is equipped with a collet attached to the end of a piston rod which holds the stud. Air pressure keeps the stud away from the plate until the weld is to be made. At the proper time air pressure drives the stud against the workpiece. As soon as the tip of the stud touches the workpiece, a high amperage, low voltage discharge takes place. The current creates an arc which melts the entire face of the stud and a similar area of the work. The stud is then driven at a high velocity into the molten pool.

Studs

Most commonly used stud diameters range from $\frac{1}{8}''$ to $1\frac{1}{4}''$. Stud stock can be round, square, or rectangular. However, rectangular studs are not usually recommended for welding when the width of the stock is more than five times its thickness.

Positioning of the studs on the workpiece is done in several ways. The simplest method consists of center-punching the location of the studs and then placing the studs over the center punch marks. When close tolerance location is required, the practice is to use a template to hold the ferrules.

Weld Time and Current

Welding time varies with the diameter of the stud. Time is expressed

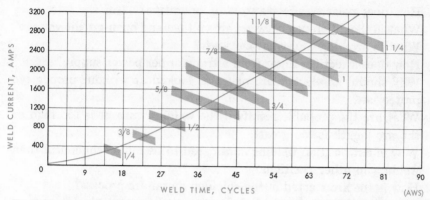

Fig. 2-22. Relationship between welding current and times for various diameter studs.

as a fractional part of a cycle (1/60 cycle) necessary to complete the weld. The cycle time may range from a few cycles for a small cross sectional point to 45 or more cycles for a ¾" diameter stud.

The amount of current required also depends on the size of the stud. Thus, it takes less current to weld a ⅜" diameter stud than one with a ⅞" diameter. In any successful operation there must be a close relationship between weld time, current, and stud diameter. Fig. 2–22.

REVIEW QUESTIONS

1. Why is a constant-current power supply with a drooping volt-amp characteristic considered the most adequate for shielded metal-arc welding?
2. When a power supply unit is rated at a 60 percent duty cycle, what does this mean?
3. Why are most welding machines rated at only 60 percent duty cycle? When is the exception?
4. In a DC welding supply unit what is the direction of current flow in a straight and reverse polarity circuit?

5. How does polarity affect a welding operation?
6. What is the difference between arc voltage and open-circuit voltage?
7. What is arc blow and what causes it?
8. How does a rectifier differ from a transformer power supply?
9. Why should an electric arc never be looked at without proper eye protection?
10. What are the probable results if the welding current is too high or the arc length is too great?
11. Why is flat position welding considered to be more economical than welding in other positions?
12. How is the arc started in the shielded metal-arc process?
13. What are the functions of the various substances which are used as coatings on electrodes?
14. What effects do oxides and nitrides have on a molten puddle?
15. What are some of the factors which must be considered in selecting an electrode for shielded metal-arc welding?
16. What are some of the characteristics of electrodes classified as fast-freeze, fill-freeze, and fast-fill?
17. What two techniques are used in stud welding?

Chapter 3 | Gas Tungsten-Arc Welding

There are two main types of gas shielded-arc welding processes. One is called gas tungsten-arc welding and the other gas metal-arc welding. Gas tungsten-arc welding is often referred to as Tig and gas metal-arc welding as Mig. The Mig process is described in Chapter IV. Both employ a shielding gas to protect the weld zone from the atmosphere. Since the shielding gas excludes the atmosphere from the molten puddle welded joints will more nearly possess the same chemical, metallurgical, and physical properties as the base metal. As a consequence welded joints are stronger, more ductile and have less distortion than welds made by most other welding processes. Gas shielded-arc welding, see Fig. 3–1, simplifies joining of nonferrous metals since no flux is needed and there is less danger of corrosion due to flux entrapment.

Fig. 3-1. Tig welding of wire cloth.

Inasmuch as the shielding gas is transparent, the operator can clearly observe the weld as it is being made. Moreover, there is very little smoke, fumes, or sparks, all of which contribute to making a neater and sounder weld. Postweld cleaning is virtually eliminated because there is no slag. This in itself results in considerable cost saving especially where multipasses are required on a joint.

Gas tungsten-arc welding. This is a welding process where coalescence is achieved by heating with an electric arc produced by a virtually nonconsumable tungsten electrode. During the welding cycle a shield of inert gas expels the air from the welding area and prevents oxidation of the electrode, weld puddle, and surrounding heat-affected zone. See Fig. 3–2. The electrode creates the arc only and is not consumed in the weld as in shielded metal-arc welding. For joints where additional weld metal is needed a filler rod is fed into the puddle in a manner similar to the oxyacetylene process.

The actual welding operation can be semi-automatic or fully automatic. In semi-automatic the current and gas flow are preset and function automatically. However, the operator must manually manipulate the torch and filler rod, if required, over the weld seam. When the

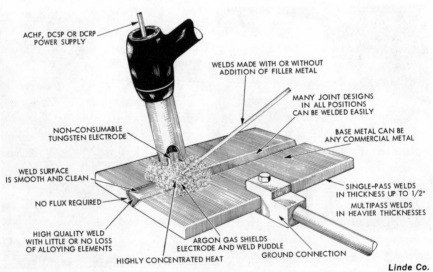

ACHF, DCSP OR DCRP POWER SUPPLY

WELDS MADE WITH OR WITHOUT ADDITION OF FILLER METAL

MANY JOINT DESIGNS IN ALL POSITIONS CAN BE WELDED EASILY

NON-CONSUMABLE TUNGSTEN ELECTRODE

BASE METAL CAN BE ANY COMMERCIAL METAL

WELD SURFACE IS SMOOTH AND CLEAN

SINGLE-PASS WELDS IN THICKNESS UP TO 1/2"

NO FLUX REQUIRED

MULTIPASS WELDS IN HEAVIER THICKNESSES

HIGH QUALITY WELD WITH LITTLE OR NO LOSS OF ALLOYING ELEMENTS

ARGON GAS SHIELDS ELECTRODE AND WELD PUDDLE

HIGHLY CONCENTRATED HEAT

GROUND CONNECTION

Linde Co.

Fig. 3-2. Tig welding uses a non-consumable tungsten electrode to produce an arc which is shielded by an inert gas.

equipment is fully automatic the travel of the arc, arc distance, gas flow, and filler rod are mechanically controlled. The operator simply sets the conditions for welding and then observes the welding operation.

Gas tungsten-arc welding is frequently known by such manufacturing trade names as Heliarc (Linde), and Heliwelding (Airco).

Application

Gas tungsten-arc welding originally was developed to weld magnesium which always was somewhat difficult to weld because of its tendency to oxidize rapidly. Later the gas tungsten-arc process was found to be very suitable for welding aluminum, stainless steel, carbon steel, copper and its alloys, and nickel and its alloys. In addition to being very effective for welding a wide range of commercial types of metals, this process is now being used extensively to weld various combinations of dissimilar metals. It also has wide application for hardfacing damaged or worn steel dies and high speed cutting tools.

The gas tungsten-arc process is especially adaptable for welding light gage materials where the utmost in weld quality or finish is required. However, heavy steel plates ⅛″ or more in thickness can be welded successfully.

Equipment

Equipment for gas tungsten-arc welding consists of a power supply, an electrode holder (torch), a non-consumable electrode, and a flowmeter to regulate the supply of shielding gas. See Fig. 3–3.

Power Supply. Modern gas tungsten-arc power supply units are constant-current AC or DC machines. The choice of an AC or DC welder depends on certain distinct weld characteristics that may be required. Some metals are joined more easily with AC current while with others better results are obtained when DC current is used. See Table I. To understand the effects of the two different currents an explanation of their behavior in a welding process is necessary.

Direct Current Reverse Polarity (DCRP). With direct current the welding circuit may be either straight or reverse polarity. When the machine is set for straight polarity the flow of electrons from the electrode to the plate exert considerable heat on the plate. In reverse polarity the flow of electrons is from the plate to the electrode, thus causing a greater concentration of heat at the electrode. See Fig. 3–4. The intense heat at

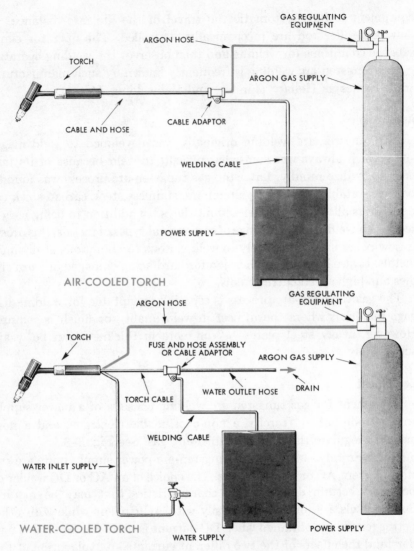

GAS REGULATING EQUIPMENT

ARGON HOSE

TORCH

ARGON GAS SUPPLY

CABLE AND HOSE

CABLE ADAPTOR

WELDING CABLE

POWER SUPPLY

AIR-COOLED TORCH

GAS REGULATING EQUIPMENT

ARGON HOSE

TORCH

FUSE AND HOSE ASSEMBLY OR CABLE ADAPTOR

ARGON GAS SUPPLY

TORCH CABLE

WATER OUTLET HOSE

DRAIN

WELDING CABLE

WATER INLET SUPPLY

WATER-COOLED TORCH

WATER SUPPLY

POWER SUPPLY

Fig. 3-3. A schematic layout of Tig welding.

the electrode tends to melt off the end of the electrode and contaminate the weld. Hence, for any given current DCRP requires a larger diameter electrode than DCSP. For example, a $\frac{1}{16}''$ tungsten electrode normally can handle about 125 amperes in a straight-polarity circuit. However, if

Table I. Current Selection for Inert-Gas-Shielded Arc Welding.

METAL	AC CURRENT WITH HIGH FREQUENCY STABILIZATION	DC CURRENT STRAIGHT POLARITY	DC CURRENT REVERSE POLARITY
Magnesium up to 1/8 in. thick	1	N.R.	2
Magnesium above 3/16 in. thick	1	N.R.	N.R.
Magnesium Castings	1	N.R.	2
Aluminum	1	N.R.	N.R.
Aluminum Castings	1	N.R.	N.R.
Stainless Steel up to 0.050 in.	1	2	N.R.
Stainless Steel 0.050 in. and up	2	1	N.R.
Brass Alloys	2	1	N.R.
Silver	2	1	N.R.
Hastelloy Alloys	2	1	N.R.
Silver Cladding	1	N.R.	N.R.
Hard–Facing	1	2	N.R.
Cast Iron	2	1	N.R.
Low Carbon Steel 0.015 in. to 0.030 in.	2	1	N.R.
Low Carbon Steel 0.030 in. to 0.125 in.	N.R.	1	N.R.
High Carbon Steel 0.015 in. to 0.030 in.	2	1	N.R.
High Carbon Steel 0.030 in. and up	2	1	N.R.
Deoxidized Copper up to 0.090 in.	N.R.	1	N.R.

Key: 1. Excellent Operation — best recommendation.

2. Good Operation — second recommendation.

N.R. Not recommended.

reverse polarity is used with this amount of current the tip of the electrode would melt off. Consequently a 1/4" diameter electrode would be required to handle 125 amperes of welding current.

Polarity also affects the shape of the weld. DCSP produces a narrow deep weld whereas DCRP with its larger diameter electrode and lower current forms a wide and shallow weld. See Fig. 3–5. For this reason DCRP is rarely used in gas tungsten-arc welding except for welding aluminum and magnesium. These metals have a heavy oxide coating which is more readily removed by the greater current cleaning action of DCRP. The same cleaning action is present in the reverse-polarity half of the A-C welding cycle. No other metals actually require the kind of cleaning action that is normally needed on aluminum and magnesium. The cleaning action develops because of a bombardment of

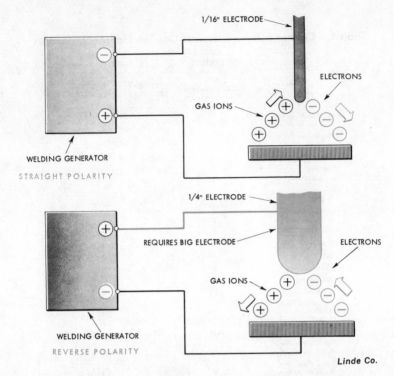

Fig. 3-4. Straight and reverse polarity in direct current welding.

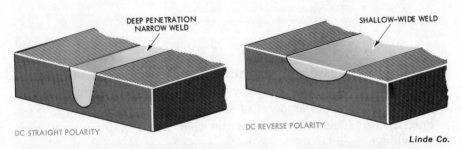

Fig. 3-5. Effect of polarity on weld shape.

positive charged gas ions that are attracted to the negative charged workpiece. These gas ions when striking the metal have sufficient power to break the oxide and dislodge it from the surface.

Direct Current Straight Polarity (DCSP). Direct current straight polarity is used for welding most metals because better welds are achieved. With the heat concentrated at the plate the welding process

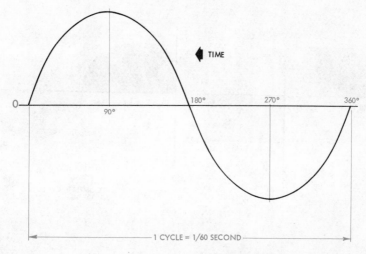

Miller Electric Manufacturing Co.
Fig. 3-6. Flow of AC current.

is more rapid, there is less distortion of the base metal, and the weld puddle is deeper and narrower than with DCRP. Since more heat is liberated at the puddle smaller diameter electrodes can be used.

Alternating Current. A characteristic of alternating current is that the current flows first in one direction and then in the other. A complete change of flow is referred to as a cycle. As shown in Fig. 3–6, the AC sine curve forms 360°. The curve starts at 0, and ascends to its maximum height at 90°, creating the first quarter cycle. The curve next decreases from maximum to 0 completing a half cycle. At this point the current alternates direction. The third quarter builds up to maximum strength in the opposite direction and finally it decreases again to 0 completing the full 360 cycle. The cycling process is repeated as long as the current is flowing.

AC welding is actually a combination of DCSP and DCRP. Notice in Fig. 3–7A that half of each complete AC cycle is DCSP and the other half is DCRP. Unfortunately oxides, scale and moisture on the workpiece often tend to prevent the full flow of current in the reverse-polarity direction. If no current whatsoever flowed in the reverse-polarity direction the current wave would resemble something as shown in Fig. 3–7B. During a welding operation the partial or complete stoppage of current flow (rectification) would cause the arc to be unstable and sometimes even go out. To prevent such rectification AC welding

Gas Tungsten-Arc Welding 51

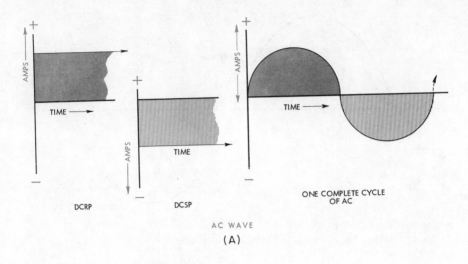

DCRP   DCSP

ONE COMPLETE CYCLE
OF AC

AC WAVE

(A)

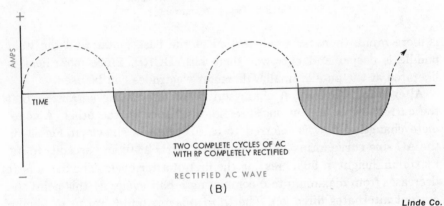

TWO COMPLETE CYCLES OF AC
WITH RP COMPLETELY RECTIFIED

RECTIFIED AC WAVE

(B)

Linde Co.

Fig. 3-7. Characteristic of an AC wave.

machines incorporate a high frequency current flow unit. The high-frequency current is able to jump the gap between the electrode and the workpiece piercing the oxide film and forming a path for the welding current to follow.

A comparison of weld contours obtained with high-frequency stabilized AC current is illustrated in Fig. 3–8.

Torches. Manually operated welding torches are constructed to conduct both the welding current and the inert gas to the weld zone. See Fig. 3–9A-E. These torches are either air- or water-cooled. Air-cooled torches are designed for welding light gage materials where low current

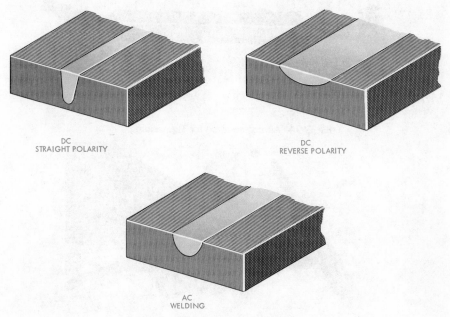

DC
STRAIGHT POLARITY

DC
REVERSE POLARITY

AC
WELDING

Fig. 3-8. Comparison of weld contours.

values are used. Water-cooled torches are recommended when the welding requires amperages over 200. A circulating stream of water flows around the torch to keep it from overheating. The tungsten electrode which supplies the welding current is held rigidly in the torch by means of a collet that screws into the body of the torch. A variety of collet sizes are available so different diameter electrodes can be used. Gas is fed to the weld zone through a nozzle which consists of a ceramic cup. Gas cups are threaded into the torch head to provide directional and distributional control of the shielding gas. The cups are interchangeable to accommodate a variety of gas flow rates. Some torches are equipped with a "gas lens" to eliminate turbulence of the gas stream which tends to pull in air and cause weld contamination. Gas lenses have a permeable barrier of concentric fine-mesh stainless steel screens that fit into the nozzle. See Fig. 3–9C-E.

Pressing a control switch on the torch starts the flow of both current and gas. On some equipment the flow of current and gas is energized

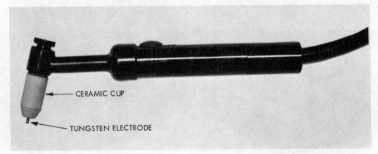

Fig. 3-9A. Air-cooled torch for Tig welding.

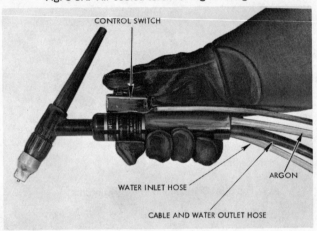

Fig. 3-9B. Water-cooled torch for Tig welding.

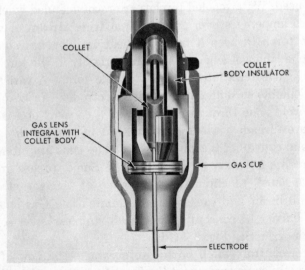

Fig. 3-9C. Section of torch showing gas lens.

Fig. 3-9D. Gas stream of conventional Tig torch.

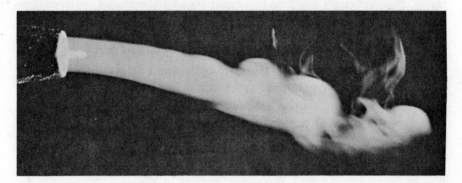

Fig. 3-9E. Gas stream of Tig torch with gas lens.

by a foot control. The advantage of the foot control is that the current flow can be better controlled as the end of the weld is reached. By gradually decreasing the current it is less likely for a cavity to remain in the end of the weld puddle and less danger of cutting short the shielding gas.

Types of Electrodes. Basic diameters of non-consumable electrodes are $\frac{1}{16}''$, $\frac{3}{32}''$, and $\frac{1}{8}''$. They are either pure tungsten or alloyed tungsten. The alloyed tungsten usually have one to two percent thorium or zirconium. The addition of thorium increases the current capacity and electron emission, keeps the tip cooler at a given level of current, minimizes movement of the arc around the electrode tip, permits easier arc starting, and the electrode is not as easily contaminated by accidental contact with the workpiece. The two percent thoria electrodes normally maintain their formed point for a greater period than the one percent type. The higher thoria electrodes are used primarily for critical sheet metal weldments in aircraft and missile industries. They have little advantage over the lower thoria electrode for most steel welds. The introduction of the "striped" electrode combines the advantage of the pure, low, and high thoriated tungsten electrodes. This electrode has a solid stripe of two percent thoria inserted in a wedge the full length of the electrode.

The diameter of the electrode selected for a welding operation is governed by the welding current to be used. Larger diameter tungsten electrodes are required with reversed polarity than with straight polarity. See Appendix for recommended sizes of electrodes, current, and material thickness for Tig welding.

Gas Regulators. Gas flow is regulated by a single or dual stage pressure reducing regulator and flow meter. See Fig. 3–10. The flowmeter is calibrated in cubic feet per hour (cfh) or liters per minute (lpm). The correct gas flow to the torch is set by turning the adjusting screw on the regulator. The rate of flow required depends on the kind and thickness of the metal to be welded. See Appendix.

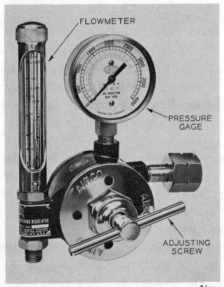

Airco

Fig. 3-10. An inert-gas regulator with flowmeter.

Shielding Gas

Shielding gas for gas tungsten-arc welding can be argon, helium, or a mixture of argon and helium. Argon is used more extensively because it is less expensive than helium. Argon is 1.4 times as heavy as air and 10 times as heavy as helium. There is very little difference between the viscosity of these two gases. Since argon is heavier than air it provides a better blanket over the weld. Moreover, there is less clouding during the welding process with argon and consequently it permits better control of the weld puddle and arc.

Argon normally produces a better cleaning action especially in welding aluminum and magnesium with alternating current. With argon there is a smoother and quieter arc action. The lower arc voltage characteristic of argon is particularly advantageous in welding thin material because there is less tendency for burning through the metal. Consequently, argon is used most generally for shielding purposes when welding materials up to $\frac{1}{8}''$ in thickness both for manual welding and low-speed machine welding.

The use of argon also permits better control of the arc in vertical and overhead welding. As a rule, the arc is easier to start in argon than in helium and for a given welding speed the weld produced is narrower with a smaller heat-affected zone.

Where welding speed is important especially in machine welding or in welding heavy materials and metals having high heat conductivity helium is sometimes used. Higher welding speeds are possible with helium because higher-arc voltage can be obtained at the same current. Since the arc voltage in helium is higher, a lower current is possible to get the same arc power. Hence, welds can be made at higher speeds inasmuch as the increase in power comes from the increase in voltage rather than in current. A mixture of argon and helium is often used in welding metals that require a higher heat input.

Welding Technique

A weld is made by applying the arc heat on the abutting edges of the workpiece. Before the arc is generated the electrode must be adjusted so that it extends about $\frac{1}{8}''$ to $\frac{3}{16}''$ beyond the end of the gas cup for butt welding and approximately $\frac{1}{4}''$ to $\frac{3}{8}''$ for fillet welding. See Fig. 3–11.

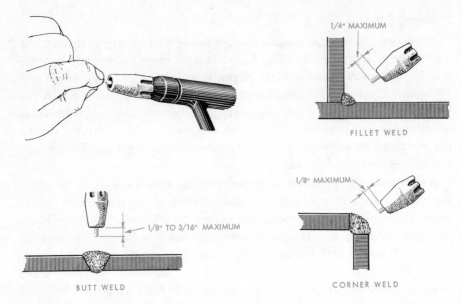

1/4" MAXIMUM

FILLET WELD

1/8" MAXIMUM

1/8" TO 3/16" MAXIMUM

BUTT WELD

CORNER WELD

Fig. 3-11. Adjust the electrode so it extends beyond the edge of the gas cup.

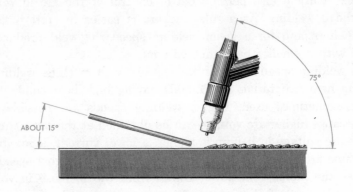

75°

ABOUT 15°

Fig. 3-12. Position of the torch for Tig welding.

The arc may be struck in two ways: with a DC machine the electrode is brought downward to touch the plate. As soon as the arc is struck, the electrode is raised so it is about ⅛″ above the workpiece. On an AC machine the electrode does not have to touch the metal to start the arc. The electrode is simply held about ⅛″ above the metal and the high frequency current will jump the gap as soon as the power

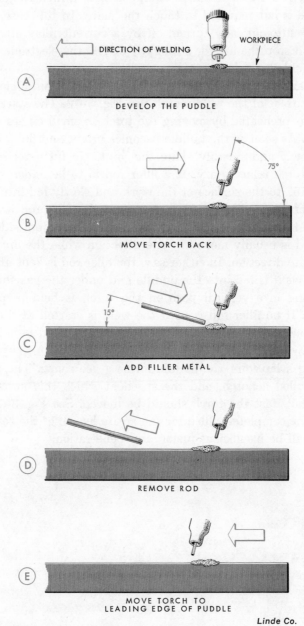

DIRECTION OF WELDING

WORKPIECE

(A)

DEVELOP THE PUDDLE

75°

(B)

MOVE TORCH BACK

15°

(C)

ADD FILLER METAL

(D)

REMOVE ROD

(E)

MOVE TORCH TO
LEADING EDGE OF PUDDLE

Linde Co.

Fig. 3-13. Forming a molten puddle with a Tig torch and performing the weld in a flat position.

is turned on. If a DC machine is equipped with a high frequency unit the electrode is not required to touch the plate. In this case, the high frequency is automatically turned off by a current relay after the arc is started. To stop the arc for either AC or DC the electrode is merely snapped quickly to a horizontal position.

Once the arc is initiated the torch is positioned at an angle of about 75° to the surface of the weld metal. See Fig. 3–12. The starting point of the weld is preheated by moving the torch in small circles as shown in Fig. 3–13. As soon as the puddle becomes bright and fluid, the torch is moved slowly and steadily along the joint. No further circular motion of the torch is necessary. If a filler rod is to be added the rod is held about 15° to the surface of the work and slowly fed into the weld puddle. When large weld beads are required the filler rod is often fed into the weld puddle by oscillating the rod and arc from side to side. The filler rod is usually moved in one direction while the arc is moved in the opposite direction. In either case the filler rod is kept at all times near the arc as it is fed into the puddle and under the gas shield.

For welding in a vertical position the torch is held perpendicular to the work. If no filler rod is used the weld is started at the top and moved downward. When filler rod is required it should be added from the bottom or leading edge of the puddle. See Fig. 3–14.

In welding a lap joint the puddle forms a vee shape. The center of the vee is called a notch, and the speed at which this notch travels determines how fast the torch should be moved. See Fig. 3–15A. The notch must be completely filled for the entire length of the seam otherwise there will be insufficient fusion and penetration.

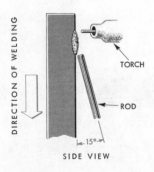

TORCH

ROD

15°

DIRECTION OF WELDING

SIDE VIEW

Fig. 3-14. Vertical Tig welding.

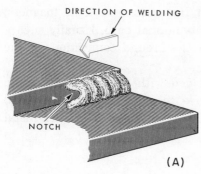

DIRECTION OF WELDING

NOTCH

(A)

Fig. 3-15. Welding a lap joint with the Tig process.

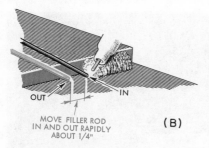

OUT

IN

MOVE FILLER ROD
IN AND OUT RAPIDLY
ABOUT 1/4"

(B)

If a filler rod is to be used on the lap joint the end of the rod is dipped in and out of the puddle about every ¼" travel of the puddle as shown in Fig. 3–15B. Care must be taken to avoid laying bits of filler rod on the cold, unfused base metal. The addition of the right amount of rod at the correct moment will produce a uniform bead of the proper proportions.

During the welding process there may be occasions when the arc has a tendency to wander. This may be due to (1) low electrode current density, (2) contamination of the electrode, (3) magnetic effects, and (4) air drafts. The first two causes are distinguished by a rapid movement of the arc from side to side. Magnetic effects will usually displace the arc to one side or the other along the entire length of the weld. The fourth cause will make the arc wander in various ways depending on the amount of air draft present. These difficulties may be remedied by first insuring that the correct current is used. Contamination of the electrode often results when the tip of the electrode touches the work or filler rod. The electrode must then be cleaned by grinding or using a new one. Magnetic effects are the results of the magnetic field established by the current flowing through the workpiece. This magnetic field will attract or repel the arc from its normal path. Usually such a condition can be overcome by shifting the ground connection on the workpiece.

All standard type joints may be welded with the gas tungsten-arc process, such as square groove, vee groove, tee, and lap. Edges of metal that are ⅛″ or less in thickness need not be beveled. Heavier metal usually requires beveling.

Linde Co.

Fig. 3-16. Automatic Tig welding unit.

Automatic Gas Tungsten-Arc Welding

Much of the gas tungsten-arc welding is done manually. However, machine welding is frequently used where standard production welding is feasible and practical. Automatic or machine welding does involve a greater investment of equipment and requires closer control of joint tolerances. On the other hand labor costs are reduced since fewer operators are needed and welding speeds increased.

Mechanized gas tungsten-arc welding equipment usually necessitates a 100 percent duty cycle power supply. The remaining equipment is relatively simple. The torch is either mounted on a carriage which moves over the workpiece or the torch is stationary and the part to be welded moves under the torch. See Fig. 3–16.

TIG SPOT WELDING

For many years spot welding was confined to the conventional process commonly referred to as resistance spot welding. See Chapter V. With this method the material to be joined is placed between two copper electrodes which are brought together and when the current is turned on sufficient heat is liberated to fuse the pieces together. The limitation of the process is that pressure must be applied on both sides of the pieces and the work has to be of a size so it can be conveniently fed into the spot welding machine.

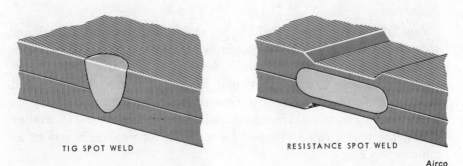

TIG SPOT WELD   RESISTANCE SPOT WELD

Airco

Fig. 3-17. A comparison of welds made by Tig spot welding and conventional resistance welding.

Fig. 3-18. Spot welds made with the Tig process.

The development of Tig spot welding now makes it possible to produce localized fusion similar to resistance spot welding without requiring accessibility to both sides of the joint. A special tungsten-arc gun is applied to one side of the joint only. Heat is generated from resistance of the work to the flow of electrical current in a circuit of which the work is a part. A comparison of weld cross sections made by resistance spot welding and Tig spot welding is shown in Fig. 3–17.

The Tig spot welding process has a wide range of applications in fabricating sheet metal products involving joints which are impractical to resistance spot welding because of the location of the weld or the size of parts or where welding can be made only from one side. See Fig. 3–18.

Equipment

Any DC power supply providing up to 250 amperes with a minimum open circuit voltage of 55 can be adapted for spot welding. The gun has a nozzle with a tungsten electrode. See Fig. 3–19. Various shape nozzles are available to meet particular job requirements. The standard nozzle can also be machined to permit access in tight corners or its diameter reduced to weld items such as small holding clips. As a matter of fact the nozzle can be shaped for a variety of welding functions as indicated in Fig. 3–20.

For most operations a ⅛″ diameter electrode is used. The end of the electrode should normally be of the same diameter as the electrode.

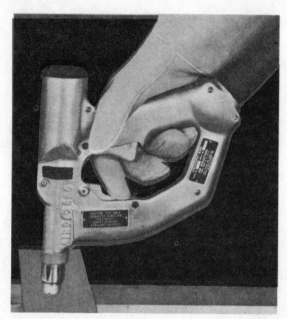

Fig. 3-19. Tig spot welding gun.

Airco

However, when working at low amperage settings (100 amperes or less) better results will be obtained if the end of the electrode is tapered slightly to provide a blunt point approximately one half the diameter of the electrode. This will prevent the arc from "wandering."

Whenever the end of the electrode "balls" excessively after only a few welds have been made, it is usually an indication of excessive amperage, dirty material, or insufficient shielding gas.

Making a Weld

To make a spot weld the end of the gun is placed against the work and the trigger is pulled. Squeezing the trigger starts the flow of cooling water and shielding gas and also advances the electrode to touch the work. The electrode automatically retracts establishing an arc which is extinguished at the end of a preset length of time. The electrode is usually set at the factory to provide an arc length of $\frac{1}{16}''$ which has been found to be generally satisfactory for practically all welding applications.

Gas Tungsten-Arc Welding 65

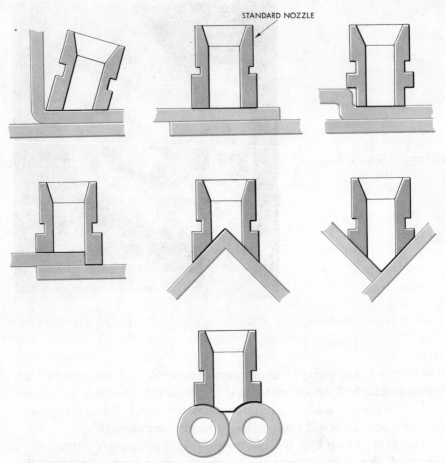

Fig. 3-20. Tig spot welding gun nozzle can be shaped for a variety of welding jobs.

Amperage

The amperage required for a weld will naturally be governed by the thickness of the metal to be welded. The major effect of increasing the amperage, when both pieces are approximately the same thickness, is to increase the penetration. However, in doing so it also tends to increase the weld diameter somewhat as shown by the dotted line in Fig. 3–21A. Increasing the amperage when the bottom part is considerably heavier

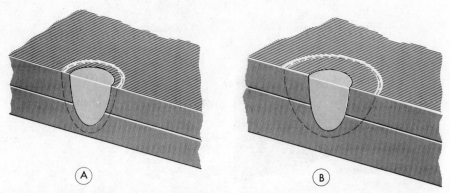

Fig. 3-21. Increasing the amperage will affect weld penetration and weld diameter.

than the top part will result in an increase in weld diameter with little or no increase in penetration as shown in Fig. 3–21B.

Weld Time

Weld time is set on the dial in the control cabinet. The dial is calibrated in 60ths of a second and is adjustable from 0 to 6 seconds. The effect of increasing the weld time is to increase the weld diameter. But in so doing it also increases the penetration somewhat as shown by the dotted lines in Fig. 3–22.

Fig. 3-22. Increasing the weld time increases the weld diameter and penetration.

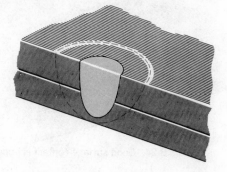

Shielding Gas

Helium will produce greater penetration than argon although argon will produce a larger weld diameter. See Fig. 3–23. Gas flow should be set at approximately 6 cfh.

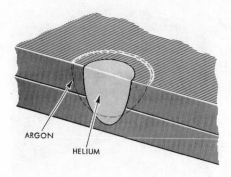

Fig. 3-23. The effects of shielding gas in making a spot weld.

ARGON

HELIUM

Surface Condition and Surface Contact

Mill scale, oil, grease, dirt, paint, and other foreign materials on or between the contacting surfaces will prevent good contact and reduce the weld strength. The space between the two contacting surfaces resulting from these surface conditions or poor fit-up acts as a barrier to heat transfer and prevents the weld from breaking through into the bottom piece. Consequently, good surface contact is important for sound welds. See Fig. 3-24.

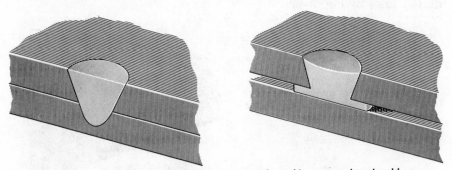

Fig. 3-24. Good surface contact is important in making a sound spot weld.

Backing

Although Tig welding can be done from one side only it is obvious that the bottom part must have sufficient rigidity to permit the two parts to be brought into contact with pressure applied by the gun. If the thickness, size or shape of the bottom part is such that it does not provide this rigidity then some form of backing support or jigging will be required. Backing may be either of steel or copper.

REVIEW QUESTIONS

1. What are some of the specific features of the Tig welding process?
2. What effects does polarity have in welding with the Tig process?
3. Why are AC machines for Tig welding designed to produce a high frequency current?
4. Why is an AC current preferred for welding some metals with the Tig process?
5. What factor determines whether a torch should be air or water cooled for Tig welding?
6. When is a filler rod used in Tig welding?
7. Why are alloyed tungsten electrodes often used instead of pure tungsten electrodes?
8. How does helium and argon affect the characteristics of a weld?
9. How is the arc started and stopped in Tig welding?
10. What are some of the probable reasons that may cause an arc to wander during the welding process?
11. What determines the size of the electrode to be used for welding?
12. How far should the electrode extend beyond the end of the gas cup on the torch?
13. How does Tig spot welding differ from the conventional resistance spot welding?
14. Increasing the amperage and weld time in Tig spot welding will have what effects on the resulting weld spot?
15. What effect will the shielding gas have in making a spot weld?

Chapter 4

Gas Metal-Arc Welding

Gas metal-arc welding is sometimes referred to as Mig or metallic-inert gas welding. It is also called by manufacturers' trade names such as Aircomatic (Airco), and Sigma (Linde). In this process coalescence is achieved by striking an electric arc between the workpiece and a continuous consumable wire electrode which is fed through a torch at controlled speeds. A shielding gas flows through the torch and forms a blanket over the weld puddle to protect it from atmospheric contamination. The welding can be semi-automatic or fully mechanized. With semi-automatic, the operator concerns himself only with torch-to-work distance, torch manipulation, and welding speed. Wire-feed rates, electrical settings, and gas flow are preset. When the equipment is completely mechanized all of these variables and welding functions are performed automatically. See Fig. 4–1.

Fig. 4-1. Fabricating sprocket wheels by automated Mig welding.

There are several variations of the basic Mig process, each best suited for a particular purpose. Spray arc, short arc, cored-wire, and CO_2 welding are all, basically, Mig processes.

Application

Originally, gas metal-arc welding was considered a high current density, small diameter filler wire process. Since then other developments have made this technique equally practical for welding at lower current densities. Hence gas metal-arc welding is now just as effective for welding light gage materials as for welding heavy structural plates. Because of the versatility of the Mig process it has found wide applications in practically every industry—automotive, construction, electrical, aerospace, container, piping, transportation, etc. See Fig. 4–2. Most metals can be welded with comparative ease including aluminum, carbon steels, low alloy steels, stainless steels, nickel, copper, magnesium, titanium, and zirconium.

The outstanding feature of gas metal-arc welding is the production of

Fig. 4-2. Fabrication of heat transfer surfaces.

high quality welds at high welding speeds without having to use flux of any kind or the necessity for any slag removal. The increased welding speed of Mig over Tig in heavier materials is shown by the curves in Fig. 4–3. Notice that as the material thickness increases the effective speed of Tig welding decreases.

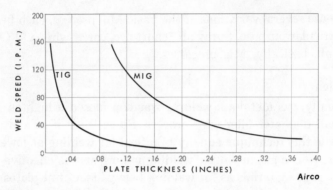

Fig. 4-3. Comparison of typical welding speeds for Tig and Mig welding of square butt joints in a flat position.

Manual Mig Welding

Manual Mig welding is actually semi-automatic. Welding current and wire feed speed are electrically interlocked so that the welding arc is self-correcting. The operator first sets the equipment to give him the correct length of welding arc for the job at hand. During welding, if he holds the torch too close to the work, the current automatically increases and the wire is burned off faster than it is fed until the correct arc length is re-established. If he raises the torch too high over the work, the current automatically decreases and the wire feeds faster than it is burned off, shortening the arc to the correct length. This self-regulating feature of Mig welding equipment enables the operator to concentrate his attention on the weld.

Welding Current

Different welding currents have a profound effect on the results obtained in gas metal-arc welding. Optimum efficiency is achieved with direct current reverse polarity (DCRP). See Fig. 4–4. The heat in this instance is concentrated at the weld puddle and therefore provides deeper penetration at the weld. Furthermore, with DCRP, there is greater surface cleaning action which is important in welding metals having heavy surface oxides such as aluminum and magnesium.

Straight polarity (DCSP) is very impractical with Mig welding because weld penetration is wide and shallow, spatter is excessive, and there is no surface cleaning action. The ineffectiveness of straight polarity largely results from the pattern of metal transfer from the electrode

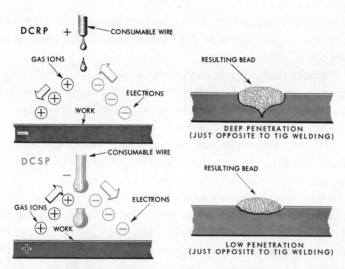

DCRP + CONSUMABLE WIRE

GAS IONS

ELECTRONS

WORK

RESULTING BEAD

DEEP PENETRATION
(JUST OPPOSITE TO TIG WELDING)

CONSUMABLE WIRE

DCSP

GAS IONS

ELECTRONS

WORK

RESULTING BEAD

LOW PENETRATION
(JUST OPPOSITE TO TIG WELDING)

Fig. 4-4. Effects of polarity in Mig welding.

to the weld puddle. Whereas in reverse polarity the transfer is in the form of a fine spray, with straight polarity the transfer is largely of the erratic globular type. The use of AC current is never recommended since the burn-offs are unequal on each half-cycle.

Spray Arc Welding

This is a high heat method with rapid deposition of weld metal. It is for welding all common metals from about $\frac{3}{32}''$ to over $1''$ in thickness. "Spray arc" describes the way in which molten metal is transferred to the work. The metal transfer is a spray of very fine droplets. The transition point at which the current level causes molten metal to spray is determined by the type of welding current, type and size of wire, and choice of shielding gas mixture.

Types of Metal Transfer

Three types of metal transfer are associated with this form of welding: spray, globular, and short arc. The manner in which metal transfer occurs depends on electrode wire size, shielding gas, arc voltage, and welding current.

Spray Transfer. In spray transfer very fine droplets or particles of the electrode metal are rapidly projected through the arc from the end of the electrode to the workpiece in the direction in which the electrode is pointed. While in the process of transfering through the welding arc the metal particles do not interrupt the flow of current and there is virtually a constant spray of metal.

Spray transfer requires a high current density. With high current the arc becomes a steady quiet column with a well defined narrow incandescent cone-shape core within which metal transfer takes place. See A, Fig. 4–5. The use of argon or a mixture of argon and oxygen is necessary for spray transfer. Argon produces a pinching effect on the molten tip of the electrode permitting only small droplets to form and transfer during the welding process.

With high heat input, heavy wire electrodes will melt readily and deep weld penetration becomes possible. Since the individual drops are small the arc is stable and can be directed where required. The fact that the metal transfer is produced by an axial force which is stronger than gravity makes spray transfer effective for out-of-position welding. It is particularly adapted for welding heavy gage metals. It is not too practical for welding light gage materials because of the resulting burn through.

Globular Transfer. This type of transfer occurs when the welding current is low or below what is known as the transition current. The transition range extends from the minimum value where the heat melts the electrode to the point where the high current value induces spray transfer. Notice in Fig. 4–6 only a few drops are transferred per second at low current values, while many small drops are transferred at high current values.

In globular transfer the molten ball at the tip of the electrode tends to grow in size until its diameter is two or three times the diameter of the wire before it separates from the electrode and transfers across the arc to the workpiece. See B, Fig. 4–5. As the globule moves across

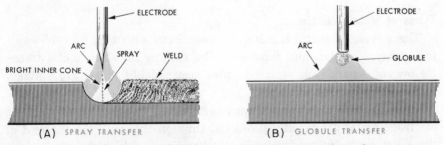

Fig. 4-5. Characteristics of spray and globule types of transfer.

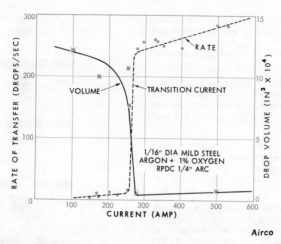

Fig. 4-6. Effects of current on the size and frequency of drops transferred in an arc shielded by inert gas.

the arc it assumes an irregular shape and rotary motion because of the physical forces of the arc. This frequently causes the globule to reconnect with the electrode and workpiece causing the arc to go out and then re-ignite. The result is poor arc stability, shallow penetration, and excessive spatter. For this reason a globular transfer is avoided in welding except in the buried arc process which will be described later.

Short Arc Welding (Short Circuiting Transfer). This is a reduced heat technique, with pinpoint arc, for all common metals. Short arc welding was developed specifically for use with thin-gage metals to eliminate the problems of distortion, burnthrough, and spatter experienced by stick electrode. Its relatively low heat input has since extended the use of the short arc technique to the welding of heavy thicknesses of quenched and tempered steels of the 100,000 psi yield strength class. See illustrations in Fig. 4–7, A-E. It is generally considered to be the most practical at current levels below 200 amperes with fine wires of .045" or less in diameter. The use of fine wire produces weld pools that remain relatively small and are easily managed, making all-position welding possible.

As the molten wire is transferred to the weld each drop touches the weld puddle before it has broken away from the advancing electrode wire. The circuit is "shorted," and the arc is extinguished.

Electromagnetic "pinch force" squeezes the drop from the wire. The short circuit is broken and the arc re-ignites. "Shorting" occurs from 20

Start of the short arc cycle. High temperature electric arc melts advancing wire electrode into a drop of liquid metal. Wire is fed mechanically through the welding torch. Arc heat is regulated by the power supply. (A)

Molten electrode moves toward workpiece. Note cleaning action. Argon gas mixture, developed for short arc, shields molten wire and seam, insuring regular arc ignition, preventing spatter, and weld contamination. (B)

Electrode makes contact with workpiece, creating short circuit. Arc is extinguished, allowing to cool. Frequency of arc extinction in short arc varies from 20 to 200 times per second, according to job requirements. (C)

Drop of molten wire breaks contact with electrode, causing arc to reignite. Electrode is broken by pinch force, a squeezing power common to all current carriers. Amount and suddenness of pinch is controlled by power supply. (D)

With arc renewed, short arc cycle begins again. Because of precision control of arc characteristics and cool, uniform operation, short arc produces perfect welds on metals as thin as .030-In. carbon or stainless steel. (E)

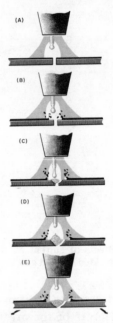

Fig. 4-7. Typical mode of metal transfer in the short arc.

to 200 times a second according to preset controls. "Shorting" of the arc pinpoints the effective heat. The result is a small, relatively cool weld puddle which reduces burn-through. Intricate welds are possible in most all positions.

In short-arc welding the shielding gas mixture consists of 25 percent carbon dioxide, which provides increased heat for higher speeds, and 75 percent argon which controls spatter. However, considerable usage is now being made of straight CO_2 or a mixture of CO_2 and argon for welding carbon and low alloy steels.

Mig Welding Equipment

Gas metal-arc welding equipment consists of four major units: power supply, wire feeding mechanism, welding gun, and gas supply. See Figs. 4–8 and 4–9.

Power Supply. The recommended machine for Mig welding is a rectifier or motor generator supplying direct current with normal limits of 200-250 amperes for all-position welding. Direct current reverse polarity

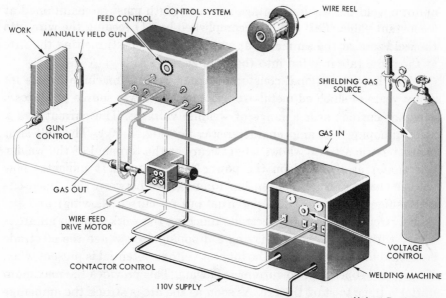

Fig. 4-8. Schematic layout of a semi-automatic Mig welder.

Hobart Brothers Co.

Linde Co.

Fig. 4-9. A fully automatic Mig welder.

(DCRP) is used for greatest efficiency since DCRP contributes to better melting, deeper penetration, and better cleaning action.

Constant Current versus Constant Potential Power Supply. In Mig welding heat is generated by the flow of current through the gap between the end of the wire electrode and the workpiece. A voltage forms across this gap which varies with the length of the arc. To produce a uniform weld the welding voltage and arc length must be maintained at a constant value. This can be accomplished by (1) feeding the wire into the weld zone at the same rate at which it melts, or (2) melting the wire at the same rate it is fed into the weld zone.

With the conventional constant current welding machine in use for many years in shielded metal-arc welding, the power source produces a constant current over a range of welding voltages. The current has a steep drooping volt-amp characteristic. See Fig. 4–10A. The volt-amp characteristic actually shows what occurs at the terminal of the welder (electrode) as the load on the power source varies. It indicates how voltage changes in its relationship to amperage between the "open circuit" stage (static electrical potential but no current flowing) and the "short circuit" condition (electrode touching the work). When an arc is struck with a power source having a drooping arc voltage the electrode is shorted to the workpiece. The highest voltage potential is present when the circuit is open and no current is flowing. This provides the maximum initial voltage to start the arc. As soon as the arc is struck the amperage shoots up to maximum and the voltage drops to minimum. Then as the electrode is moved the voltage rises to maintain the arc and the amperage drops to its normal working level.

During welding the voltage automatically varies directly and amperage inversely with the length of the welding arc. Consequently, the operator can keep reasonable control over the heat input to the work.

When a conventional power source is used for Mig welding the wire

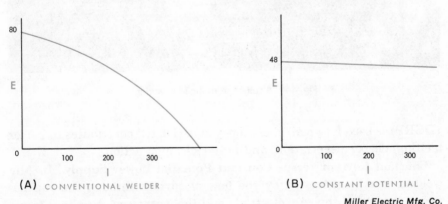

Miller Electric Mfg. Co.

Fig. 4-10. Comparison of volt-amp curves for a conventional constant current welder and a constant potential welder.

feed speed must be adjusted to narrow limits to prevent the wire from burning back to the nozzle or plunging into the weld plate. Although the operator can, by means of electronic speed controls, adjust the wire speed for a predetermined arc length, nevertheless whenever the nozzle to work distance changes the arc length (voltage) changes. Thus, if the nozzle to work distance increases the arc length increases. The result is a non-uniform weld.

With the need for better arc control the *constant voltage (potential) power supply* was developed. The constant potential welding power supply has a nearly flat volt-ampere characteristic. See Fig. 4–10B. This means that the preset voltage level can be held throughout its range. Although its static voltage potential at open circuit is lower than a machine with drooping characteristic, it maintains approximately the same voltage regardless of the amount of current drawn. Accordingly there is unlimited amperage to melt the consumable wire electrode. The power supply becomes self-correcting with respect to arc length. The operator can change the wire feed speed over a considerable range without affecting the stubbing or burning back the wire. In other words, the arc length can be set on the power supply and any variations in nozzle to work distance will not produce changes in the arc length. For example, if the arc length becomes shorter than the pre-selected value there is an automatic increase of current and the wire speed automatically adjusts itself to maintain a constant arc length. Similarly, if the arc becomes too long the current decreases and the wire begins to feed faster.

Stated in another way, when the wire is fed into the arc at a specific rate a proportionate amount of current is automatically drawn. The constant potential welder therefore provides the necessary current required by the load imposed on it. When the electrode wire is fed faster the current increases; if it is fed slower the current decreases.

Because of this self-correcting feature less operator skill is necessary to achieve good welds. There are only two basic controls: a rheostat on the welding machine to regulate the voltage and a rheostat on the wire feed mechanism to control the speed of the wire feed motor.

Slope Control. Some power supply units designed for Mig welding have provisions for controlling the slope. The incorporation of slope control gives the machine greater versatility. Thus, by altering the flat shape of the slope it is possible to control the pinch force on the consumable wire which is particularly important in the short circuiting transfer method of welding. With better control of the short circuit, the weld puddle can be kept more fluid with better resulting welds. Slope control also helps to decrease the sudden current surge when the electrode makes its initial contact with the workpiece. By slowing down the rate of current rise the amount of spatter can be reduced.

Wire Feeding Mechanism. The wire feeding mechanism automatically drives the electrode wire from the wire spool to the gun and arc. See Fig. 4–11. Control on the panel can be adjusted to vary the wire feeding speed. In addition, the control panel usually includes a welding power contactor and a solenoid to energize the gas flow. On units designed for welding with a water-cooled gun a control is also available to turn on and shut off the water flow.

The wire feeder can be mounted on the power supply machine or it can be separate from the welding machine and mounted elsewhere to facilitate welding over a large area.

Welding Gun. The function of the welding gun is to deliver the wire, shielding gas, and welding current to the arc area. The manually operated gun is either water- or air-cooled. An air-cooled gun is especially designed to weld light gage metals that require less than 200 amperes

Miller Electric Mfg. Co.

Fig. 4-11. Typical wire feeding unit for Mig welding.

with argon as a shielding gas. However, such a torch can usually function at higher amperage (300) with CO_2 because of the cooling effects of this gas. A water-cooled gun is best when welding with currents that are higher than 200 amperes.

Guns are either of the push or pull type. The pull gun uses drive rolls in the gun that pulls the welding wire from the wire feeder, and the push gun has the wire pushed to it by drive rolls in the wire feeder itself. The pull gun handles small diameter wires; the push gun moves heavier diameter wires. The pull type is also used to weld with soft wires such as aluminum and magnesium while the push gun is considered more suitable for welding with hard wires such as carbon and stainless steels and where currents are often in excess of 250 amperes. Both guns have a trigger switch that controls the wire feed and arc as well as the shielding gas and water flow. When the trigger is released the wire feed, arc, shielding gas, and water, if a water-cooled torch is used, stop immediately. With some equipment a timer is included to permit the shielding gas to flow for a predetermined time to protect the weld until it solidifies.

Guns are available with a straight or a curved nozzle. See Fig. 4–12. Curved nozzles provide easy access to intricate joints and patterns.

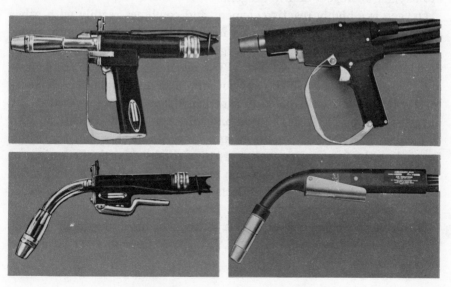

Fig. 4-12. Types of guns used for Mig welding.

Shielding Gas[1]

In any gas shielded-arc welding process the shielding gas can have an appreciable effect upon the properties of a weld deposit. Therefore, welding is done in a controlled atmosphere. In shielded metal-arc welding this is accomplished by placing a coating on the electrode which produces a non-harmful atmosphere when it disintegrates in the welding arc. In the case of Mig welding the same effect is accomplished by surrounding the arc area with gas supplied from an external source. See Table I.

The air in the arc area is displaced by the shielding gas. The arc is then struck under the blanket of shielding gas and the welding is accomplished. Since the molten weld metal is exposed only to the shielding gas it is not contaminated and strong dense weld deposits are obtained. The reason for shielding the arc area is to prohibit air from coming in contact with the molten metal.

By volume, air is made up of 21 percent oxygen, 78 percent nitrogen, 0.9 percent argon, and 0.04 percent other gases (primarily carbon dioxide). The atmosphere will also contain a certain amount of water depending upon its humidity. Of all of the elements that are in the air, the three which cause the most difficulty as far as welding is concerned are oxygen, nitrogen, and hydrogen.

Oxygen is a highly reactive element and combines readily with other elements in the metal or alloy to form undesirable oxides and gases. The oxide-forming aspect of the oxygen can be overcome with the use of deoxidizers in the steel weld metal. The deoxidizers, such as manganese and silicon, combine with the oxygen and form a light slag which floats to the top of the weld pool. If the deoxidizers are not provided the oxygen will combine with the iron and form compounds which can lead to inclusions in the weld material and lower its mechanical properties. On cooling the free oxygen in the arc area combines with the carbon of the alloy material and forms carbon monoxide. If this gas is trapped in the weld metal as it cools it collects in pockets which cause pores or hollow spaces in the weld deposit.

Of all of the elements in the air nitrogen causes the most serious problems in welding steel materials. When iron is molten, it is able to

---

[1]Courtesy Hobart Brothers Co.

Table I.   Shielding Gas for M. Welding.

MATERIAL	PREFERRED GAS	REMARKS
Aluminum Alloys	1. Argon	With DC Reverse Polarity removes oxide surface on work piece.
Magnesium Aluminum Alloys	1. 75% He. 25% A.	Greater heat input reduces porosity tendencies. Also cleans oxide surface.
Stainless Steels	1. Argon + 1% O_2	Oxygen eliminates under-cutting when DC Reverse Polarity is used.
	2. Argon + 5% O_2	When DC straight polarity is used 5% O_2 improves arc stability.
Magnesium	1. Argon	With DC straight polarity removes oxide surface on work piece.
Copper (Deoxidized)	1. 75% He. 25% A.	Good wetting and increased heat input to counteract high thermal conductivity. Light gages.
	2. Argon	
Low Carbon Steel	1. Argon + 2% O_2	Oxygen eliminates under-cutting tendencies also removes oxidation.
Low Carbon Steel	1. Carbon Dioxide (spray Transfer)	High quality low current out of position welding low spatter.
	2. Carbon Dioxide (Buried Arc)	High speed low cost welding accompanied by spatter loss.
Nickel	1. Argon	Good wetting decreases fluidity of weld metal.
Monel	1. Argon	Good wetting decreases fluidity of weld metal.
Inconel	1. Argon	Good wetting decreases fluidity of weld metal.
Titanium	1. Argon	Reduces heat-affected zone, improves metal transfer.
Silicon Bronze	1. Argon	Reduces crack sensitivity of this hot short material.
Aluminum Bronze	1. Argon	Less penetration of base metal. Commonly used as a surfacing material.

NOTE: 1. — First Choice
2. — Second Choice

(Airco)

take a relatively large amount of nitrogen into solution. At room temperature, however, the solubility of nitrogen in iron is very low. Therefore, in cooling, the nitrogen precipitates out in the form of iron nitrites. These nitrites cause high yield strength, tensile strength, hardness, but a pronounced decrease in the ductility and impact resistance of the steel materials. The loss of ductility due to the presence of iron nitrites often leads to cracking in and adjacent to the weld metal. As stated above, air contains approximately 78 percent nitrogen by volume and, therefore, if the weld metal is not protected from the air during welding very pronounced decreases in weld quality will occur. In excessive amounts nitrogen can also lead to gross porosity in the weld deposit.

Hydrogen is also harmful to welding. Very small amounts of hydrogen in the atmosphere produce an erratic arc. Of more importance is the effect that hydrogen has on the properties of the weld deposit. As in the case of nitrogen, iron can hold a relatively large amount of hydrogen when it is molten but upon cooling it has a low solubility for hydrogen. As the metal starts to solidify it rejects the hydrogen. Hydrogen that becomes entrapped in the solidifying metal collects at certain points and causes large pressures or stresses to occur. These pressures lead to minute cracks in the weld metal which can later develop into large cracks. Hydrogen also causes defects known as "fish eyes" and under-bead cracking.

The effects of oxygen, nitrogen, and hydrogen make it essential that they are excluded from the weld area during welding. This is done by using inert gases for shielding. The inert gases consist of atoms which are very stable and do not react readily with other atoms. In nature it is found that there are only six elements possessing this stability and each of these elements exists as a gas. The six inert gases are helium, neon, argon, krypton, xenon and radon. Since the inert gases are very stable and do not readily form compounds with other elements they are very useful as shielding atmospheres for arc welding. Of the six inert gases only helium and argon are important to the welding industry. This is because these are the only two inert gases which can be obtained in quantities at an economical price.

Carbon dioxide gas can also be used for shielding the weld area. Although it is not an inert gas compensations can be made for its oxidizing tendencies and it can readily be employed for shielding the weld. Characteristics of this gas will be explained in detail later.

Argon. Argon gas has been used for many years as a shielding medium for fusion welding. Argon is obtained by the liquification and distillation of air. Air contains approximately 0.94 percent argon by volume or 1.3 percent by weight. This seems like a small quantity but calculations show that the amount of air covering one square mile of the earth's surface contains approximately 800,000 pounds of argon.

In manufacturing argon, air is put under great pressure and refrigerated to very low temperatures. Then, the various elements in the air are boiled off by raising the temperature of the liquid.

Argon boils off from the liquid at a temperature of $-302.4°$ F. For welding the purity of the argon is approximately 99.995 percent.

Argon has a relatively low ionization potential. This means that the welding arc tends to be more stable when argon is used as the shielding gas. For this reason argon is often used in conjunction with other gases for arc shielding. The argon gives a quiet arc and thereby reduces spatter. Since argon has a low ionization potential the arc voltage is reduced when argon is added to the shielding gas. This results in lower power in the arc and therefore lower penetration. The combination of lower penetration and reduced spatter makes the use of argon desirable when welding sheet metal.

Straight argon is seldom used for arc shielding except in welding such metals as aluminum, copper, nickel, and titanium. When welding steel the use of straight argon gas leads to undercutting and poor bead contour. Also, the penetration pattern obtained with pure argon is shallow at the bead edges and has a deep portion at the center of the weld. This can lead to lack of fusion at the root of the weld if the arc is not directed exactly over the center of the weld.

Argon Plus Oxygen. In order to reduce the poor bead contour and penetration pattern obtained with argon gas when welding on mild steel, it has been found that the addition of oxygen to the shielding gas is desirable. Small amounts of oxygen added to the argon produce significant changes. Normally, the oxygen is added in amounts of 1, 2 or 5 percent. Using gas metal-arc welding wires the amount of oxygen which can be employed is limited to 5 percent. Additional oxygen might lead to the formation of porosity in the weld deposit.

Oxygen improves the penetration pattern by broadening the deep penetration finger at the center of the weld bead. It also improves bead contour and eliminates the undercut at the edge of the weld that is obtained with pure argon when welding steel. Argon-oxygen mixtures are very common for welding low alloy steels, carbon steels, and stainless steel.

Carbon Dioxide. Unlike argon or helium gases which are made up of single atoms the carbon dioxide gas is made up of molecules. Each molecule contains one carbon atom and two oxygen atoms.

At normal temperatures carbon dioxide is essentially an inert gas. However, when subjected to high temperatures carbon dioxide will disassociate into carbon monoxide and oxygen. In the high temperature of the welding arc this disassociation takes place to the extent that 20 to 30 percent of the gas in the arc area is oxygen. Because of this oxidizing characteristic of CO_2 gas, the wires used with this gas must contain deoxidizing elements. The deoxidizing elements have a great affinity for the oxygen and readily combine with it. This prevents the oxygen atoms from combining with the carbon or iron in the weld metal and producing low quality welds. The most common deoxidizers used in wire electrodes are manganese, silicon, aluminum, titanium, and vanadium.

The purity of carbon dioxide gas can vary considerably depending upon the process used to manufacture it. However, standards have been set up for the purity that must be obtained in carbon dioxide gas that is to be used for arc welding. The purity specified for welding grade CO_2 is a minimum dew point of minus 40° F. This means that gas of this purity will contain approximately 0.0066 percent moisture by weight.

Carbon dioxide gas eliminates many of the undesirable characteristics that are obtained when using argon for arc shielding. With the carbon dioxide a broad, deep penetration pattern is obtained. This makes it easier for the operator to eliminate weld defects such as lack of penetration and lack of fusion. Bead contour is good and there is no tendency towards undercutting. Another advantage of CO_2 shielding is its relatively low cost when compared to other shielding gases.

The chief drawback of the CO_2 gas is the tendency for the arc to be somewhat violent. This can lead to spatter problems when welding on thin materials where appearance is of particular importance. However, for most applications this is not a major problem and the advantages of CO_2 shielding far outweigh its disadvantages.

Helium[2]. Helium (He) has an ionization potential of 24.5 volts. It is lighter than air and has high thermal conductivity. The helium arc plasma will expand under heat (thermal ionization) reducing the arc density. With helium there is a simultaneous change in arc voltage where the voltage gradient of the arc length is increased by the discharge of heat from the arc stream or core. This means that more arc energy is lost in the arc itself and is not transmitted to the work. The

[2]Courtesy Miller Electric Mfg. Co.

result is that, with helium, there will be a broader weld bead with relatively shallower penetration than with argon. (For Tig welding, the opposite is true.) This also accounts for the higher arc voltage, for the same arc length, that is obtained with helium as opposed to argon. See Fig. 4–13.

Helium is derived from natural gas. The process by which it is obtained is similar to that of argon. First the natural gas is compressed and cooled. The hydrocarbons are drawn off, then nitrogen, and finally the helium. This is a process of liquifying the various gases until at −452° F., the helium is produced.

Helium has sometimes been in short supply due to governmental restrictions and, therefore, has not been used as much as it might have been for welding purposes. It is difficult to initiate an arc in a helium atmosphere with the tungsten arc process. The problem is less acute with the gas metal-arc process.

Helium is used primarily for the nonferrous metals such as aluminum, magnesium, and copper. It is also used in combination with other shielding gases.

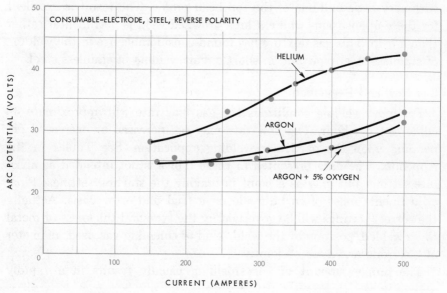

Fig. 4-13. Arc characteristics of various shielding gas.

Argon-CO_2. For some applications of mild steel welding, welding grade CO_2 does not provide the arc characteristics needed for the job. This will usually manifest itself, where surface appearance is a factor, in the form of intolerable spatter in the weld area. In such cases a mixture of argon-CO_2 has usually eliminated the problem. Some welding authorities believe that the mixture should not exceed 25 percent CO_2. Others feel that mixtures with up to 80 percent CO_2 are practical.

The reason for wanting to use as much CO_2 as possible in the mixtures is primarily cost. By using a cylinder of each type of gas, argon and CO_2, the mixture percentages may be varied by the use of flowmeters. This method precludes the possibility of gas separation such as may occur in pre-mixed cylinders. When it is considered that pre-mixed argon-CO_2 gas is sold at the price of pure argon then it makes good sense to mix your own. The price of CO_2 is approximately 15 percent that of argon in most areas of the country.

Argon-CO_2 shielding gas mixtures are employed for welding mild steel, low alloy steel, and in some cases for stainless steels.

Argon-Helium-CO_2[3]. This mixture of shielding gases is used primarily for welding austenitic stainless steels. The combination of gases provides a unique characteristic to the weld. It is possible to make a weld with very little building of the top bead profile. The result is excellent for those applications where a high crowned weld is detrimental rather than a help. This gas mixture has found considerable use in the welding of stainless steel pipe, and in short circuit welding of stainless steel.

Gas Flow and Regulation

For most welding conditions the gas flow rate will approximate 35 cubic feet per hour. The flow rate may be increased or decreased depending upon the particular welding application. See Tables in the Appendix. The data presented in these tables is not intended as absolute settings but only as a point in making the starting settings. Final adjustments must often be made on a trial and error basis. Actually the correct settings will be governed by the type and thickness of metal to be welded, position of the weld, kind of shielding gas used, diameter of electrode and type of joint.

The proper amount of gas shielding usually results in a rapidly

---

[3]Courtesy Miller Electric Mfg. Co.

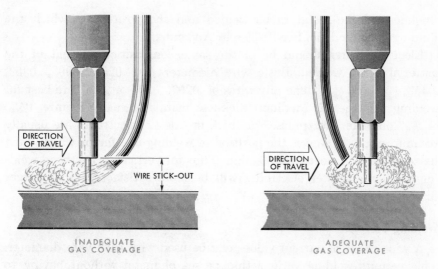

DIRECTION
OF TRAVEL

WIRE STICK-OUT

DIRECTION
OF TRAVEL

INADEQUATE
GAS COVERAGE

ADEQUATE
GAS COVERAGE

Fig. 4-14. The soundness of a weld depends on adequate gas coverage.

crackling or sizzling arc sound. Inadequate gas shielding will produce a popping arc sound with resultant weld discoloration, porosity, and spatter. An excessive amount of gas is equally objectionable because it causes turbulence which pulls in outside air and results in weld contamination.

"Gas drift" may occur from high weld travel speeds or from unusually drafty or windy conditions in the weld area. Since one or more of these factors may cause the gas to "drift" away from the arc, the result is inadequate gas coverage. See Fig. 4–14. The gas nozzle should be adjusted for proper coverage and outside influences should be eliminated by proper windbreakers or shields.

Correct positioning of the nozzle with respect to the work is determined by the nature of the weld. The gas nozzle may be placed up to 2″ from the work. Too much space between nozzle and work reduces the effectiveness of a gas shield while too little space may result in excessive weld spatter which collects on the nozzle and shortens its life.

Wire Size

Best results are obtained by using the proper diameter wire for the thickness of the metal to be welded and the position in which the welding is to be done. See Tables in Appendix.

Electrode wires should be of the same composition as that of the material being welded. Basic wire diameters are .020″, .030″, .035″, .045″, ¹⁄₁₆″ and ⅛″. Generally wires of .020″, .030″ or .035″ are best for welding thin metals. Medium thickness metals normally require .045″ or ¹⁄₁₆″ diameter electrodes. For thick metals ⅛″ electrodes are usually recommended. However, the position of welding is a factor which must be considered in electrode selection. Thus for vertical or overhead welding smaller diameter electrodes will be more satisfactory than larger diameter wires.

Welding Current

A wide range of current values can be used with each wire diameter. This permits welding various thicknesses of metal without having to change wire diameter. The correct current to use for a particular joint must often be determined by trial. The current selected should be high enough to secure the desired penetration without cold lapping but low

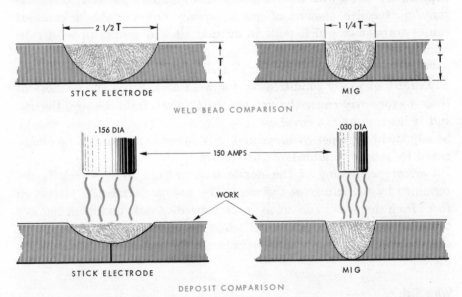

Fig. 4-15. The high current density of Mig welding produces deeper penetration and narrower beads.

enough to avoid undercutting and burn through. See Tables in Appendix.

The term current is often related to current density. Current density is the amperage per square inch of cross sectional area of the electrode. Thus at a given amperage the current density of .030" diameter electrode is higher than with .045" diameter electrode. Current density is calculated by dividing the welding current by the electrode area.

Each type and size of electrode wire has a minimum and maximum current density. For example, if the welding current falls below this minimum, a satisfactory weld cannot be made.

The success of Mig welding depends on the concentration of a high current density at the electrode tip. Whereas the arc stream of Mig is sharp and deeply penetrating, metallic arc (stick electrode) is soft and widespread. Consequently, the width-to-depth ratio of gas metal arc will be less than with stick electrode. See Fig. 4–15.

Wire Feed

The amperage of the welding current used limits the speed of the wire feed to a definite range. However, it is possible to make adjustments of the wire feed within the range. For a specific amperage setting a high speed of wire feed will result in a short arc. A low speed contributes to a long arc. Also a higher speed must be used for overhead welding than for flat position welding.

Wire Stickout

Wire stickout refers to the distance the wire projects from the contact tip of the gun. See Fig. 4-16. Stickout influences the welding current

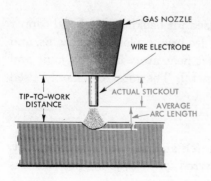

Fig. 4-16. Correct wire stickup is important to achieve sound welds.

since it changes the preheating in the wire. When the stickout increases the preheating increases which means that the power source does not have to furnish as much welding current to melt the wire at a given feed rate. Since the power source is self-regulating the current output is automatically decreased. Conversely, if the stickout decreases the power source is forced to furnish more current to burn off the wire at the required rate.

For most Mig welding application the wire stickout should measure up to 1″. An excessive amount of wire stickout results in increased wire preheating; this tends to increase the deposit rate. Too much wire stickout may cause a "ropy" appearance in the weld bead. Too little stickout results in the wire fusing to the contact tip which decreases the life of the tip. As the amount of wire stickout increases, it may become more difficult to follow the weld seam, particularly with a small diameter wire.

The wire, in a near plastic-state between the tip and arc, tends to move (whip) around, describing a somewhat circular pattern. Decreasing the amount of wire stickout and straightening the welding wire tend to decrease the amount of wire whip.

Joint Edge Preparation and Weld Backing

Preparation of the edge of each member to be joined is recommended to aid in the penetration and control of weld reinforcement. For Mig welding beveling the edges is usually necessary for butt joints thicker than ¼″ if complete root penetration is desired. For thinner sections a square butt joint is best.

In the Mig process weld backing is helpful in obtaining a sound weld at the roots. Backing prevents molten metal from running through the joint being welded especially when complete weld penetration is required.

There are several types of material used for backing: steel and copper blocks, strips, and bars; carbon blocks, fire bricks, plastics, asbestos, and fire clay. Some of these serve to conduct heat away from the joint and also to form a mold or dam for the metal. The most commonly used backing for Mig welding is copper or steel.

Positioning Work and Welding Wire

The proper position of the welding torch and weldment is important. In Mig welding the flat position is preferred for most joints because this position improves the molten metal flow, bead contour, and gives better gas protection. However, on gage material it is sometimes necessary or advantageous to weld with the work inclined 10 to 20 degrees. The welding is done in the downhill position. This has a tendency to flatten the bead and increase the travel speed.

The alignment of the welding wire in relation to the joint is very important. The welding wire should be on the center line of the joint for most butt joints if the pieces to be joined are of equal thickness. If the pieces are unequal in thickness the wire may be moved toward the thicker piece. The recommended position of the welding gun for fillet and butt welds is shown in Fig. 4–17.

Either a pulling or pushing technique may be used with little or no

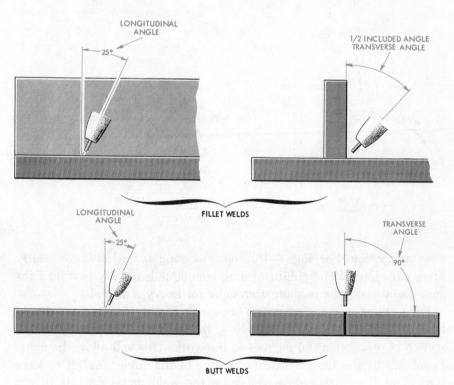

Fig. 4-17. Correct nozzle angle for Mig welding.

Fig. 4-18. Either a pushing or pulling technique is used in Mig welding.

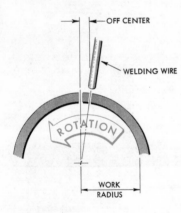

Fig. 4-19. Position of wire electrode in making a circular weld.

weaving motion. See Fig. 4–18. Some weaving is desirable for poorly fitted edge joints. The pulling or drag technique is usually best for light gage metals and the pushing technique for heavy materials.

For welding circular seams, as shown in Fig. 4–19, the wire should be shifted off-center approximately ⅓ the work radius opposite the direction of rotation, at 90 degrees to the work. This will allow the metal to solidify by the time it reaches the top of the circle. A shift of more than ⅓ of the work radius will cause the weld metal deposit to run ahead of the weld bead.

Arc Starting

Starting an electrical arc for a welding process involves three major factors: electrical contact, arc voltage, and time. To assure good arc starts it is necessary for the electrode wire to make good electrical contact with the work. The electrode must exert sufficient force on the workpiece to penetrate impurities.

Arc initiation becomes increasingly more difficult as wire stickout increases. A reasonable balance of volts and amperes must be maintained in order to assure the proper arc and to deposit the metal at the best electrode melting rate.

The response time of a power source and circuit is generally fixed by equipment design. Most power units have an optimum response time for wires.

The arc may be generated by the *run-in start method* or *scratch start method*, depending on the type of equipment used. Some welding units provide circuiting for both by incorporating separate toggle switches on the feeder panel.

In the run-in method the gun is aimed at the work piece without touching the workpiece. The gun trigger is depressed and this immediately energizes the wire and starts the arc.

With the scratch method the end of the welding wire must be scratched against the workpiece to start the arc.

Once the arc is started the gun is simply held at the correct angle and moved at a uniform speed. When reaching the end of a weld the trigger is released which stops the wire feed and interrupts the welding current.

BURIED ARC CO₂ WELDING PROCESS

Buried arc CO_2 welding is a high-energy, fast-weld method in which the end of the wire electrode is held either level or below the surface of the work with practically a zero arc length. See Figs. 4–20 and 4–21. This process is designed for high speed welding of mild steel. Although this process is designed for high speed welding on carbon and low alloy steels, usually in the range of $\frac{3}{32}''$ to $\frac{1}{4}''$ in thickness, it can be employed for manual welding. Its greatest application is in mechanized welding. The process is widely used in many industries

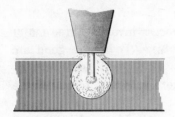

Fig. 4-20. Buried arc CO_2.

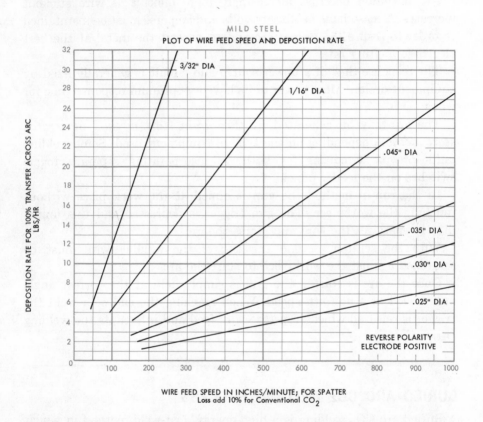

MILD STEEL
PLOT OF WIRE FEED SPEED AND DEPOSITION RATE

DEPOSITION RATE FOR 100% TRANSFER ACROSS ARC LBS./HR

3/32" DIA

1/16" DIA

.045" DIA

.035" DIA

.030" DIA

.025" DIA

REVERSE POLARITY
ELECTRODE POSITIVE

WIRE FEED SPEED IN INCHES/MINUTE; FOR SPATTER
Loss add 10% for Conventional CO_2

Fig. 4-21. Buried arc CO_2 plot of wire feed speed and deposition rate for mild steel welding.

where the fabrication of parts requires deep penetration and fast deposition of weld metal without critical control of bead contour. In most instances welding wire diameters range from 0.045" to ⅛".

Regular Mig welding equipment is utilized for the buried arc process. The shielding gas is pure carbon dioxide which provides additional economy over the more expensive argon. The metal transfer is globular, but since the wire is buried and a high density current is used, a deep cavity is formed. This deep cavity traps the molten globules that normally would be ejected sideways through the arc. Thus, splatter which otherwise would be severe is minimized and does not affect the welding process.

PULSED-SPRAY ARC WELDING[4]

The pulsed-spray process is an extension of spray-transfer welding to a current level much below that required for continuous spray transfer. The pulsing current used may be considered as having its peak current in the spray-transfer current range and its minimum value in the globular transfer current range.

The need for current values less than the transition level becomes apparent when attempting to weld under heat transfer conditions which are inadequate relative to the minimum heat input râte with spray transfer. For example, when welding out-of-position the high current will result in a molten pool which cannot be retained in position unless the material being welded has an adequate thermal-conductivity (coupled with the joint type and plate thickness).

The same factors explain the burn-through obtained when a weld on thin material is attempted with too high a welding current. While smaller diameter electrodes have lower transition currents the basic limitation cannot be avoided. The net result is that the spray transfer process is very applicable to flat position use but is rather limited in its use for out-of-position and thin material welding.

Pulsed-spray transfer is achieved by pulsing the current back and forth between the spray transfer and globular transfer current ranges. Fig. 4–22 illustrates, on the left, the current-time relationships for two power sources, A and B, with A putting out a current in the globular transfer range and B putting out a current in the spray-transfer range. Fig. 4–22 also shows, on the right, the two outputs combined to produce a simple pulsed output by electrically switching back and forth between them.

Transfer is restricted to the spray mode. Globular transfer is sup-

[4]Courtesy Airco.

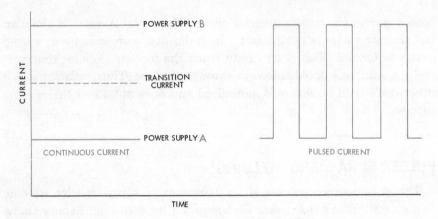

Fig. 4-22. Illustration of how a switching system can correct two steady-state DC output currents in a simple pulsing-current output wave form.

pressed by not allowing sufficient time for transfer by this mode to occur. Conversely, at the high current level, spray transfer is ensured by allowing more than sufficient time for transfer to occur.

For the given electrode deposited by the pulsed-spray method all the advantages of the spray-transfer process are available at average current levels from the minimum possible with continuous spray transfer down to values low in the globular transfer range.

The Power Source

To suppress globular transfer the time period between consecutive pulses must be less than sufficient for a transfer by the globular mode to occur. The time period between pulses produced by the use of the positive half-cycles from a 60 hertz (cps) power line is short enough to suppress transfer at all current levels in the globular transfer current range. Conversely, the pulse duration is long enough to ensure that transfer by the spray mode will occur at an appropriate current in the spray transfer range.

As Fig. 4–23 illustrates the power source combines a standard, three phase, full wave unit with a single phase, half wave unit, both of the constant-potential type. The three phase unit is termed the "background" unit and the single phase unit is termed the "pulsing" unit. These units are connected in parallel but commutate in operation. The pulsing current output is schematically illustrated in Fig. 4–24.

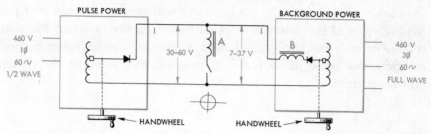

Fig. 4-23. Block diagram of the essential features of the pulsed-current power supply.

Airco

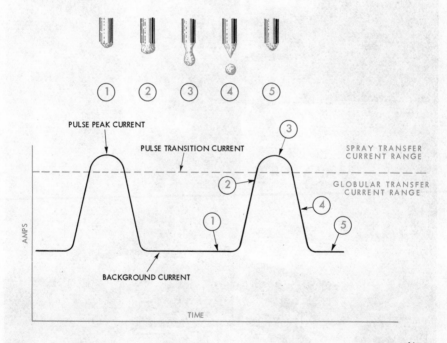

Airco

Fig. 4-24. Illustration of the output current wave form of the pulsed-current power supply; also showing the metal transfer sequence.

The units are made to switch back and forth in operation by means of the varying output voltage of the pulsing unit. The diode rectifiers in each unit alternately permit or block the passage of current depending upon whether there is a positive or negative voltage difference across their terminals. When the pulse is off or its voltage is less than the background voltage the diode rectifiers of the background unit pass the full value of the instantaneous current. Conversely, when the pulse voltage exceeds the background voltage blocking the background diode rectifiers, the pulse diode rectifiers pass the full value of the instantaneous current.

Airco

Fig. 4-25. Pulsed power Mig welder.

Two chokes are used. See Fig. 4–24. The choke labeled "A" performs a commutation function. When the pulse voltage drops below the background voltage it sustains the welding current momentarily, giving the background unit time to respond to the demand for current. Choke "B", in series with the output of the background unit, filters the background current and prevents undesirable arc outages at low background current levels. The pulsed-current power source is shown in Fig. 4–25.

Table II.   Typical Pulsed-Current Power Supply Settings.

ELECTRODE TYPE	ELECTRODE DIAMETER, IN	PULSE PEAK RANGE, V	AVERAGE CURRENT RANGE, AMP	AVERAGE VOLTAGE
MILD AND LOW ALLOY STEEL	0.035	34–36	55–130	18–20
	0.045	37–39	90–180	19–23
	1/16	42–44	110–250	20–25
STAINLESS STEEL	0.035	33–35	55–130	18–20
	0.045	36–38	90–180	19–23
	1/16	41–43	110–250	20–25
ALUMINUM	1/16	34–36	80–250	20–30

Table III.   Average Current vs. Electrode Feed Speed for Typical Diameters of Steel and Aluminum Electrodes when Using Pulsed-Current Unit.

ELECTRODE FEED SPEED (IPM)	CORRESPONDING AVERAGE CURRENT, AMP. MILD STEEL AND STAINLESS STEEL (a)			
	0.035 IN (DIA)	0.045 IN (DIA)	1/16 IN (DIA)	ALUMINUM (b) 1/16 IN (DIA)
70	....	70	115	70
90	....	90	175	90
105	50	105	215	105
125	60	125	....	125
135	70	135	....	135
155	80	155	....	155
185	90	185	....	185
220	110	220	....	220
235	120			
255	130			
275	140			
300	150			
325	160			
345	170			
365	180			
380	190			
425	210			
500	....			

(a) ARGON + 290 O_2
(b) ARGON

Operation of the pulsed-current power source is similar to that of conventional constant potential sources. With the arc off, the value of the pulse peak voltage, which depends upon the electrode type and diameter (Table II), is selected and remains constant. This setting is made by rotating the pulse peak voltage handwheel while pressing a button which converts the average voltage meter to a peak voltage meter. The electrode feeder is set at the value which will produce the required current and is determined from Table III for the type and diameter of electrode to be used. The arc is then initiated and the background voltage handwheel rotated to produce the proper arc length. So long as the type and diameter of electrode remain the same all further power source adjustments are made by means of the background voltage handwheel. The meters on the power supply read the average voltage and the current which are the values familiar to every welding operator.

Features of Pulsed-spray Welding

Pulsed-spray welding provides many features not previously available.

1. The heat range bridges the gap between, and laps over into, the heat input ranges available from the spray and short-circuiting arc processes. In its lower heat input range the pulsed-spray process brings the advantages of the continuous spray-transfer process. Also, due to lower heat input, the use of spray transfer is extended greatly into poor heat transfer areas, mainly related to welding out-of-position and on thinner materials.

2. The area of overlay with the spray-transfer process occurs because, having a higher transition current, a larger diameter electrode leaves the continuous spray and enters the pulsed-spray range at a higher current than a smaller electrode. Further, the use of a larger diameter electrode can continue down to a current considerably below the transition current of the smaller diameter electrode.

Large diameter electrodes are lower in cost. They have a lower surface-to-volume ratio which is important considering the correlation between material found on the electrode surface and weld porosity and, in certain cases, weld cracking. Also, electrode feeding problems are more often experienced with the smaller diameters of electrodes.

3. The pulsed-spray and the short-circuiting arc processes are different. In general, where applicable, the minimum heat inputs obtainable are approximately equal but, conversely, the maximum heat input of the globular transfer range is achieved by the pulsed spray but not by the short-circuiting arc process. The pulsed-spray method produces the higher ratio of heat input to metal deposition, permits the use of a completely inert gas shield where necessary, and is free from spatter.

The pulsed-spray process will not displace the short-circuiting arc process in those areas where the short-circuiting arc process is properly applicable and more economical.

4. The pulsed-spray process is characterized by a uniformity of root penetration which approaches that possible with the gas tungsten-arc process; because of this feature, the process may permit deletion of weld backing in some cases.

TUBULAR WIRE WELDING (CORED-WIRE WELDING)

Tubular wire welding is a gas-metal arc welding process in which a continuous fluxed core wire instead of a solid wire serves as the electrode. The wire can be used on any automatic or semi-automatic Mig welding equipment and is employed principally in combination with CO_2 as a shielding gas. The wire is frequently referred to by the manufacturer's trade name such as "Fluxcor" (Airco) and "FabCo" (Hobart).

The flux ingredients in the wire include ionizers to stabilize the arc, deoxidizers to purge the deposit of gas and slag, and elements to produce high strength, ductility and toughness in weld deposits. The flux generates a gas shield, which is also augmented by the regular CO_2 shield, and a slag blanket that retards the cooling rate and protects the weld deposit as it solidifies.

Tubular wire is designed for high current densities and deposition rates which when combined with high duty cycles result in sharply increased production speeds. It is especially intended for application in large fillet single and multi-pass welds in either a horizontal or flat position using DCRP current. Because of its deep penetrating qualities into the weld root tubular wire fillet welds of smaller leg size will have the same strength as stick fillet welds of large size. Double welded

butt joints up to ¾″ thick can be welded without edge penetration.
The actual operation of tubular wire welding is similar to other Mig
welding processes.

VAPOR-SHIELDED ARC WELDING[5]

Vapor-shielded arc welding, known as Innershield, is a welding process
introduced by Lincoln Electric Company. This process uses a vapor

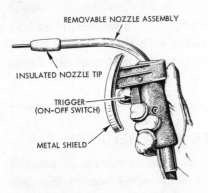

REMOVABLE NOZZLE ASSEMBLY
INSULATED NOZZLE TIP
TRIGGER (ON-OFF SWITCH)
METAL SHIELD

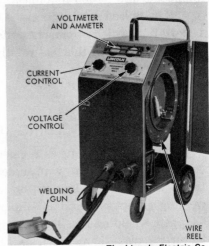

VOLTMETER AND AMMETER
CURRENT CONTROL
VOLTAGE CONTROL
WELDING GUN
WIRE REEL

The Lincoln Electric Co.
Semi-automatic welder.

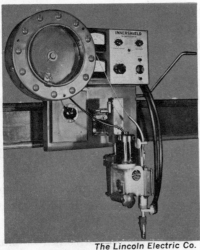

The Lincoln Electric Co.
Automatic welder.

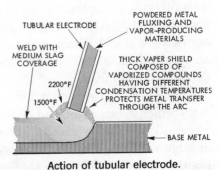

TUBULAR ELECTRODE
WELD WITH MEDIUM SLAG COVERAGE
2200°F
1500°F
POWDERED METAL FLUXING AND VAPOR-PRODUCING MATERIALS
THICK VAPER SHIELD COMPOSED OF VAPORIZED COMPOUNDS HAVING DIFFERENT CONDENSATION TEMPERATURES PROTECTS METAL TRANSFER THROUGH THE ARC
BASE METAL

Action of tubular electrode.

Fig. 4-26. Innershield welder.

[5]Lincoln Electric Co.

instead of a gas to shield the arc and molten metal. (A vapor in its natural state is a liquid or a solid, as distinct from a gas, which is in its natural state.) The vapor is generated by a continuous tubular electrode containing vapor-producing materials. The electrode also serves as a filler rod.

Equipment

Equipment for the vapor-shielded arc consists of a DC power source, a control station for adjusting amperage and voltage, and a continuous wire feed mechanism. The equipment can be fully automatic or semi-automatic. See Fig. 4–26.

With semi-automatic equipment the operator moves a gun-type nozzle along the weld seam. To weld he simply pulls the trigger which starts the welding current and wire feed. When he wants to stop welding he merely releases the trigger.

How Does Vapor Shielding Work?

The tubular mild steel wire which serves as a filler rod contains all of the necessary ingredients for shielding, deoxidizing, and fluxing. Some of the ingredients are metallic salts and oxides which melt before the electrode melts, thereby vaporizing at a temperature lower than the melted electrode metal. These metallic salts vaporize and expand when they meet the intense heat of the arc. The vaporized salts expand until they reach a distance surrounding the arc where the temperature corresponds to their condensing temperature. Then the vaporized salts recondense forming a thick vapor shield around the arc and molten metal. This shield prevents contamination by the surrounding atmosphere. Other materials in the electrode have alloying functions in creating high grade weld metal.

SUBMERGED ARC WELDING

Submerged arc welding is a process where an electric arc is submerged or hidden beneath a granular material. See Fig. 4–27. The electric arc provides the necessary heat to melt and fuse the metal. The granular material, called flux, completely surrounds the electric arc

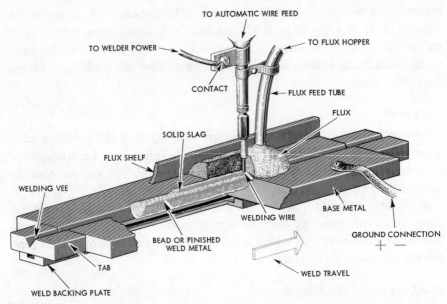

TO AUTOMATIC WIRE FEED

TO WELDER POWER

TO FLUX HOPPER

CONTACT

FLUX FEED TUBE

FLUX

SOLID SLAG

FLUX SHELF

WELDING VEE

WELDING WIRE

BASE METAL

BEAD OR FINISHED WELD METAL

GROUND CONNECTION
+ −

WELD TRAVEL

TAB

WELD BACKING PLATE

Linde Co.

Fig. 4-27. Cutaway view of submerged arc welding.

shielding the arc and the metal from the atmosphere. A metallic wire is fed into the welding zone underneath the flux.

The welding process can be either semi-automatic or fully automatic. In the semi-automatic, a special hand welding gun is used. See Fig. 4–28. Any regular Mig welding DC power source can be adapted for submerged arc welding. The difference between submerged arc welding and other forms of gas-metal arc welding is that no inert shielding gas is required. The gun hopper is simply filled with flux pointed over the weld area and the gun trigger depressed. As soon as the trigger is pulled the wire is energized and the arc is started. At the same time the flux begins to flow. The actual welding operation is now carried out in the same way as in Mig welding.

As the metallic wire is fed into the weld zone the feeding hopper deposits the granulated flux over the weld puddle and completely shields the welding action. The arc is not visible since it is buried in the flux and there is no flash or spatter. That portion of the granular flux immediately around the arc fuses and covers the molten metal but after it has solidified it can be tapped off easily.

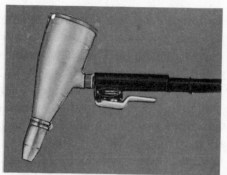

Hand gun for semi-automatic unit.

Linde Co.

Fig. 4-28. Automatic submerged arc unit.

With the fully automatic process the welding unit is arranged to move over the weld area at a controlled speed. On some arrangements the welding head moves and the work remains stationary. In others the head is stationary and the work moves.

Submerged arc welding is usually best adapted where relatively thick sections are to be joined and deep penetration is required. For example, it is possible to weld three inch thick plate in a single pass. Little, if any, edge preparation is necessary—as a rule none on material under one-half inch in thickness. Generally back-up support is essential. Welding is done on a horizontal or nearly horizontal plane.

REVIEW QUESTIONS

1. How does Mig welding differ from Tig welding?
2. In Mig welding, why is DCRP current better than DCSP current?
3. What is the general characteristic of spray transfer?
4. Why is globular transfer not very effective for most regular Mig welding processes?
5. What is meant by short circuiting transfer?
6. How does a constant potential power supply unit differ from the conventional constant current welder?
7. What is the difference between a push and pull type of Mig welding gun?
8. Why are oxygen and nitrogen harmful to a weld?
9. Why is argon generally considered a better shielding gas than helium?
10. Why are other gases sometimes mixed with argon for shielding purposes?
11. When is CO_2 used as a shielding gas?
12. What is meant by current density?
13. Why is the correct amount of wire stickout important to get good welds?
14. What methods may be used to start the arc for Mig welding?
15. What is meant by "buried" arc CO_2 welding?
16. What is pulsed-spray arc welding?
17. How does tubular wire welding differ from regular Mig welding?
18. What is submerged arc welding?

Chapter 5 | Resistance Welding

Resistance welding is a process where the required heat for fusion is produced by the resistance of the workpiece to the flow of low-voltage, high-density electric current. The generated heat must be of sufficient intensity to melt the base metal. According to Ohm's Law:

$$I = \frac{E}{R} \text{ or } R = \frac{E}{I}$$

Where I is current in amperes, E is the emf or voltage in volts, and R is the resistance in ohms. Total energy is expressed by:

$$\text{Energy} = IET$$

in which T is the time in seconds of the current flow in a circuit. If this energy is to be converted into heat then

$$H \text{ (heat energy)} = I^2 RT$$

Since H varies as I^2 the greatest heating effect is induced by using a welder which provides low voltage and high amperage. Along with heat force is required before, during, and after the application of current in order to provide a solid contact for the electrical circuit and to forge the heated parts together.

Resistance welding is strictly a machine welding process and designed for mass production of metal products. The basic resistance welding processes are: spot, seam, projection, flash, upset, and percussion.

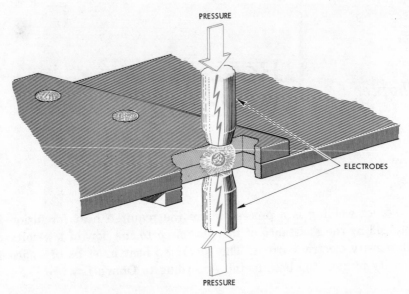

PRESSURE

ELECTRODES

PRESSURE

Fig. 5-1. In spot welding fusion is achieved by the heat resulting from the resistance of the electrical current through the workpieces held under pressure by electrodes.

SPOT WELDING

Spot welding is probably the most common type of resistance welding. The material to be joined is placed between two electrodes, pressure is applied, and the current turned on. See Fig. 5–1. In any spot welding operation three stages are involved in the welding cycle: squeeze time, weld time, and hold time. Squeeze time is the first period when the electrodes are brought together against the metal prior to the application of current. Weld time is the time when the current is flowing. Hold time is the time when the current is off but the pressure continues.

Spot welding may be done on material as low as 0.001″ in thickness and in joints having members as heavy as one inch. The bulk of resistance welding is confined to metals that are less than ¼″ in thickness. Spot welding machines are either of the single or multiple type.

Single-Spot Welding Machines

There are two types of single-spot welding machines: rocker arm and press. The rocker arm is the simplest and most common. See Fig. 5–2.

Peer Division—Landis Machine Co.
Fig. 5-2. Conventional rocker-arm spot welding machine.

It has two long horizontal horns each holding a single electrode with the upper arm providing the moving action. Because of the long horizontal horns welds can be made in areas inaccessible to other machines.

Press-type spot welders have a movable electrode and welding head which operate in a straight line. These welders have greater application in welding heavier sections. See Fig. 5–3.

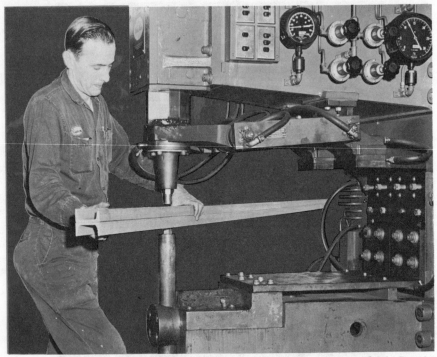

Fig. 5-3. Press type single-spot welding machine.

Although these spot welders are of stationary design, there is an increasing demand for the more maneuverable portable type. The portable, or spot-welding gun, consists of a welding head connected to the power supply unit. With this type of equipment, spot welds can be made on many irregular shaped objects. See Fig. 5–4.

Heat Application. Welding period in spot welding is controlled by electronic, mechanical, or manually operated devices. Weld time may range from one-half hertz (cycle) of a 60-hertz (cycle) frequency for very light material to several seconds for thicker plate.

Weld time is an important factor because the effectiveness of the joint depends upon the correct depth of fusion. Sufficient time is required for the current to pass through the circuit and develop enough heat to raise a small volume of metal to or above its melting point. The temperature must be high enough to insure proper fusion but not so great that it

112 *Welding Technology*

Fig. 5-4. A spot welding gun is used to weld the intake housing of a commercial fan.

will force metal from the weld zone. Excessive current may cause cavitation and weld cracks.

The amount of heat required for welding depends on the electrical resistance of the materials to be joined, the electrical resistance of the electrodes, the surface condition of the metal, and the contact resistance between the electrodes and the workpiece. The rate of temperature rise and fall has to be rapid enough to meet production welding speeds but not excessively high to produce brittle welds.

Since total heat input is a definite function of time the development of a proper weld nugget requires a definite length of time regardless of the magnitude of current flow. Insufficient time is noticeable when the heat builds up too rapidly at the contacting surfaces, resulting in excessive spitting and pitting.

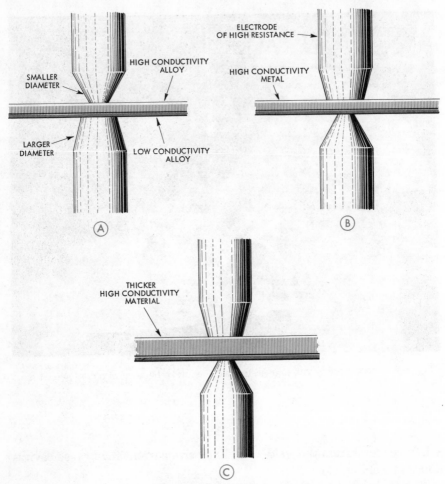

Fig. 5-5. Technique for securing better heat balance in spot welding.

It is impossible to specify welding current for spot welding because of the multiplicity of variables, such as type of materials, joint design, thickness and conductivity of metals to be joined, type of equipment, etc. Satisfactory spot welds or any other form of resistance welds can be obtained only by experimentation and testing.

Heat Balance. Uneven heat balance is a condition when materials to be welded generate an unbalanced rate of heat. When two pieces of metal are of the same thickness and composition, the heat will be the same in both pieces and a typical oval cross sectional weld nugget will be formed. However, if one piece has a higher electrical resistance than the other, more heat will be generated in the thicker piece than in the thinner section. Similarly, when two dissimilar metals are to be joined such as stainless steel to low carbon steel, their dissimilarity of composition will produce an unbalanced heat rate.

Compensation for uneven heat balance is achieved by varying the geometry of the electrode, by using different electrode materials, or by changing the thickness of the two metal pieces. For example, as shown in A of Fig. 5–5, it is possible to secure a more equitable heat balance by using a smaller electrode area on the high conductivity metal and a correspondingly larger electrode area on the lower conductivity metal. In B of Fig. 5–5 comparable results are obtained by using an electrode of high thermal resistance such as tungsten and molybdenum on the high conductivity metal. In C of Fig. 5–5 a better heat balance is secured by increasing the thickness of the higher-conductivity metal.

Electrodes. The primary functions of the electrode are to (1) conduct the required heat to the weld zone, (2) transmit the necessary force to the weld area, and (3) help dissipate the heat from the weld zone. Most electrodes for spot welding are made of low resistance copper alloy, although in some instances other high thermal resistance materials are used. Electrodes are hollow to facilitate the passage of water for cooling purposes. Cooling is an essential phase of spot welding since the heat needs to be dissipated as quickly as possible from the weld area. Electrodes are held in holders by straight or tapered fits or threads.

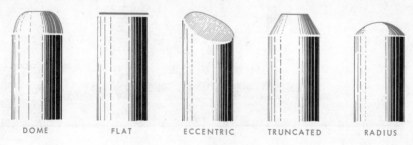

DOME   FLAT   ECCENTRIC   TRUNCATED   RADIUS

Fig. 5-6. Types of electrode faces.

The shape and size of the electrode face are governed by the thickness and geometry of the assembly to be welded. The average diameter of a spot weld is usually three times the thickness of the thinnest outer section plus 0.06″. The face may be dome, flat, eccentric, truncated, or radius. See Fig. 5–6. The eccentric is used in corners or where a centered-face electrode is impractical to function.

The proper maintenance of the correct shaped face is important for effective welding. Thus if a ¼″ diameter electrode face is allowed to increase to ⅜″ by wear or mushrooming the contact area is practically doubled with a corresponding decrease in current density. Unless this is compensated for by an increase in current setting the result will be weak welds. Poor welds are also due to misalignment of electrodes, improper electrode pressure, and convex or concave electrode faces. As a rule, a study of each job will establish the number of welds which can be made before a slight dressing is necessary. The dressing is best done with an abrasive paddle or abrasive cloth. Heavy dressing with a coarse file is not recommended since it wastes too much electrode material and destroys the true contour of the face. When the shape of the face has become deformed and a major redressing is required the electrode is replaced with a new one and the old electrode reshaped by turning in a lathe.

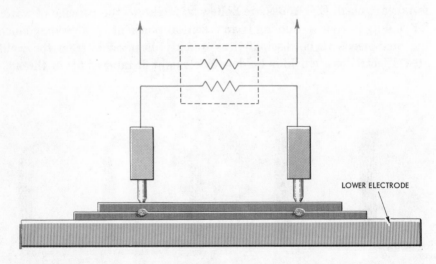

LOWER ELECTRODE

Fig. 5-7. In multiple-spot welding the lower electrode consists of a mandrel or bar.

Regular spot welding often leaves slight depressions on the metal which are usually undesirable on the "show side" of the finished product. These depressions can be minimized by using an electrode of greater area on the side of the material that is to have minimum marking.

Multiple-Spot Welders

Multiple-spot welders have a series of hydraulically or air operated welding guns mounted in a framework or header but using a common mandrel or bar for the lower electrode. See Fig. 5–7. The guns are connected by flexible bands to individual transformers or to a common buss bar attached to the transformer. See Fig. 5–8. With some machines two or four guns are often attached to a transformer.

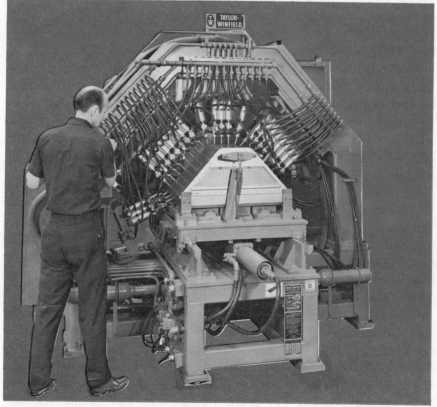

The Taylor-Winfield Corp.

Fig. 5-8. A multiple-spot welding machine.

In operation all guns make contact with the workpiece simultaneously under pressure. The guns are synchronized to fire in a certain sequence with welding current fed to only one gun at a time or to several guns. Multiple-spot welding machines are used extensively in the automotive industries.

SEAM WELDING

Seam welding is similar to spot welding except that the spots overlap each other, or the spots are spaced at short intervals making a continuous weld seam. In this process the metal pieces pass between roller type electrodes as shown in Fig. 5–9. As the electrodes revolve the current is automatically turned on and off at intervals corresponding to the speed at which the parts are set to move.

Electrodes

Electrodes for seam welders are disk shaped rollers with different face contours. See Fig. 5–10. The face of the roller may be straight, beveled, or concave. The straight and double beveled edge is used where there is sufficient clearance on both sides of the work. A single beveled edge electrode is often required where the weld must be made close to an obstruction such as a flange. The radius or concave face frequently will produce the best weld appearance.

Diameter of electrodes may vary from 2″ to 2′. The material is usually a copper alloy of the heat-treated, precipitation-hardening kind. The contacting face varies in width depending on the thickness of the material to be welded. Normally the width of a seam weld will range from 1½ to 3 times the thickness of the thinner piece to be welded. Weld widths are basically greater on thin materials than on thick materials. This is done to minimize excessive electrode wear.

Cooling of electrodes is accomplished either by internally circulating water or by external spray of water over the electrode rollers. Clean tap water ordinarily is used for cooling nonferrous materials and stainless steel. When cooling steels that are very susceptible to rusting in a wet condition a common practice is to use a 5 percent borax solution. With adequate cooling warpage can be kept to a minimum. Warpage becomes critical especially when several parallel seams are to be welded.

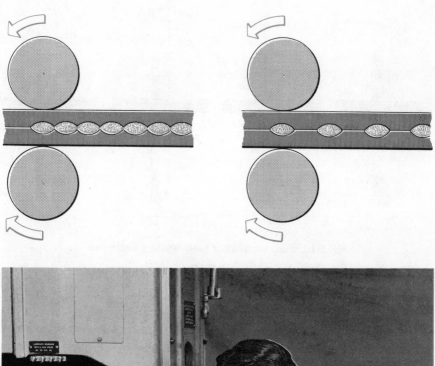

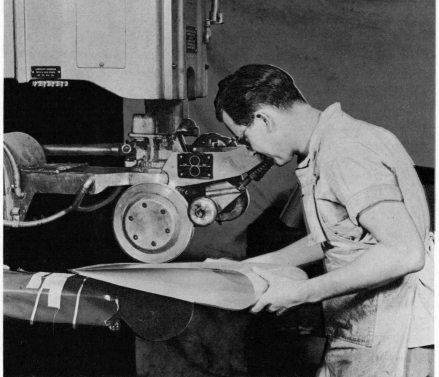

Fig. 5-9. In seam welding the metal pieces pass between roller type electrodes.

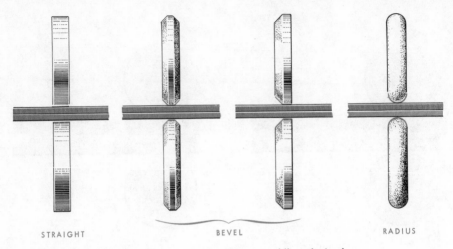

STRAIGHT   BEVEL   RADIUS

Fig. 5-10. Face contours of seam welding electrodes.

Current

The current value for seam welding is contingent on several variables such as type of material, welding speed, thickness of the joint, and even the amount of water used for cooling purpose. As a rule, correct current setting is determined by "trial and error" until the desired quality weld is obtained. Improper current setting will often result in cracked and porous welds.

Welding Speed and Force

The speed with which seam welding is to be accomplished will vary with the thickness and type of material as well as the joint geometry. The usual practice is to utilize the fastest speed which is within the capacity of the machine and is consistent with variables of the welding operation.

Electrode force for seam welding is generally greater than that regularly used for spot welding. However, the nature of the joining surface is always an important consideration. Surfaces of joints must fit closely together and be free from wrinkles to insure a gas-tight or liquid-tight seam. Surface indentations on one side of the joint can be minimized by using a wider face electrode on the side where such markings are to be avoided.

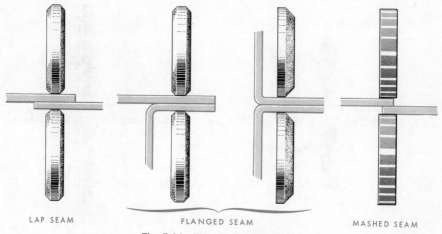

LAP SEAM   FLANGED SEAM   MASHED SEAM

Fig. 5-11. Types of weld seams.

Types of Weld Seams

Several types of seams are used in seam welding. See Fig. 5–11. The simplest and most common is the lap seam where the two edges are lapped sufficiently to permit a sound weld. The flange joint is frequently employed in fastening tops and bottoms to containers. The mash seam is adaptable for joining light gage sheet metal with a thickness of less than $\frac{1}{16}''$. As shown in Fig. 5–11, the mash seam is made by forcing the overlapping edges together. An electrode with a sufficiently wide face to cover the overlap is required. The electrode is operated at a slower speed and with higher pressure. The width of the lap for a strong quality weld is usually held at about $1\frac{1}{2}$ times the thickness of the sheet.

PROJECTION WELDING

Projection welding involves the joining of parts by a resistance welding process which closely resembles spot welding. This type of welding is widely used in attaching fasteners to structural members. The localization of current and heat is predetermined by one or more projections

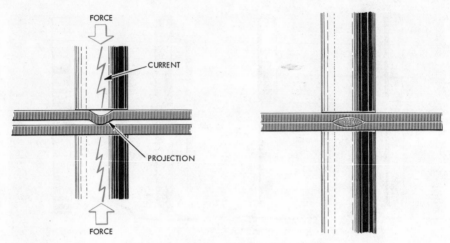

Fig. 5-12. Example of projection welding.

on the parts to be joined. These projections are formed by embossing, stamping, casting, or machining. Thus the welding is localized by the preparation of the workpiece and is not contingent on the shape and size of the electrode. See Fig. 5–12.

The height of the projections may vary from a few thousandths of an inch to ⅛″ or more. Since the major heat develops in the part with the projections the projections are usually produced on the heavier of the two pieces to be joined. In welding dissimilar metals the projections are located on the piece having the higher conductivity. Projections may be square, round, oblong, or diamond shape.

Electrodes

One feature of projection welding is that numerous welds can be made simultaneously without complicated electrode configuration. Multiple welds are limited only by the capacity of the equipment in controlling pressure and current. See Fig. 5–13.

When areas to be joined are flat, except for the projections, the electrodes are flat and large enough to contact a wide area. Where the area is irregular the electrode is shaped to fit the surface. The welding process consists of placing the projections in contact with the mating fixtures and aligning them between the electrodes.

Fig. 5-13. Projection welding a mild steel air impeller. Thirty projection welds are made in one weld stroke.

There are many variables involved in projection welding such as stock thickness, kind of material, and number of projections, that make it impossible to predetermine the correct current and required pressure. Only by trial runs followed by careful inspection can proper settings be established.

Metals Adaptable for Projection Welding

Not all metals can be projection welded. Brass and copper as a rule do not lend themselves to projection welding because the projections collapse too easily under pressure. Aluminum projections welding is generally limited to extruded parts. Galvanized sheet-steel, tin plate, and stainless steel as well as most other thin gage steels can be successfully projection welded.

FLASH WELDING

In flash welding the two pieces of metal to be joined are clamped in dies which conduct the electric current to the work. The ends of the

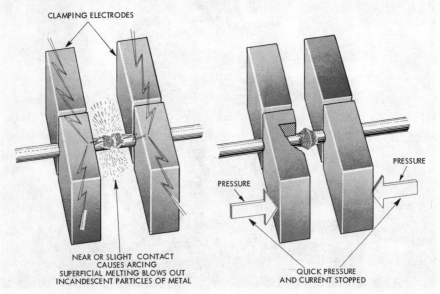

CLAMPING ELECTRODES

NEAR OR SLIGHT CONTACT
CAUSES ARCING
SUPERFICIAL MELTING BLOWS OUT
INCANDESCENT PARTICLES OF METAL

PRESSURE

PRESSURE

QUICK PRESSURE
AND CURRENT STOPPED

Fig. 5-14. In flash welding the two edges are brought into proximity, causing intense arcing which results in melting the metal.

Fig. 5-15. Flash-butt welding application showing welding of aluminum bands for luggage.

two metal pieces are moved together until an arc is established. The flashing action across the gap melts the metal and as the two molten ends are forced together fusion takes place. See Figs. 5–14 and 5–15. The current is cut off as soon as the forging action is completed.

Flash welding is used to butt or miter weld sheet, bar, rod, tubing, and extruded sections. It has unlimited application for both ferrous and nonferrous metals. It is not generally recommended for welding cast iron, lead, or zinc alloys.

Alignment of Parts

Parts to be welded are clamped by copper alloy dies shaped to fit each piece. For some operations the dies are water cooled to dissipate the heat from the welded area.

The most important factor in flash welding is the precision alignment

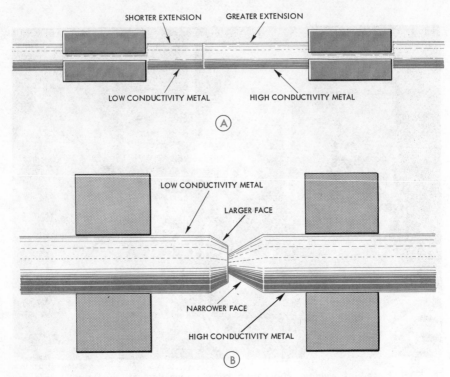

SHORTER EXTENSION · GREATER EXTENSION

LOW CONDUCTIVITY METAL · HIGH CONDUCTIVITY METAL

(A)

LOW CONDUCTIVITY METAL

LARGER FACE

NARROWER FACE

HIGH CONDUCTIVITY METAL

(B)

Fig. 5-16. Method of achieving heat balance in flash welding.

of the parts. Misalignment not only results in a poor joint but also produces uneven heat and telescoping of one piece over another.

If the same plasticity is to be obtained in both pieces a correct heat balance is necessary. Proper heat balance can be achieved in several ways. One method is to extend the part with the higher heat conductivity a greater distance out of the die. See Fig. 5–16A. Another way is to resort to unequal beveling of the two pieces as shown in Fig. 5–16B. In very unusual situations one piece is preheated before the welding action is started.

The only serious problem encountered in flash welding is the resultant bulge or increased size left at the point of the weld. If the finished area of the weld is important, then it becomes necessary to grind or machine the joint to the proper size.

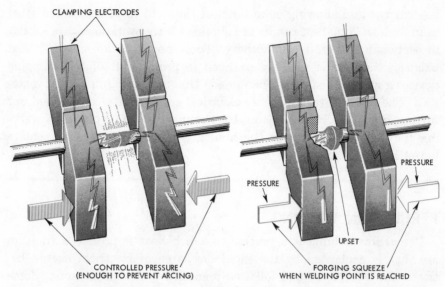

CLAMPING ELECTRODES

PRESSURE

PRESSURE

PRESSURE

UPSET

CONTROLLED PRESSURE
(ENOUGH TO PREVENT ARCING)

FORGING SQUEEZE
WHEN WELDING POINT IS REACHED

Fig. 5-17. In upset welding the two ends are butted together and a high current is passed through which melts the metal. Continuous pressure fuses the ends.

UPSET WELDING

Upset welding is sometimes referred to as butt welding. In this process the metals to be welded are brought into contact under pressure, an electric current is passed through them, and the edges are softened and fused together. See Fig. 5–17. Flash welding has an arcing action at the contacting surfaces, whereas in upset welding no flashing occurs.

Principles of Operation

As the parts are brought into contact force is applied even before the heating is started. This force is maintained throughout the heating cycle. Current is passed through the contact area until the temperature is sufficient to melt the interface between the two workpieces. The upset force then fuses the two sections together at which time both the current and pressure are cut off.

Although the operation and control of the butt welding process is almost identical to flash welding the basic difference is in the use of less current and allowing more time for the weld to be completed. Just as in flash welding best results are obtained if the workpieces are similar in sectional area and conductivity. To compensate for unequal heat balances the thicker piece is arranged to project out of the clamping devices a greater distance. The same is true in welding dissimilar materials. The metal with the higher electrical conductivity is extended out further from the clamping jaws. Restricting the contact area by varying the beveling surface of the abutting ends also serves to achieve better heat balance.

PERCUSSION WELDING

Percussion welding is a process in which heat is produced from an arc that is generated by the rapid discharge of electrical energy between the workpieces and followed immediately by an impacting force which weld the pieces together. The process has somewhat restricted applications because it is suitable for joining pieces that normally cannot be welded by the flash or upset method. It is generally confined to butt welded joints where the total area to be welded at any one time does not exceed $\frac{1}{2}$ sq in. and where the abutting surfaces are in close alignment. If used on larger sections the arc fails to distribute itself evenly over the entire surface and unwelded spots result in the joint.

The advantage of the percussion process is that heat penetration is shallow around 0.010″. This makes it feasible to join heat-treated metals without destroying the effects obtained by heat treating. Thus it is possible to weld stellite tips to bronze or stainless steel valve stems or an aluminum fixture to stainless steel or copper without affecting their metallurgical properties. The process is often used in welding wire, bar, and tubing.

Principle of Operation

The arc is started by bringing the workpieces into light contact with each other. As soon as the flow of current is established, the pieces are separated a distance to maintain the arc. On some equipment the arc is started by superimposing an auxiliary high frequency AC voltage on a low-voltage direct current. The high frequency voltage ionizes the gap which then makes it possible for the DC current to maintain the arc.

With the arc established the parts are brought together by percussion force. The force is supplied by pneumatic cylinders or electromagnets. The heat liberated by the arc and the subsequent application of force brings about the required fusion. Since the heat is generated by the arc and does not emanate from the electrical resistance of the materials being welded the fusion temperature of the metals is not an important factor. Hence, dissimilar metals can be welded with comparative ease.

HIGH FREQUENCY RESISTANCE WELDING

High frequency resistance welding employs a high frequency current to generate the required heat for bonding edges of metal. The process has wide application in fabricating tubes, pipes, structural shapes, heat exchangers, cable sheathing, and other parts where high speed is required. Both ferrous and nonferrous materials can be welded in thicknesses ranging from 0.004″ to ½″ in thickness.

Principle of Operation

The process is based on the use of current that reverses at a high rate—450,000 hertz (cycles) as compared with the line frequency of

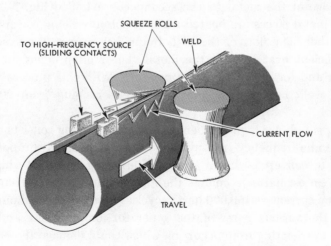

Fig. 5-18. Example of high frequency resistance welding.

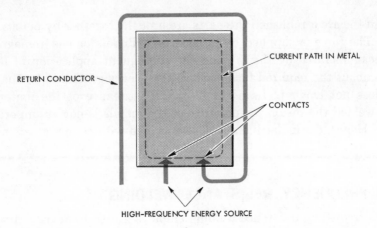

RETURN CONDUCTOR

CURRENT PATH IN METAL

CONTACTS

HIGH-FREQUENCY ENERGY SOURCE

Fig. 5-19. Path of current flow in high frequency resistance welding.

60 hertz (cycles). The current instead of following a direct low resistance path between two contacts follows a long low inductance path. The current is carried to the workpiece by relatively small water-cooled sliding contacts placed on either side of the two abutting edges of the metal. See Fig. 5–18. The current flows from one contact, moves along the surface of the weld seam and penetrates to the root of the weld metal. It then completes the circuit by returning along the opposite edge of the metal to the second contact. See Fig. 5–19. Thus, the two metal pieces are heated without actually being in contact with one another. The flow of the high frequency current resistance generates sufficient heat to melt the metal. The molten edges are brought together and forge welded under pressure. With this process it is possible to weld butt or lap joints at speeds ranging from 20 to over 1000 fpm.

Power supply for high frequency resistance welding consists of a 60 to 560-kw radio frequency transmitter. The transmitter is equipped with a rectifier to convert 60 hertz (cycle) alternating current to high direct current, an oscillator to change the high voltage direct current to high frequency current at 450,000 hertz (cycles) and a water pumping system to cool the various parts of the generator and contacts and prevent harmful overheating from occurring which could damage the equipment and cause a hazardous condition.

REVIEW QUESTIONS

1. How is the heat required for fusion generated in resistance welding?
2. What is the difference between the single and multiple spot welder?
3. In the multiple spot welder, what serves as the bottom electrode?
4. What are the variables which affect the amount of heat required for effective spot welding?
5. Why is correct heat balance important in spot welding?
6. What can be done to achieve proper heat balance?
7. What are some of the probable causes for poor spot welds?
8. How does seam welding differ from spot welding?
9. How are electrodes for seam welding cooled?
10. In what ways does projection welding differ from spot welding?
11. Why are some metals not recommended for projection welding?
12. What is the difference between flash welding and upset welding?
13. What is the basic principle of percussion welding?
14. What are some of the advantages and limitations of percussion welding?
15. What is the operating principle of high-frequency resistance welding?

| Chapter 6 | # Special Welding Processes |

In order to meet special needs of aerospace, nuclear, and other types of industries new metal alloys have come into common usage. At the same time different fabricating techniques had to be developed to cope with the more complex mechanical properties of these metals. One of the critical problems was to find ways to produce virtually perfect weldments. As a rule, most so-called exotic metals are highly susceptible to atmospheric contamination during a welding process. Even with gas-shielded arc where high purity inert gases are used there is often a minute amount of remaining contamination. Since these new metals have such a low tolerance to impurities different welding techniques had to be employed. Some of the new processes are: electron beam welding, laser beam welding, ultrasonic welding, inertia welding, and plasma-arc welding.

ELECTRON BEAM WELDING

Electron beam welding is essentially a fusion welding process. Fusion is achieved by focusing a high power density beam of electrons on the area to be joined. Upon striking the metal the kinetic energy of the high velocity electrons changes to thermal energy causing the metal to melt and fuse.

The electrons are emitted from a tungsten filament heated to approximately 2000°C. Since the filament would quickly oxidize at this temperature, if it were exposed to normal atmosphere, the unit as well as the focusing devices and workpiece are placed in a vacuum. Using a vacuum chamber also prevents the electrons from colliding with molecules of air which otherwise would make them scatter and lose their kinetic energy. Although welding in a chamber has certain limitations the important advantage is that metal can be welded without fear of chemical contamination.

Electron beam welding can be used to join materials ranging from thin foil to 2 inches in thickness. It is particularly adaptable to the welding of refractory metals such as tungsten, molybdenum, columbium, tantalum, and metals which oxidize readily, such as titanium, beryllium, and zirconium. It also has wide application in joining dissimilar metals, aluminum, standard steels, and ceramics.

Advantages

Electron beam welding has several distinct advantages:

1. Welds are nearly straightsided, narrow, and deep. See Fig. 6–1.

2. Welding can be done with a rapid but low energy input. This reduces workpiece distortion and confines the heat-affected zones to narrow limits.

3. The weld size and locations can be controlled producing welds with better metallurgical properties.

4. Multiple welds can be made in a single pass without danger of contamination.

Limitations

One of the major limitations of electron beam welding is that the work must be done in a vacuum chamber. Consequently, the piece

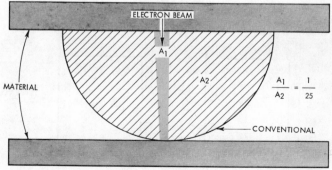

The Buehler Corp.

Fig. 6-1. This illustrates the depth-to-width ratio of a typical electron beam weld compared to a conventional fusion weld.

must be small enough to fit into the chamber. This limitation is being reduced to some extent because larger chambers are now manufactured to accommodate a vast variety of products. Another limitation is that when the workpiece is in the chamber and in a vacuum it becomes inaccessible and therefore must be manipulated by some special device. However, engineers believe that in the future electron beam welding, with all of its advantages, can be done without a vacuum chamber and applied to any normal atmospheric requirement.

Major Units of the System

Electron beam welding equipment includes the following basic modules:

Electron Gun. The gun consists of a filament, cathode, anode, and

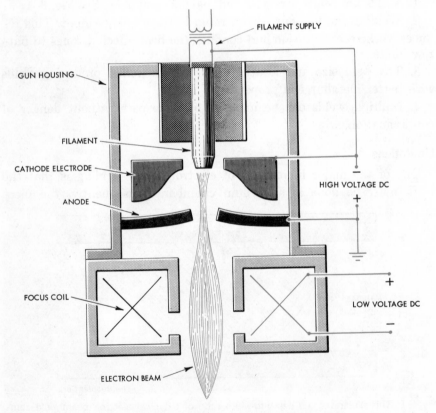

Fig. 6-2. Basic elements of an electron beam gun.

focusing coil mounted above the work chamber. See Fig. 6–2. The electrons emitted from the heated filament carry a negative charge and are repelled by the cathode and attracted by the anode. The electrons pass through an aperture in the anode and then through a magnetic field generated by the electromagnetic focusing coil. An optical viewing system provides a line of sight down the path of the electron beam system provides a line of sight down the path of the electron beam centerline to the weld area when the beam is off. See Figs. 6–3 and 6–4. By varying the current to the focusing coil the operator can focus the beam for gun-to-work distances ranging from ½ inch to 25 inches. The electron beam can be controlled to produce a spot less than .005 of an inch in diameter.

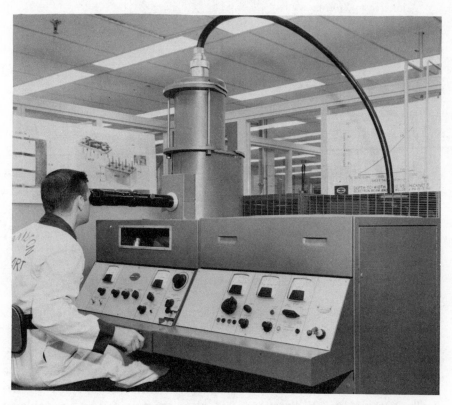

Fig. 6-3. Electron beam welder.

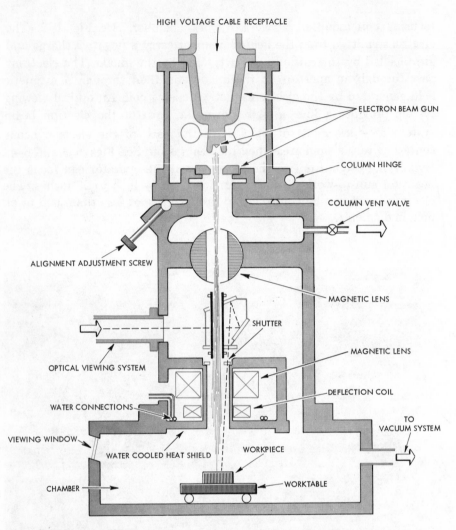

HIGH VOLTAGE CABLE RECEPTACLE

ELECTRON BEAM GUN

COLUMN HINGE

COLUMN VENT VALVE

ALIGNMENT ADJUSTMENT SCREW

MAGNETIC LENS

SHUTTER

OPTICAL VIEWING SYSTEM

MAGNETIC LENS

DEFLECTION COIL

WATER CONNECTIONS

TO VACUUM SYSTEM

VIEWING WINDOW

WORKPIECE

WATER COOLED HEAT SHIELD

CHAMBER

WORKTABLE

Fig. 6-4. Schematic of electron beam gun column.

Vacuum Chamber. This is usually rectangular in shape with heavy glass windows or ports to permit viewing the work. A work table in the chamber is arranged so it can be operated either manually or electrically in the X and Y directions. "T" slots are provided on the table to attach fixtures or workpieces for welding.

Vacuum Pumping System. This is designed to provide a clean, dry vacuum chamber in a relatively short time. The capacity of the pump required is governed by the volume and area of the chamber. The pumping equipment is usually completely automatic.

Electrical Controls. Controls include set-up controls and operating controls. The set-up controls are the instruments required for the initial setup of the welding operation, such as meters for beam voltage, beam current, focusing current, and filament current.

The operating controls consist of stop-and start sequence, high voltage adjustment, focusing adjustment, filament activation, and work table motion. These controls are mounted so they are easily accessible to the operator while he is observing the welding action through the viewing window.

Power Unit. This furnishes the main high voltage supply up to 150 kv and a low filament power up to 6 volts.

LASER WELDING

Laser welding is like welding with a white-hot needle. Fusion is achieved by directing a highly concentrated beam to a spot about the diameter of a human hair. The word "laser" means light amplification by stimulated emission of radiation. The light beam has even a higher energy concentration than the electron beam. The highly concentrated beam generates a power intensity of one billion or more watts per square centimeter at its point of focus. Because of its excellent control of heat input the laser can fuse metal next to glass or even weld near varnish coated wires without damaging the insulating properties of the varnish.

Since the heat input to the workpiece is extremely small in comparison to other welding processes the size of the heat affected zone and the thermal damage to the adjacent parts of the weld are minimized. Thus, it is possible to weld heat-treated alloys without affecting their heat-treated condition. As a matter of fact, the weldment can be held in the hand immediately after the weld is completed.

The laser can be used to join dissimilar metals and other difficult to weld metals such as copper, nickel, tungsten, aluminum, stainless steel, titanium, and columbium. Furthermore, the laser beam can pass through transparent substances without affecting them, thereby making it possible to weld metals that are sealed in glass or plastic. The fact that the heat source is a light beam, the effect of atmospheric contamination on the weld joint is not a problem.

The current application of laser welding is largely in aerospace and electronic industries where extreme control in weldments is required. Its major limitation is the shallow penetration. Present equipment restricts it to metals not over 0.020 inches in thickness.

The duration of the beam is usually about 0.002 seconds with a pulse rate of one to ten times per second. As each point of the beam hits the metal a spot is melted but solidifies in microseconds. The line of weld thus consists of a series of round solidified puddles each overlapping the other. The workpiece is either moved beneath the beam or the energy source is moved across the line of weld.

The principal limitation of the laser for welding is that it cannot be used for continuous welding. The heat generated from the high intensity external light source for pumping the atoms to a high energy state permits the light source to stay operative for only short periods of time. Approximately 100 to 1000 units of heat energy are released for every unit of light energy. This limits the repetition rate with which pulses can be produced.

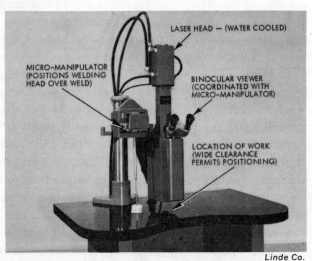

LASER HEAD — (WATER COOLED)

MICRO–MANIPULATOR
(POSITIONS WELDING
HEAD OVER WELD)

BINOCULAR VIEWER
(COORDINATED WITH
MICRO-MANIPULATOR)

LOCATION OF WORK
(WIDE CLEARANCE
PERMITS POSITIONING)

Linde Co.

Fig. 6-5. Laser beam welder.

Focusing the beam onto the workpiece is accomplished with an optical system and the actual control of the welding energy by means of a switch. See Fig. 6–5. An important feature of the laser is that the weld does not have to be done in a vacuum.

Theory of the Laser Beam.[1] Atoms have been made to generate energy by exciting them in such common devices as fluorescent lights and television tubes. Fluorescence refers to the ability of certain atoms to emit light when they are exposed to external radiation of shorter wave lengths.

In the Laser Welder, the atoms that are excited to produce the laser light beam are locked in a man-made ruby rod 3/8" in diameter. See Fig. 6–6. The ruby is identical to a natural ruby but has a more perfect crystal structure. About .05 percent of its weight consists of chromium oxide.

The chromium atoms give the ruby its red color because they absorb green light from external light sources. When the atoms absorb this light energy, some of their electrons are excited. Thus, green light is said to pump the chromium atoms to a higher energy state.

The atoms eventually return to their original state. In doing so, they give up a portion of the extra energy they previously absorbed (as green light) in the form of red fluorescent light.

When the red light emitted by one excited atom hits another excited atom, the second atom gives off red light which is in phase with the colliding red light wave. In other words, the red light from the first atom is amplified because more red light exactly like it is produced.

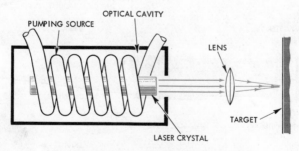

Fig. 6-6. Schematic diagram of laser ruby.

[1]Courtesy Union Carbide Corp.—Linde Division

By using a very intense green light to excite the chromium atoms in the ruby rod, a larger number of atoms can be excited and the chances of collisions are increased.

To further enhance this effect, the parallel ends of the rod are mirrored to bounce the red light back and forth within the rod. When a certain critical intensity of pumping is reached (the so-called threshold energy), the chain reaction collisions become numerous enough to cause a burst of red light. The mirror at the front end of the rod is only a partial reflector, allowing the burst of light to escape through it.

ULTRASONIC WELDING

If two metal pieces with perfectly smooth surfaces were brought into intimate contact the metal atoms of one piece conceivably could unite with the atoms of the other piece to form a permanent bond. However, regardless of how smooth such surfaces could be produced a sound metallurgical bond normally would not occur because it is virtually impossible to prepare surfaces that are absolutely smooth. No matter what means were used to smooth surfaces they would still possess peaks and valleys as measured on an atomic scale. As a result only the peaks of both workpieces which came into intimate contact would unite leaving the countless atoms of valleys without the power of forming a bond. Furthermore, smooth surfaces are never actually clean. Oxygen molecules from the atmosphere react with the metal to form oxides.

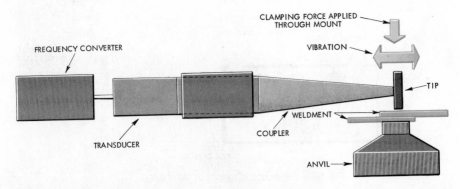

Fig. 6-7. Schematic of an ultrasonic welder.

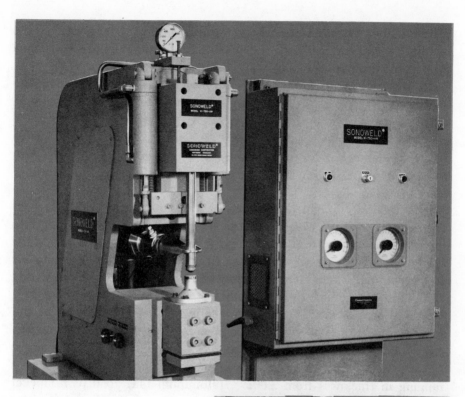

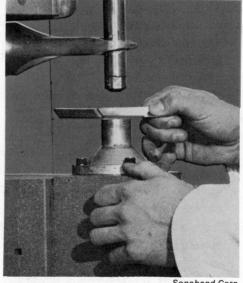

Fig. 6-8. Ultrasonic spot welder.

Sonobond Corp.

These oxides attract water vapor forming a film of moisture on the oxidized metal surface. Both the moisture and oxide film act as barriers to prevent intimate contact.

In the ultrasonic welding process these three existing barriers are broken down by plastically deforming the interface between the workpieces. This is done by means of vibratory energy which disperses the moisture, oxide, and irregular surface to bring the areas of both pieces into close contact and form a solid bond. Vibratory energy is generated by a transducer. A schematic of a typical ultrasonic welding unit is shown in Fig. 6–7.

The welding equipment consists of two units: a power source or frequency converter which converts 60 hertz (cycle) line power into high-frequency electrical power, and a transducer which changes the high-frequency electrical power into vibratory energy. The components to be joined are simply clamped between a welding tip and supporting anvil with just enough pressure to hold them in close contact. See Fig. 6–8. The high frequency vibratory energy is then transmitted to the joint for the required period of time. The bonding is accomplished without applying external heat, filler rod, or melting metal. Either spot-type welds or continuous-seam welds can be made on a variety of metals ranging in thickness from .00017″ (aluminum foil) to 0.10″. Thicker sheet and plate can be welded if the machine is specifically designed for them. High strength bonds are possible both in similar and dissimilar metal combinations.

Ultrasonic welding is particularly adaptable for joining electrical and electronic components, hermetic sealing of materials and devices, splicing metallic foil, welding aluminum wire and sheet, and fabricating nuclear fuel elements.

Welding variables such as power, clamping force, weld time for spot welds, or welding rate for continuous seam welds can be preset and the welding cycle completed automatically. A foot switch or some other triggering mechanism lowers the welding head, applies the necessary clamping force, and starts the flow of ultrasonic energy.

Successful ultrasonic welding depends on the proper relationship between these welding variables which is usually determined experimentally for a specific application. Thus, clamping force may vary from a few grams for very light materials to several thousand pounds for heavy pieces. Weld time can range from 0.005 to 1.0 seconds for spot welding and a few feet per minute to 400 fpm for continuous-seam welding. The high-frequency electrical input to the transducer may be anywhere from a fraction of a watt to several kilowatts.

PLASMA WELDING

Plasma welding is a process which utilizes a central core of extreme temperature surrounded by a sheath of cool gas. The required heat for fusion is generated by an arc heating a gas to such a high temperature that the gas becomes ionized. The tip of the electrode is located within a torch nozzle having a small opening that constricts the arc. As gas is fed through the arc it becomes heated to what is known as the plasma temperature range. The plasma jet is forced through the constricting orifice and is accelerated to sonic velocity (4000 ft per sec) with an intense heat (30,000°F), which is hotter than any flame or conventional electric arc. When this super-heated columnar arc is directed on metal it makes possible butt welds up to one-half inch in thickness or more in a single pass without filler rod or edge preparation. See Fig. 6–9.

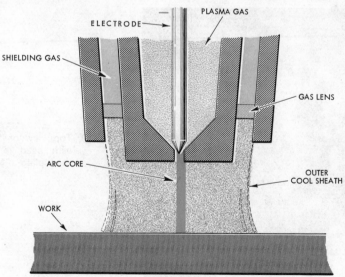

Fig. 6-9. Plasma welding uses a central core of extreme temperature surrounded by a sheath of cool gas.

Plasma is often considered the fourth state of matter. The other three are gas, liquid, and solid. Plasma results when a gas is heated to a high temperature and changes into positive ions, neutral atoms, and negative electrons. When matter passes from one state to another latent heat is generated. Thus latent heat is required to change water into steam, and similarly, the plasma torch supplies energy to a gas to change it into plasma. When the plasma changes back to a gas the heat is released.

In some respects plasma welding may be considered as an extension of the conventional gas tungsten-arc welding. The main difference is that in plasma welding the arc column is constricted and it is this constriction that produces the much higher heat transfer rate.

As the arc gas strikes the metal it cuts or "keyholes" entirely through the piece producing a small hole which is carried along the weld seam. See Fig. 6-10. During this cutting action the melted metal in front of the arc flows around the arc column, then is drawn together immediately behind the hole by surface tension forces and reforms in a weld bead.

The specially designed torch (Fig. 6-11) for plasma welding can be hand held or mounted for stationary or mechanized applications. See Fig. 6-12. The process can be used to weld stainless steels, carbon

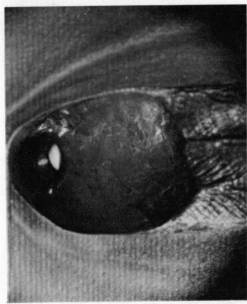

Fig. 6-10. Top view of "keyhole" with bead in ½ inch thick plate of titanium.

Thermal Dynamics Corp.

Fig. 6-11. Torch for plasma welding.

Thermal Dynamics Corp.

Thermal Dynamics Corp.

Fig. 6-12. Plasma welding torch can be mounted for stationary or mechanized applications.

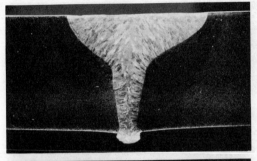

Single-pass butt weld in ¼ in. thick Type 304 stainless steel. Welds produced by "keyholing" have this typical "wine glass" shape. No edge preparation or filler metal.

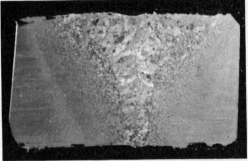

Single-pass butt weld in ½ in. thick titanium. "Wine glass" configuration bounds coarse grain structure. No edge preparation or filler metal.

Single-pass butt weld in ¼ in. thick mild steel using "keyhole" technique. No edge preparation or filler metal.

Thermal Dynamics Corp.

Fig. 6-13. Types of weld penetration on different metals produced with plasma welding.

steels, Monel, Inconel, titanium, aluminum, copper and brass alloys. See Fig. 6–13. Although for many fusion welds no filler rod is needed a continuous filler wire can be added for various fillet types of weld joints.

Equipment. A regular heavy duty DC rectifier is used as the source of power for plasma welding. A special control console is required to

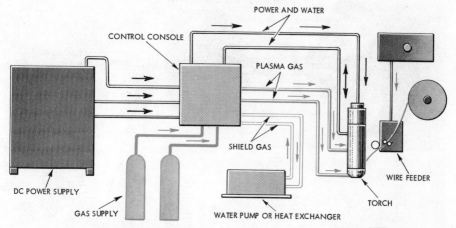

Fig. 6-14. Schematic of a plasma welding installation.

provide the necessary operating controls. A water cooling pump is usually needed to assure a controlled flow of cooling water to the torch at a regulated pressure. Proper cooling prolongs the life of the electrode and nozzle. See Fig. 6–14.

Gas supply is either argon or helium. In some application argon is used as the plasma gas and helium as the shielding gas. However, in many operations argon is used for both shielding and generating the plasma arc.

INERTIA WELDING

Inertia or friction welding is a process where stored kinetic energy is used to generate the required heat for fusion. The two workpieces to be joined are axially aligned. One is held stationary by means of a chuck or fixture and the other is securely clamped in a rotating spindle.

The rotating member is brought up to a certain speed so as to develop sufficient energy. Then the drive source is disconnected and the pieces brought into contact under a pre-computed thrust load. At this point the kinetic energy contained in the rotating mass converts to

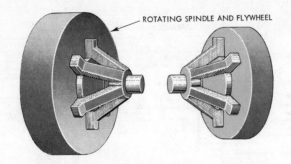

ROTATING SPINDLE AND FLYWHEEL

1. PIECES ARE ALIGNED
   AND CLAMPED

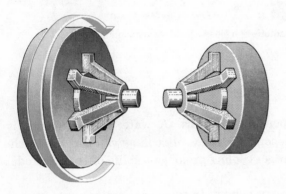

2. FLYWHEEL IS ROTATED BY
   AN EXTERNAL ENERGY SOURCE

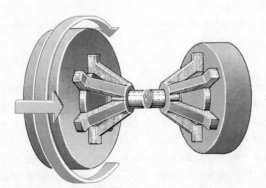

3. MEMBERS BROUGHT INTO CONTACT

Fig. 6-15. In inertia welding heat resulting from stored kinetic energy is used to forge the pieces together.

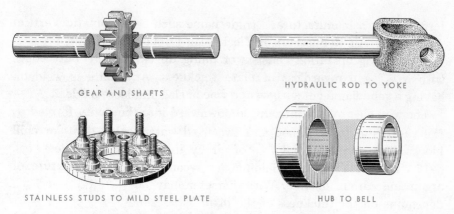

GEAR AND SHAFTS

HYDRAULIC ROD TO YOKE

STAINLESS STUDS TO MILD STEEL PLATE

HUB TO BELL

Fig. 6-16. Examples of parts welded by inertia welding.

frictional heat. The metal at and immediately behind the interface is softened, permitting the workpiece to be forged together. See Figs. 6–15 and 6–16.

Inertia welding has several advantages over conventional flash or butt welding. It produces improved welds at higher speed and lower cost, less electrical current is required, and costly copper fixtures for hold parts are eliminated. With inertia welding there is less shortening of the components, which often occurs in flash or butt welding. Also the heat-affected zone adjacent to the weld is confined to a narrow band and therefore does not draw the temper of the surrounding area.

The inertia welding process is applicable for welding many dissimilar or exotic metals as well as similar metals. Weld strength is normally equal to that of the original metals.

ELECTRO-GAS WELDING

Electro-gas welding is a process which uses a gas-shielded metal arc and is designed for single-pass welding of vertical joints in steel plates ranging in thickness from ⅜″ to 1½″. The process is frequently referred to by a manufacturer's trade name such as Automatic Vertical Aircomatic (Air Reduction) or Electroslag (Hobart Brothers).

The joining operation consists of lining up two square butt edges with a ½″ gap regardless of plate thickness. Any plate is weldable having a good flame-cut surface and free of slag and loose oxide.

The wire electrode is introduced downward into the cavity formed by the two plates to be joined and two fixed water-cooled dams or chill blocks. The cavity is kept free of air by the shielding gas. See Figs. 6–17A and 6–17B. The shielding gas recommended is a mixture of argon and carbon dioxide. Wire diameter may be $\frac{1}{16}″$, $\frac{5}{64}″$, or $\frac{7}{32}″$, depending on the thickness of the plate.

The welding head is suspended from an elevator mechanism which provides automatic control of the vertical travel speed during welding. This mechanism raises the welding head automatically at the same rate as the advancing weld metal. The welding head is self-aligning and can follow any alignment irregularity in plate or joint.

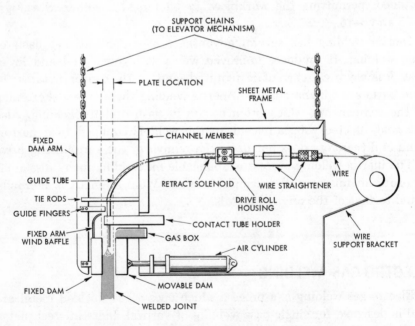

Fig. 6-17A. Schematic of an electro-gas welding head.

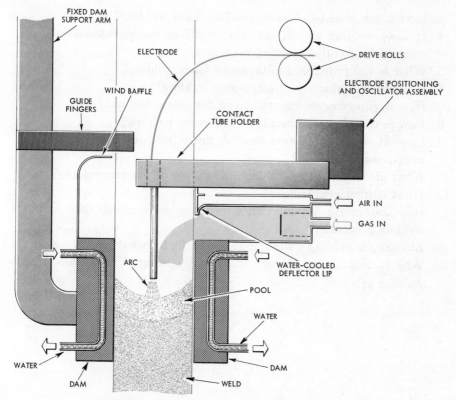

FIXED DAM
SUPPORT ARM

ELECTRODE

DRIVE ROLLS

ELECTRODE POSITIONING
AND OSCILLATOR ASSEMBLY

WIND BAFFLE

GUIDE
FINGERS

CONTACT
TUBE HOLDER

AIR IN

GAS IN

ARC

WATER-COOLED
DEFLECTOR LIP

POOL

WATER

WATER

WATER

DAM

DAM

WELD

Fig. 6-17B. Enlarged schematic of welding portion of head.

Once the equipment is positioned on the joint welding is completely automatic. Wire feed and current are constant. At the end of the weld the process automatically stops.

This welding technique is especially recommended for use in ship-yards, and in the fabrication of storage tanks and large diameter pipe.

REVIEW QUESTIONS

1. Why is a vacuum chamber required for electron beam welding?
2. What are some of the advantages and limitations of electron beam welding?

3. How is the laser beam generated for laser welding?
4. In laser welding why is the heat input to the workpiece small in comparison to other welding processes?
5. What is the principal limitation of laser welding?
6. What is the principle of ultrasonic welding?
7. How is vibratory energy produced for ultrasonic welding?
8. How is the plasma generated for plasma welding?
9. In what way does plasma welding differ from the conventional gas tungsten-arc welding?
10. What are some of the specific advantages of plasma welding?
11. What is inertia welding?
12. How does inertia welding differ from the conventional flash or butt welding process?
13. Electro-gas welding is designed specifically for what kind of welding?
14. Why is the electro-gas welding usually referred to as a vertical welding process?

Chapter 7

Metallurgy Of Welding

In theory as well as in practice welding is a vital phase of the general area of metallurgy and it encompasses the same disciplines of physics, chemistry, mechanics, stress analysis, and testing as well as the practical art of making sound castings. Fusion welding has frequently been referred to as a high quality casting process. All of the problems that are normally associated with metal casting are present in fusion welding, that is, expansion, contraction, shrinkage cracks, segregation of elements, formation of compounds, grain growth, and response to changes in temperature (hardening and resulting brittleness).

To successfully weld metals a basic knowledge of the metallurgical nature of different metals is essential. Actually metals are like people. In people, mental and physical behavior is largely determined through heredity and learning. Certain limitations may exist but there is also the capability of improving these limitations to some degree. The elements composing metals correspond to heredity and they influence the limit or optimum behavior of materials.

Welding affects any material. Sometimes the effects are slight with no deleterious results. At other times the effects are very harmful even including changes in the properties of the metal.

Each material known to man is classified as an element, a compound, or a mixture.

Elements. An element is a material made only of itself. It is a pure substance and does not contain other ingredients. Typical elements are iron, copper, gold, oxygen, hydrogen, and sulfur.

Compounds. A compound is a material composed of two or more elements joined together in definite proportions. For example, water (H_2O) is a compound of one part oxygen and two parts hydrogen. Some compounds may have three elements; others may have four or more elements.

Mixtures. A mixture is a material composed of two or more elements or compounds in which the ingredients are mixed together without any fixed proportions. Each ingredient retains its own identity.

The Atom

The atom is the smallest particle of matter that exhibits the characteristics of an element. Physicists believe that the atom is not a solid body, but consists of neutrons, protons, and electrons held together by the attraction of positive and negative electrical charges. The center of the atom is made of neutrons and protons and is where practically all the mass or weight is concentrated. The outer portion of the atom contains the electrons. The neutron has an equal number of positive and negative charges and is electrically neutral. The proton is a positive charged particle whereas the electron is a negative charged particle. Fig. 7-1 is a pictorial representation of an atom for one of the simplest

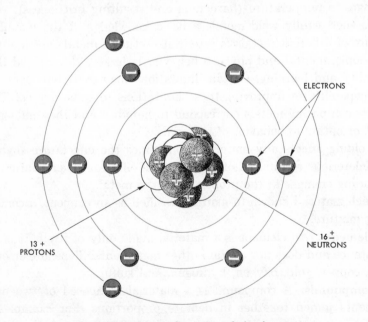

Fig. 7-1. Atomic configuration of aluminum.

metals, aluminum. Notice that in this illustration the atom looks like a miniature solar system in which the nucleus is the sun and the electrons the orbiting planets. However, the atom should be pictured as a sphere rather than a plane as it is shown on paper.

METALS AND NONMETALS

The 103 known elements are divided into two main groups—metals and nonmetals. Scientists usually determine in which group an element belongs by conducting an electrolysis test. To do this they dissolve

Table I. Properties of Metals

Metal or Composition	Chemical Symbol	Specific Gravity	Weight per Cubic Foot, Pounds	Weight per Cubic Inch, Pound	Melting Point, Deg. F.
Aluminum	Al	2.70	168.5	0.0975	1220
Antimony	Sb	6.618	413.0	0.2390	1167
Barium	Ba	3.78	235.9	0.1365	1562
Bismuth	Bi	9.781	610.3	0.3532	520
Boron	B	2.535	158.2	0.0916	4172
Brass	..	8.60	536.6	0.3105	1823
Bronze	..	8.78	547.9	0.3171	1841
Cadmium	Cd	8.648	539.6	0.3123	610
Calcium	Ca	1.54	96.1	0.0556	1490
Chromium	Cr	6.93	432.4	0.2502	2939
Cobalt	Co	8.71	543.5	0.3145	2696
Copper	Cu	8.89	554.7	0.3210	1981
Gold	Au	19.3	1204.3	0.6969	1945
Iron, Cast	..	7.03–7.73	438.7–482.4	0.254–0.279	1990–2300
Iron, Wrought	Fe	7.80–7.90	486.7–493.0	0.282–0.285	2750
Lead	Pb	11.342	707.7	0.4096	621
Magnesium	Mg	1.741	108.6	0.0628	1204
Manganese	Mn	7.3	455.5	0.2636	2300
Mercury (68° F.)	Hg	13.546	845.3	0.4892	−38
Molybdenum	Mo	10.2	636.5	0.3683	4748
Nickel	Ni	8.8	549.1	0.3178	2651
Platinum	Pt	21.37	1333.5	0.7717	3224
Potassium	K	0.870	54.3	0.0314	144
Silver	Ag	10.42–10.53	650.2–657.1	0.376–0.380	1761
Steel, Carbon	..	...	489.0–490.8	0.283–0.284	2500
Tantalum	Ta	16.6	1035.8	0.5998	5162
Tellurium	Te	6.25	390.0	0.2257	846
Tin	Sn	7.29	454.9	0.2633	449
Titanium	Ti	4.5	280.1	0.1621	3272
Tungsten	W	18.6–19.1	1161–1192	0.672–0.690	6098
Uranium	U	18.7	1166.9	0.6753	3362
Vanadium	V	5.6	394.4	0.2022	3110
Zinc	Zn	7.04–7.16	439.3–446.8	0.254–0.259	788

an element in an acid and then pass an electric current through the solution. If the element is a metal, its atoms will show a positive charge and will be attracted to the negative pole where the electric current enters the solution. Hence metals are scientifically defined as elements which, when in solution and in a pure state, carry a positive charge and seek the negative pole in an electrical cell. Over three-fourths of the known elements are metals. See Table I for a partial list of these elements. Insofar as welding is concerned, the metallic elements are the most significant.

Structure of Metals

Each metallic element is made up of a vast number of atoms which are arranged in certain patterns in the solid state. Atoms of elements that arrange themselves in an orderly geometric pattern upon solidification are said to be crystalline. Atoms of nonmetallic elements do not solidify into any specific pattern.

Crystals grow in all directions upon solidification until they meet other crystals also growing in the melt. When crystals meet and produce irregular boundaries they are called *grains*. See Fig. 7–2.

In the solid or crystalline state the protons in the nucleus attract the electrons of adjacent atoms as well as their own orbiting electrons. The

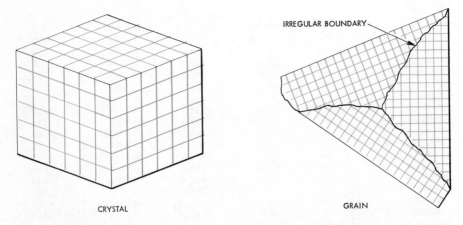

CRYSTAL   GRAIN

Fig. 7-2. A regular perfect shape is called a crystal and an irregular boundary makes it a grain.

protons of each atom repel the protons of other atoms. Likewise the electrons of each atom repel one another. When the forces of attraction and repulsion are equal or in balance there is a state of equilibrium. These interatomic forces give metals their high strength and rigidity.

Structure refers to the arrangement of the atoms in a metal according to a microstructural and a macrostructural scale. (Microstructure is the appearance of a polished and etched metal specimen under a microscope; macrostructure is the appearance of a fractured metal specimen as viewed by the naked eye.) The mechanical properties of metals are governed by their structures.

During welding temperatures range from room temperature to well above the melting points of pure metals and alloys. These temperature differences within the metal cause many structures to form and various grain size changes to take place, all of which have specific effects on welded areas.

Space Lattice

A space lattice is a pictorial representation of the orderly geometric arrangement of atoms of metals in a solid state. The atoms of metals that are particularly important to welders are described in terms of three different types of crystals: (1) the body-centered cubic lattice (BCC), (2) the face-centered cubic lattice (FCC), and (3) the hexagonal close-packed lattice (HCP). See Fig. 7–3.

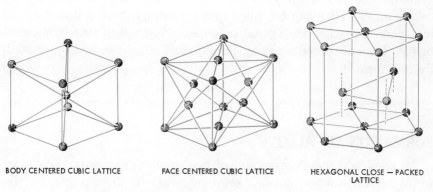

BODY CENTERED CUBIC LATTICE   FACE CENTERED CUBIC LATTICE   HEXAGONAL CLOSE — PACKED LATTICE

Fig. 7-3. The most common crystal patterns.

In the *body-centered cubic* lattice there are 9 atoms with one atom at each corner of the cube and one in the center of the cube. The *face-centered cubic* structure has 14 atoms with one atom at each corner of the cube and one in the center of each face of the cube. In the *hexagonal close-packed* arrangement the ends of the crystal each require 7 atoms with one atom in the center and one at each corner. The two ends are separated by three equally spaced atoms. There are a total of 17 atoms in this lattice.

Most metals always solidify upon cooling in only one lattice form at room temperature. However, a few metals can exist in two lattice forms at different temperatures. Metals of this kind are said to be *allotropic*, of which iron is an example. The following are the lattice type of some of the more common metals:

Lattice Type	Metal
FCC	Aluminum
BCC and FCC at 1670°F	Iron
FCC	Copper
FCC	Lead
BCC	Tin
HCP	Magnesium
HCP	Zinc
HCP	Titanium

Metals which solidify into the body-centered cubic structure are strong with high yield strength and limited workability. Face-centered cubic metals have lower yield strength but possess great capacity for cold working. They can be rolled into exceedingly thin foil and drawn into very fine wire. Hexagonal close-packed structures are somewhat brittle. The only exceptions are titanium and zirconium which are readily cold formed.

FORMATION OF ALLOYS

An alloy is a metal composed of two or more elements. Alloys are formed in three ways: as solid solutions, mechanical mixtures, and intermetallic compounds.

Solid Solutions

Solid solution alloys are those whose elements completely dissolve in each other. There are two types of solid solutions known as substitutional and interstitial.

Substitutional. In the substitutional solid solution the solute atoms replace atoms of the solvent in the space lattice. For example, if nickel is added to copper in Monel metals the copper atoms will replace some of the nickel atoms.

Interstitial. In the interstitial solid solution the solute atoms instead of replacing the solvent atoms will merely occupy positions between the solvent atoms. Since most atoms are too large to fit easily into interstitial positions only partial solubility is possible. An example of an interstitial solid solution in a metal is the carbon in iron at elevated temperatures or gamma iron, and in stainless steels where chromium is dissolved in iron.

General Characteristics of Solid Solution Alloys. Solid solution alloys have several specific characteristics.

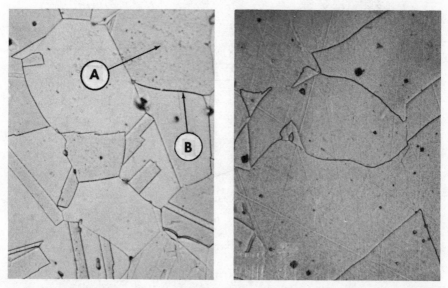

Fig. 7-4. Solid solution alloys, left, resemble pure metals, right. The solid solution shown is Austenite. "A" marks the solution while "B" indicates the boundary.

1. When viewed under a microscope the solute is not visible. The grain structure looks the same as the grain structure of the solvent or pure metal. See Fig. 7–4.

2. The difference in size between the atoms of the solvent and the atoms of the solute creates strains in the crystal structure. The distortion caused by these strains increases the resistance of the metal to being deformed. This increases the yield strength, the tensile strength, and the hardness of the metal. It also slightly lowers the ductility. See Fig. 7–5.

3. Electrical and thermal conductivity of solid solutions are lower than the conductivity of pure metals. Any irregularity of crystal structure always lowers conductivity.

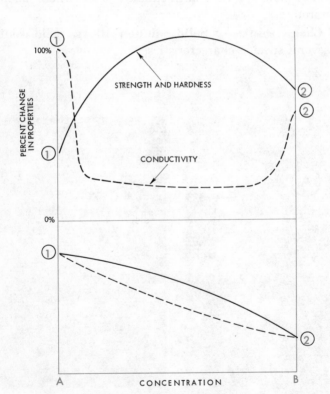

Fig. 7-5. Behavior that might be expected from a solid solution alloy.

Mechanical Mixtures

Metals which are soluble in one another in the liquid state but freeze into separate crystals which are insoluble in one another in the solid state form mechanical mixtures called *eutectics*.

A eutectic composition is fixed. There is a definite ratio of elements which never varies. Typical eutectic alloys contain such elements as cadmium-bismuth, antimony-lead, copper-silver, aluminum-copper.

An examination of any eutectic composition diagram such as the one shown in Fig. 7–6 for cadmium-bismuth, will indicate how the eutectic is formed. Notice that point C on this diagram designates the eutectic composition and temperature at which this alloy melts or solidifies. The left hand ordinate represents 100 percent cadmium and the right hand ordinate 100 percent bismuth. Adding cadmium to bismuth lowers its melting point; also bismuth added to cadmium reduces its melting point.

When an alloy of 60 percent bismuth and 40 percent cadmium is allowed to cool from a liquid, it remains liquid until it reaches the temperature indicated at point C. In the liquid state cadmium and bismuth are completely soluble in one another in all proportions. At a tempera-

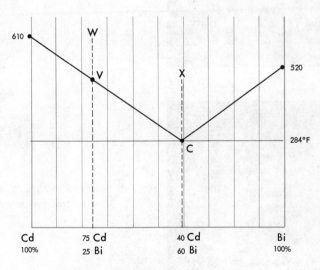

Fig. 7-6. Cadmium-Bismuth diagram.

ture of 284°F, cadmium and bismuth become completely insoluble in one another. At this stage they instantly change from a liquid to a solid state.

The grains of this alloy are composed of alternate layers of pure cadmium and pure bismuth. See Fig. 7–7. If there is more cadmium or bismuth than is needed to form the eutectic, then the excess constituent is precipitated from the liquid as pure metal before the eutectic forms and solidifies over a range of temperatures. For example, if the alloy in the liquid state contains 75 percent cadmium and 25 percent bismuth there will be more cadmium than needed to form all eutectic. When the melt is cooled to the temperature at point "V" (Fig. 7–6), pure cadmium is precipitated from the liquid. The remaining liquid

Fig. 7-7. Eutectic Cadmium-Bismuth alloy.

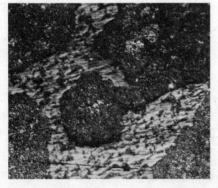

Fig. 7-8. Structure of 75% Cd and 25% Bi. Dark masses are pure cadmium.

becomes richer in bismuth. This continues as the temperature drops until the eutectic composition is reached at which time the eutectic solidifies at a constant temperature. The microstructure of such an alloy will contain grains of pure cadmium and grains of eutectic. See Fig. 7–8.

Intermetallic Compounds

Metals may combine with metals or nonmetals to form compounds. These compounds are called intermetallic compounds and are similar to such combinations as $NaCl$, H_2SO_4, etc.

In general, intermetallic compounds are the principal hardeners in alloys. Compounds are never used by themselves because of their low strength and brittleness but they strengthen the metals to which they are added.

In an alloy of this type, metals may combine in a very definite ratio, such as Fe_3C in which three atoms of iron react with one atom of carbon to form the compound iron carbide, which is usually referred to as *cementite*. Aluminum and copper react to form the compound $CuAl_2$, where two atoms of aluminum reacts with one atom of copper to form the compound. See Fig. 7–9.

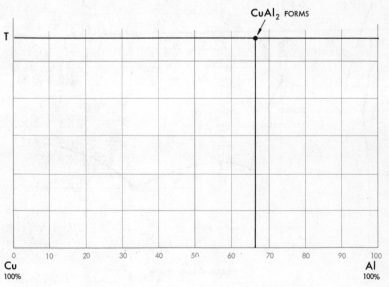

Fig. 7-9. Formation of compound, $CuAl_2$.

Compounds melt at a specific temperature of their own and this temperature cannot be predicted from the melting points of the elements which make up the compound. Examples of alloy systems involving intermetallic compounds are: aluminum-copper, beryllium-copper, lead-tin, steels, and copper magnesium. A typical system involving the formation of an intermetallic is the one for copper-magnesium shown in Fig. 7-10. The atomic percentages are plotted across the top for convenience. It enables the atom ratio to be seen at a glance. In this alloy system two compounds may be formed, Cu_2Mg or $CuMg_2$, depending upon the original alloy chosen. These compounds now behave as elements and are capable of dissolving excess constituents or reacting with them to form eutectics. The mechanisms of forming either of these reactions are the same as in the diagrams illustrated earlier for solid solutions and eutectics.

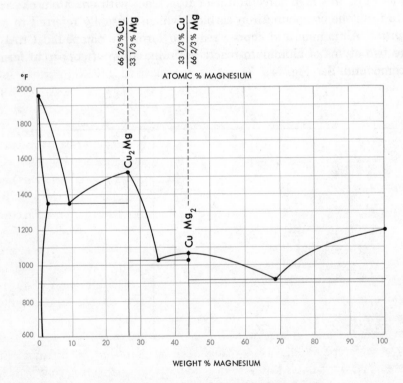

Fig. 7-10. A copper-magnesium alloy diagram.

Iron Based Alloys

Iron based alloys are among the most important of the engineering alloys. Most engineering metals are alloys because pure metals have relatively low strength. For example, commercially pure iron has a yield strength of about 30,000 pounds per square inch, a tensile strength of approximately 45,000 pounds per square inch, with a 40 percent elongation in two inches. Its strength can only be increased by cold working. On the other hand, an alloy of iron has a yield strength of 300,000 psi, a tensile strength of 325,000 psi and an elongation of 1 percent in two inches. There are thousands of alloys in between these two which have varying degrees of strength and ductility.

Equilibrium Diagram

To understand what happens to iron alloys when they are heated and cooled, a graphical chart is often used. Such a chart is usually referred to as an *equilibrium diagram.*

In studying an equilibrium diagram as shown in Fig. 7–11, it must be remembered that carbon is the most important alloying element used with iron. Iron and carbon combine to form a hard, brittle intermetallic compound. This compound, Fe_3C, is composed of three atoms of iron and one atom of carbon. Atomically 25 percent of this compound is carbon, however, it contains 6.67 percent carbon by weight.

Iron undergoes allotropic changes when it is heated to a temperature of 1670°F. At this temperature the metal changes from a body-centered cubic structure to a face-centered cubic structure. If heated still further to a temperature of 2535°F it undergoes another allotropic change. In this instance, the iron transforms from a face-centered cubic structure back to a body-centered cubic structure. When these allotropic changes take place other important changes also occur. Some of these are changes in solid solubility, formation of eutectoid structures, grain growth, and recrystallization. Changes in solid solubility and changes from one allotropic form to another can be seen on the iron-carbide diagram in Fig. 7–11.

Notice in this diagram that the left-hand ordinate is 100 percent iron and the allotropic changes are designated as points at 1670°F and at 2535°F. The right-hand ordinate is shown as 6.68 percent carbon which is 100 percent compound, Fe_3C.

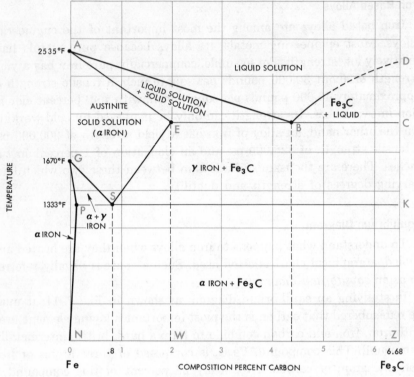

Fig. 7-11. Iron-carbide alloy diagram.

No matter what quantity of carbon is added to iron it reacts with the iron to form a compound. This intermetallic compound now behaves as though it were an element and reacts with the iron. In area AESG of the diagram iron is in a face-centered cubic structure and is capable of dissolving up to 2 percent carbon by weight. A solid solution of carbon in a face-centered cubic iron is called *austenite*. See Fig. 7–4. Thus in an austenitic structure the carbon goes into solution form and becomes evenly distributed in the iron.

Face-centered cubic iron usually is referred to as gamma iron and designated by the Greek symbol γ. At temperatures below 1670°F, body-centered iron is indicated as alpha α iron. Body-centered cubic iron is also called *ferrite*. See Fig. 7–12.

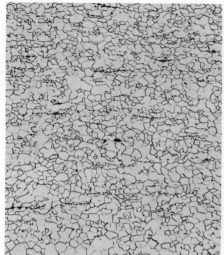

Fig. 7-12. Ferrite structure magnified.

Fig. 7-13. Example of a pearlitic structure.

The addition of carbon to iron lowers the temperature at which gamma iron transforms to alpha iron. Pure iron transforms at 1670°F and the lowest transformation occurs at a temperature of 1333°F for 0.8 percent carbon. When gamma iron changes to alpha iron there is a great change in solubility. Alpha iron has almost zero capacity to hold carbon in solution. At 1333°F austenite decomposes into alternate layers of ferrite and carbide. This structure is similar in form to the eutectic. However, the change takes place in the solid state, whereas the eutectic is a liquid to solid change. Since the structure is similar to eutectic it is called a *eutectoid* structure. The eutectoid structure is usually referred to as *pearlite* and always has the same composition. See Fig. 7–13.

If any iron carbon alloy above 0.3 percent is rapidly cooled the above changes are prevented from occurring and instead a structure called *martensite* is formed. Martensite is the constituent found in fully hardened steel which is hard and brittle. See Fig. 7–14.

Fig. 7-14. Structure of Martensite.

Iron Carbon Eutectic

In cooling from a liquid state an alloy of iron with 4.3 percent carbon solidifies at 2066°F into layers of austenite of 2 percent carbon and layers of iron carbide. Checking the diagram in Fig. 7–11, it will be found that point E indicates the percent of carbon in the austenite and point K the constituent, iron carbide. Upon further cooling from 2066°F austenite has a decreasing capacity to hold carbon in solution. The line ES shows a decreasing solubility with the lowering of temperature. Iron carbide is precipitated out of solution with decreasing temperature until a composition of 0.8 percent carbon is reached at 1333°F. At this temperature austenite decomposes into pearlite. No further change takes place upon cooling to room temperature. The microstructure of alloy now consists of grains of iron carbide and pearlite. This eutectic, which is very hard and brittle, is called white cast iron.

Plain Carbon Tool Steel

Consider an alloy of iron and carbon which contains 1 percent carbon. In cooling from the liquid state the first solid crystals formed are solid solutions of carbon in gamma iron. The composition of the solid is determined by finding the intersection of the temperature line, called an isotherm, with the solidus line AE, Fig. 7–11, and extending a perpendicular down to the percent carbon axis. As the temperature drops more solid solutions form until all of the alloy has solidified. There is now an infinite number of solid solutions similar to the solid solution alloy system. However, in this case the temperature is so high that diffusion is rapid and a uniform solid solution of 1 percent carbon is readily obtained. Upon continued cooling the solid solution is stable until the line SE is reached. This line indicates the decreasing solubility for iron carbide as in the case of the eutectic. The gamma iron precipitates the iron carbide out of solution until it reaches a temperature of 1333°F. When this temperature is attained pearlite is formed in the same manner as in the eutectic. The final structure is composed of crystals of iron carbide and pearlite. Tool steels with this structure are hard and brittle.

Structural Steel

If a structural steel containing 0.25 percent carbon is cooled slowly from above liquid it will solidify into an infinite number of solid solutions of carbon in iron. As in the previous example when the solid solutions are formed at this high temperature diffusion takes place rapidly producing a uniform solid solution of 0.25 percent carbon in iron. The composition of this alloy is less than the 0.8 percent required for the formation of all pearlite. The gamma iron now has a decreasing solubility for iron. With continued cooling alpha iron is precipitated from the solution until a temperature of 1333°F is reached. At this temperature enough ferrite has been precipitated from the solution so that the composition of the alloy is 0.8 percent carbon. Now the austenite transforms to pearlite. The alloy consists of ferrite and pearlite. Ferrite is soft and ductile, consequently structural steels have good plasticity.

No matter what the carbon content iron carbon alloys always contain pearlite. At 0.8 percent carbon the structure is all pearlite. If there is more carbon than 0.8 percent then there will be an excess of iron carbide. If the alloy contains less than 0.8 percent carbon, the structure will contain an excess of ferrite.

Hardening Steel

Steel may be changed from a comparatively soft to a very hard condition by heating it to an austenitic range and cooling it rapidly. Hardening involves dissolving carbon in gamma iron until a homogeneous solid solution is obtained followed by a suitable quench. Rapid quenching prevents the normal movement of carbon out of solution to form pearlite. When the gamma iron transforms to alpha iron the carbon which was dissolved in the austenite becomes insoluble in the alpha iron. It is, nevertheless, trapped in the structure and produces a very high state of strain in the ferrite lattice, distorting it until it becomes body-centered tetragonal in shape. This new structure which is characteristic of fully hardened steels is called martensite. See Fig. 7–14.

Fully hardened martensite is very hard but also very brittle. Because of this brittleness the metal is likely to crack when in service. Hence a fully hardened structure is often subjected to a tempering treatment.

To induce the formation of martensite three conditions must prevail. First there must be enough carbon in the steel in the form of iron carbide (cementite). If very little carbon is present the primary material is ferrite and ferrite when quenched will not become very hard. To satisfactorily form martensite a steel must have not less than 0.3 percent carbon. As the carbon content increases the hardness increases rapidly up to a certain limit. Generally, above 0.7 percent carbon the rate of

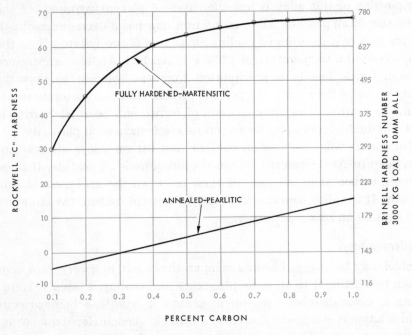

Fig. 7-15. Diagram showing effect of carbon upon structure.

hardness increases only slightly. Fig. 7–15 shows the effect of carbon on hardness.

The second condition necessary for the formation of martensite is heating to the correct temperature, known as critical point, where the carbon goes into solution and austenite is formed. The critical temperature for steel will vary according to the carbon and alloy content.

The third factor is rapid cooling. Quenching media used for this purpose are either water, brine, oil, or molten salts. The cooling rate depends on the type of alloy. As a rule plain water is recommended for fast cooling; brine for very rapid cooling, oil for slightly slower cooling, and lead baths or molten salts for slow cooling.

Tempering

Tempering involves heating a fully hardened steel to some temperature below the recrystallization temperature and holding it at this temperature to impart some toughness to it. Service requirements will determine the amount of tempering needed.

Tempering temperatures range from about 300°F to 1200°F. The first change that takes place at the selected temperature is the relief of stresses. As the tempering temperature is increased changes in the structure of the steel begin to take place which will affect its mechanical properties. Thus tensile strength, yield point, and hardness decrease while impact strength and elongation are improved. Obviously if the temperature is continued high enough austenite will again form.

Case Hardening

Case hardening is a process of hardening low-carbon or mild steels by adding carbon to the outer surface forming a hard, thin outer shell. The three principal case hardening techniques are known as carburizing, cyaniding, and nitriding.

Carburizing. Carburizing consists of heating low-carbon steel in contact with a carbonaceous material such as charcoal, coal, nuts, beans, bone, leather or a combination of these. The piece is heated to a temperature between 1650° to 1700°F where the steel as austenitic readily absorbs carbon. The length of the heating period depends on the thickness of the hardened case desired. After heating the steel is quenched which produces a material with a high-carbon surface and a relatively soft inner core.

Cyanide Hardening. Cyaniding involves heating a low-carbon steel in sodium cyanide or potassium cyanide. The cyanide is heated until it becomes liquid and the steel is placed in the liquid bath. This produces a very thin outer case which is harder than that achieved by the carburizing process.

Nitriding. Nitriding is a case hardening process used for alloys which are susceptible to the formation of chemical nitrides and where distortion must be kept to a minimum. The hard shell in this process is formed with nitrogen instead of carbon. The nitrogen producing material is ammonia. Since the heating is carried out at a low temperature no scaling occurs and consequently the piece can be finished to size before heat treatment.

Cold Working

Cold working is a process of deforming metals below the recrystallization temperature by hammering, rolling, drawing, and pressing. Metal is deformed by the slippage that occurs between any two adjacent atomic planes. This slippage is often known as strain or work hardening and results in increasing the strength and hardness of the material. The process tends to elongate and flatten the grains producing a much finer grain structure.

Many materials can only be hardened and strengthened by cold working. If metals which have been cold worked are subjected to high temperatures of welding all of the effects of cold working are lost.

During a cold working operation materials may work harden to a point where they become too hard and brittle and are likely to fracture. To continue any forming operation the material must then be softened by an annealing treatment which restores the original undistorted crystals.

Annealing

Annealing is a process used to improve toughness, remove stresses, or refine the crystal structure of a metal. The principal annealing methods are known as full annealing, spheroidizing, and normalizing.

Full Annealing. In this operation steels are heated above the critical temperature to allow the grains to recrystallize and then cooled slowly. During the soaking period at the required temperature the metal reverts

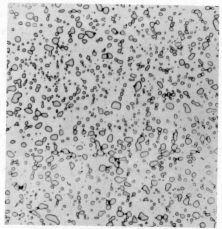

Fig. 7-16. Spheroidized annealed SAE 3250 steel.

to austenite. As it is slowly cooled in the furnace the metal breaks down in a soft ferrite and pearlite structure.

Spheroidizing. Spheroidizing is an annealing process used primarily to improve cold-working operations and machinability of high-carbon steels. The process consists of prolonged heating of the metal at temperatures slightly below the critical temperatures and cooled slowly. This treatment allows the iron carbide layers to curl up in spherical or globular carbides thereby permitting better slippage of the iron crystals. See Fig. 7–16.

Normalizing. The piece to be normalized is heated into the austenitic range until the solution is complete and then allowing it to cool in air. Cooling in air produces a finer structure than furnace cooling as in full annealing. Normalizing produces a stronger and harder structure and is often used after machining or hot working to develop a uniform grain structure.

Solution Heat Treating

Solution heat treating is a process associated with nonferrous metals and particularly alloys having aluminum, copper, or magnesium as a base metal. The heat treatment of these metals produces changes in their metallurgical structure with resultant improvement in strength and hardness. The process consists of heating the metal to the specified hardening temperature which will vary for different alloys. After proper soaking in the prescribed heat the alloy is quickly quenched in water.

Immediately after quenching the material becomes extremely ductile even more so than in an annealed state. Manufacturers often take advantage of this period to carry out certain forming operations.

During the period following the quenching operation the alloying constituent, which was in a supersaturated solid solution, slowly begins to precipitate. As the particles precipitate and coalesce hardness begins to develop. The rate of precipitation is a function of time and temperature. If the precipitation is carried out at room temperature it is referred to as *age hardening*. If it is performed at elevated temperatures it is called *artificial aging*. Usually aging at room temperature takes longer than if done at elevated temperatures. It is interesting to note that

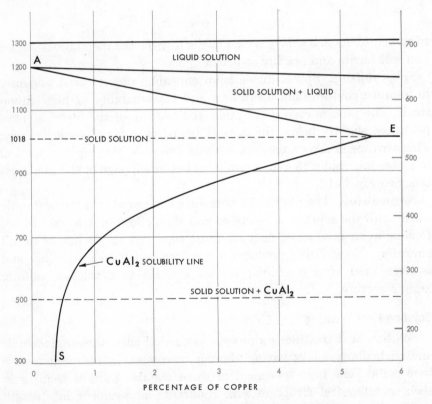

Fig. 7-17. Copper-aluminum equilibrium diagram.

precipitation can be suppressed indefinitely by holding the alloy at low temperatures.

Checking the equilibrium diagram of a copper-aluminum (2017) alloy shown in Fig. 7–17 will disclose a typical reaction of solution heat treating and precipitation of a nonferrous alloy. The actual heat treatment of this alloy depends upon the changing solubility of aluminum for the constituent $CuAl_2$ during the heating and cooling phase as indicated by the S-E curve. Thus if the copper-aluminum alloy with 4.5 percent

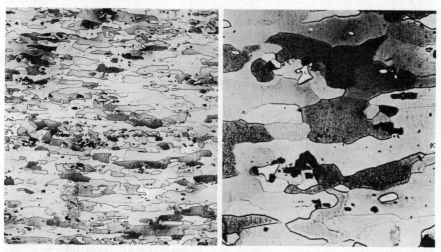

Structure of Duralumin-type alloy sheet after heat treatment. Practically all the soluble $CuAl_2$ is dissolved. (Mag. 100X)

Showing same structure magnified 500X. Light particles with black boundaries are undissolved $CuAl_2$.

Fig. 7-18. Structure of Duralumin-type alloy after heat treatment.

Cu is heated to a temperature of 950°F for a period of 14 hours, most of the $CuAl_2$ phase present in the annealed condition of the alloy is dissolved. See Fig. 7–18. If this solution is cooled rapidly the $CuAl_2$ phase will be retained in solution in the aluminum. Precipitation now occurs at the aluminum grain boundaries and along the crystallographic planes within the aluminum crystals. The very fine particles of $CuAl_2$ act as keys and build up resistance to slip greatly increasing the strength of the alloy.

Recrystallization

Recrystallization is a phenomenon associated with metals that have been cold worked. During any cold working phase the crystals are severely distorted. The process of restoring these crystals to a strain free, undistorted condition is known as recrystallization. The recrystallization process consists of heating the metal to a high enough temperature to form new unstrained crystals. Recrystallization temperature will vary for different metals and the amount of cold working. In the process of recrystallization, nuclei formed at the grain boundaries grow until they meet other new nuclei which develop into new grain boundaries. The result is an increase in ductility and decrease in hardness.

Recrystallization is also important when it is necessary to change a metal from a coarse to a fine grain structure. The transformation is made by first plastically deforming the metal while in a cold state and then heating it to bring about recrystallization.

Grain Size

A grain is a single crystal with an irregularly shaped boundary. Crystals growing simultaneously in a liquid from many directions determine the size and shape of each grain. There is no regularity of crystal structure at these boundaries. Structures which do not have a definite form or arrangement of atoms are said to be amorphous, meaning "without form." Since there is a disordered arrangement of atoms at the line of intersection they cannot seek the lowest energy level which is the crystal. The lack of order at the boundary puts a strain on the atoms in the crystal for a depth of several layers of atoms. This area is stronger and harder than the interior of the grain. As the grain size gets smaller the ratio of grain boundary to volume increases and there is greater resistance to being deformed. This is equivalent to saying that the metal is stronger.

Grain size is important in determining the response to heat treatment of metals such as steel in which the basic metal undergoes an allotropic change. In this condition the grain boundaries are at a higher energy level than the grains themselves. Consequently, the greater the number of grain boundaries the more sites for nucleation.

To fully harden a fine-grained steel it must be severely quenched otherwise a large number of nucleation sites will form pearlite and result in soft spots. Fine-grained steels have far superior mechanical properties than coarse-grained steels. Therefore, it is important that the welder preserve the grain size of the parent material. If he employs excessive heat to cause excessive grain growth the weldment will harden to a great extent and may become very brittle.

PROPERTIES OF METALS

Metals possess certain characteristics or properties which are of special significance in welding. These are broadly classified as chemical activity and mechanical properties.

Chemical Activity

Chemical activity indicates how actively an element reacts with oxygen. For example, when aluminum comes in contact with oxygen it quickly forms aluminum oxide. The formation of any oxide poses special problems to the welder because oxides affect the strength of weld deposits.

Corrosion. Corrosion is the destruction of a metal by chemical or electrochemical reactions. A metal may react with its environment to form an oxide, salt, or some other compound. When a metal is changed to one of these compounds it loses its strength, ductility, and other desirable mechanical properties.

Deterioration resulting from physical actions brought about by applied static or dynamic loads such as erosion, galling, or wear, is often referred to as corrosion-erosion, corrosive wear, fretting-corrosion, corrosion-fatigue, and stress corrosion. These conditions may be induced by an almost unlimited number of corrosive media, particularly air, water, industrial atmospheres, soils, acids, elevated temperatures, bases, and salt solutions.

Ideally a weld deposit should be identical to the base metal and thus have the same corrosion resistance. However, the intense localized heat of welding and large temperatures gradient will often cause structural changes.

Corrosion resistance is dependent upon the elements in an alloy as well as the type of structure. Elements added to produce corrosion resistance respond only when they are in a definite condition within the alloy. For example, in 18–8 stainless steel the chromium must be in solid solution to give protection. If it precipitates out of solution and combines with carbon to form chromium carbide it loses its corrosion resistance.

There are five basic chemical reactions which represent most types of corrosion.

1. Combinations of metals with nonmetals to form compounds. High temperature oxidation of iron is a reaction of this type. Chemically it is shown as:

$$2 \text{ Fe} + O_2 = 2 \text{ FeO}$$
$$\text{or}$$
$$4 \text{ Fe} + 3 \text{ } O_2 = 2 \text{ Fe}_2O_3$$

This type of reaction always occurs during fusion welding in air.

2. Reaction of a metal with oxygen in which moisture is present. Oxidation of aluminum, iron, and other active metals in moist air are examples of this type of corrosion. Expressed chemically:

$$4 \text{ Al} + 3 \text{ } O_2 = 2 \text{ Al}_2O_3$$

Metals which react in this manner always have an oxidized surface layer that must be removed prior to welding.

The above two types of corrosion are of most interest to the welder. The remaining three involve displacement of hydrogen from acids by a metal, displacement of hydrogen in impure water by a metal, and the replacement of one metal with another from a salt solution.

Mechanical Properties

Mechanical properties are measures of how materials behave under applied loads. These properties are described in terms of the kinds of forces materials have to resist and the manner in which these forces are resisted.

Common types of loads are tensile, compressive, torsional, direct shear, impact, or a combination of these. The reactions of various loads are determined by conducting tests on standardized test specimens. In these tests the desired type of load is applied and the resulting deformation measured.

Stress. Stress is defined as the internal resistance a material offers to being deformed and is measured quantitatively in terms of the applied load and the area over which it is applied. See Chapter 14.

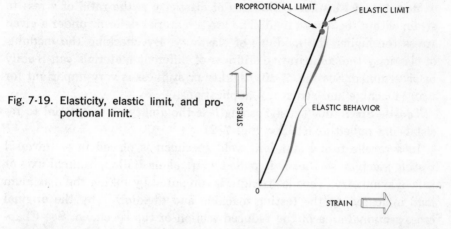

Fig. 7-19. Elasticity, elastic limit, and proportional limit.

Strain. Strain is the deformation resulting from stress and is expressed quantitatively in terms of the amount of deformation per inch.

Stress and strain are usually plotted on a graph with stress as the ordinate and strain as the abscissa. In this way the complete behavior of the material may be seen at a glance. Within the elastic limit of a material stress is proportional to strain as is evidenced by a straight line relationship. See Fig. 7–19. At any point on this straight line if stress is divided by the corresponding strain the result is a constant. The value of this constant is called *Modulus of Elasticity*.

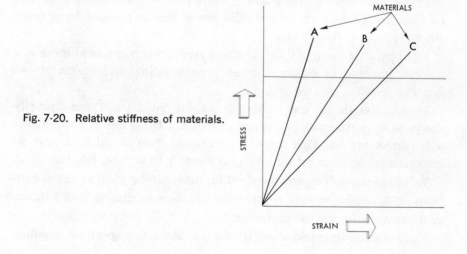

Fig. 7-20. Relative stiffness of materials.

Modulus of Elasticity. Modulus of elasticity is the ratio of stress to strain within the elastic limit. The less a material deforms under a given stress the higher the modulus of elasticity. By checking the modulus of elasticity the comparative stiffness of different materials can readily be ascertained. See Fig. 7–20. Rigidity or stiffness is very important for many machine and structural applications.

Tensile Strength. Tensile strength is the ability of a material to resist being pulled apart. See Fig. 7–21.

In a tensile test a standard weld specimen is placed in a universal testing machine so that it is pulled apart along its longitudinal axis or axis of symmetry. Tensile strength is computed by taking the maximum load indicated on the testing machine and dividing it by the original cross sectional area of the reduced section of the specimen. See Chapter 14.

Compressive Strength. Compressive strength is the ability of a material to resist being crushed. Although compression is the opposite of tension with respect to the direction of the applied load some materials react differently to each type of load. Most metals have high tensile strength and high compressive strength. However, brittle materials such as cast iron have high compressive strength but only moderate tensile strength. Very brittle materials often have high compressive strength but low tensile strength.

Yield Strength and Yield Point. Yield strength is a general term that indicates the limiting stress where some specified deformation occurs. In some materials such as the mild steels this is marked by a rapid stretching at a constant load.

Yield point is the point on the stress-strain diagram where there is a marked increase in strain with no corresponding increase in stress. See Fig. 7–22.

Most materials do not exhibit a definite yield point consequently the yield strength must be defined in terms of the stress at some allowable permanent set. This value is determined from actual field tests on materials which have proven to be satisfactory in service. See Fig. 7–23.

Values of allowable permanent set for determining yield strength vary from .0001 in./in. or 0.01 percent for brittle materials to 0.002 in./in. or 0.2 percent for ductile materials.

To determine the yield strength of a material a test specimen is pulled

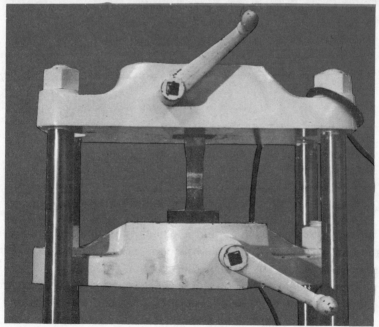

Fig. 7-21. A specimen being tested in a universal testing machine.

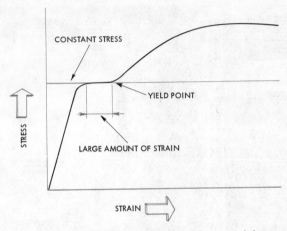

Fig. 7-22. Stress-strain diagram of a metal having a definite yield point.

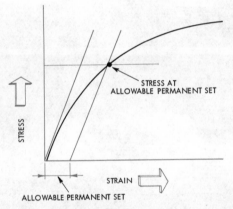

Fig. 7-23. Stress-strain diagram of a material which does not have a yield point.

on a universal testing machine and a stress-strain diagram is plotted or recorded autographically. A straight line is drawn through the point of origin and tangent to the straight line portion of the curve. The allowable permanent set is laid off on the strain axis and a line drawn through this point parallel to the line of origin. The point of intersection of this line with the stress-strain curve is read from the stress axis and is called the yield strength at the specified offset. See Fig. 7–23.

Elasticity. Elasticity is the ability of a material when deformed under a load to return to its original or undeformed condition after the load is released. See Fig. 7–19.

Elastic Limit. The elastic limit is the last point at which a material may be stretched and still return to its undeformed condition upon release of the stress. See Fig. 7–19.

Proportional Limit. The proportional limit is the final point at which stress remains proportional to strain. Within the elastic limit most materials exhibit a straight line relationship between stress and strain. See Fig. 7–19.

Fatigue Strength. Fatigue strength is the ability of a material to resist various kinds of rapidly alternating stresses. It is specified in terms of the magnitude of alternating stress for a specified number of cycles. See Fig. 7–24.

Most metals show a fatigue behavior such as seen in Fig. 7–25, where the S/N curve falls continuously with a decreasing slope up to the maximum number of cycles used in fatigue experiments. Such metals do not have a definite fatigue limit.

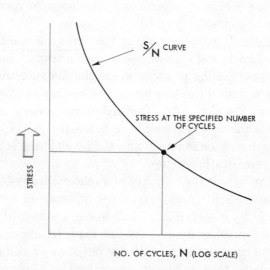

Fig. 7-24. Stress-number of cycles diagram of a material which does not have a fatigue limit.

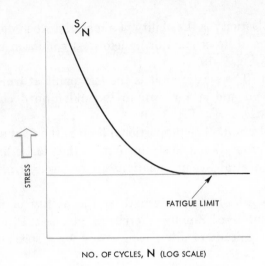

Fig. 7-25. Stress-number of cycles diagram showing the fatigue limit.

Fatigue Limit or Endurance Limit. Fatigue limit or endurance limit is the stress below which an alternating load may be repeated an indefinite number of times.

Impact Strength. Impact strength is the ability of a metal to resist suddenly applied loads and is measured in foot-pounds. The test is conducted in an impact testing machine where the distance through which its pendulum swings indicates how much total energy was used in breaking the specimen. See Fig. 7–26. The higher the impact strength of a metal the greater the energy required to break it.

Impact strength may be seriously affected by welding. It is one of the most structure-sensitive properties.

Hardness. Hardness is the property of a material to resist permanent indentation. Hardness is usually expressed in terms of the area of an indentation made by a special sized ball under a standard load, or the depth of a special indenter under a specific load. In any case the values are arbitrary and are significant only for purposes of comparison with some other property such as tensile strength. The hardness number is also used to control some industrial process such as hardening, tempering, or cold working operations.

Ductility. Ductility is the ability of a metal to deform appreciably without rupture. It is important because it permits metal to be formed and prevents localized failure that may occur because of some design or processing characteristic.

Ductility is expressed quantitatively in two different ways, as percent elongation, or as percent reduction area.

Toughness. Toughness may be considered as strength together with ductility. A tough material is one which may absorb large amounts of energy without breaking. It is found in metals which exhibit a high elastic limit and good ductility.

Cryogenic Properties. Cryogenic properties of metals represent behavior characteristics under stress in environments of very low temperatures. In addition to being sensitive to crystal structure and processing conditions metals are also sensitive to low and high temperatures. Some alloys which perform satisfactorily at room temperatures may fail completely at low or high temperatures.

In the liquefaction of air metals have to withstand temperature around $-350°$F. The changes from ductile to brittle failure occurs rather suddenly at low temperatures. The temperature at which this change takes place is called the transition temperature. To determine the transition temperature Charpy or Izod impact notched specimens are used. These specimens are immersed in liquids which are continuously lowered in temperature. The specimens are then broken in the impact machine, and temperatures plotted on a graph. See Fig. 7–27.

Materials tested at cryogenic temperatures do not undergo microstructural changes. Coarse grain size, strain-hardening, and certain embrittling elements tend to raise the transition range of temperature, whereas fine grain size, ductilizing, and refining heat treatments, and the addition of certain alloying elements tend to enhance toughness.

Creep or High Temperature Properties. Failure of a metal due to a slow but progressively increasing strain usually at high temperatures is called *creep*. Data taken over long periods of time have shown that some materials can be used at high temperatures if the stresses in them are kept low enough or if they are replaced periodically before excessive strain occurs.

At room temperature most metals and their alloys behave elastically up to a fairly high stress level. As the temperature is raised their elastic

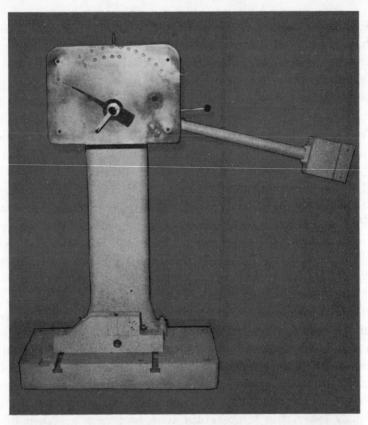

Fig. 7-26. Impact testing machine.

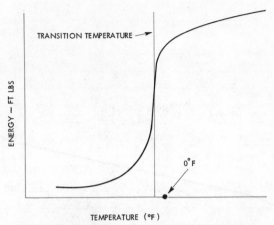

Fig. 7-27. Diagram of a material which has a sharp
well-defined transition temperature.

limit is lowered until it disappears entirely at the point of melting. Somewhere within this interval metals have their own particular useful maximum stress level for any given temperature. For high temperature applications high strength alloys are chosen because greater saving in weight can be realized which is especially important in today's jet-engines, steam-turbines, etc.

In a creep test a material being considered for a high temperature application is subjected to a stress level believed to represent operating stresses and temperatures. Stress and temperature are kept constant throughout the test and the changes in strain are measured and plotted against time. A typical strain-time curve for such a test is shown in Fig. 7–28.

It can be seen from this curve that there are three stages of strain with time. In the first stage creep begins and gradually decreases into the second stage where it remains relatively constant. It then enters the third stage where the strain begins to increase rapidly and leads to a quick fracture. The third stage is obviously useless for design application. The limit of usefulness is within the second stage. Once an operating temperature is established for a specific application an alloy is chosen on the basis of its response to this temperature.

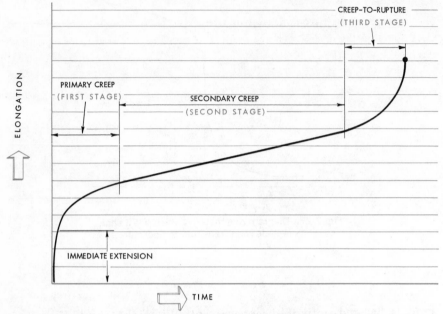

Fig. 7-28. Diagram showing the three stages of creep.

The addition of certain elements such as nickel, cobalt, and manganese will increase the creep resistance of alloys by entering into solid solution. Other elements such as chromium, molybdenum, tungsten, and vanadium increase the creep resistance by the formation of carbides. The carbide-forming elements are the most effective. Columbium and titanium in small amounts also greatly reduce creep.

WELDING DEFECTS

In the process of welding various materials precautions must be taken to prevent the development of certain defects in the weld metal otherwise these defects will severely weaken the weld. The following are some of the principal defects that are significant in any welding process.

Grain Growth

A wide temperature differential will exist between the molten metal of the actual weld and the edges of the heat-affected zone of the base metal. This temperature may range from a point far above the critical temperature down to an area unaffected by the heat. Thus the grain size can be expected to be large at the molten zone of the weld puddle and gradually reducing in size until recrystallization is reached.

Effective control of preheating and postheating will often minimize the formation of coarse grain structure. Where heavy welds require successive passes it is possible to use the heat of each succeeding pass to refine the grain of the previous pass. This can be done only if the metal is allowed to cool below the lower critical temperature between each pass. High carbon and alloy steels are particularly vulnerable to coarse grain growth if cooled rapidly. The rapid cooling develops a martensitic condition, leaving a hard and brittle weld.

Blowholes

Blowholes are cavities caused by gas entrapment during the solidification of the weld metal. They usually develop because of improper manipulation of the electrode and failure to maintain the molten pool long enough to float out the entrapped gas, slag, and other foreign matter. When gas and other matter become trapped in the grains of the solid metal small holes are left in the weld after the metal cools.

Blowholes can be avoided by keeping the molten pool at a uniform temperature throughout the welding operation. This can be done by using a constant welding speed so the metal solidifies evenly. The possibility of blowholes is especially prevalent during the stopping and starting of the weld along the seam.

Inclusions

Inclusions are impurities or foreign substances which are forced in a molten puddle during the welding process. Any inclusion tends to weaken a weld because it has the same effects as a crack. A typical example of an inclusion is slag which normally forms over a deposited weld. If the electrode is not manipulated correctly the force of the arc causes some of the slag particles to be blown into the molten pool and become lodged in the metal. Inclusions are more likely to occur in overhead welding since the tendency is not to keep the pool molten too long to prevent it from dripping off the seam.

Segregation

Segregation is a condition where some regions of the metal are enriched with an alloy ingredient while surrounding areas are actually impoverished. For example, when metal begins to solidify tiny crystals form along grain boundaries. These so called crystals or dendrites tend to exclude alloying elements. As other crystals form they become progressively richer in alloying elements leaving other regions without the benefits of the alloying ingredients. Segregation can be remedied by proper heat treating or slow cooling.

Porosity

Porosity refers to the formation of tiny pinholes generated by atmospheric contamination. Some metals have a high affinity for oxygen and nitrogen when in a molten state. Unless an adequate protective shield is provided over the molten metal gas will enter the metal and weaken it.

Residual Stresses

Any rapid heating and cooling of weld metal will produce thermal stresses which may be extremely detrimental to the strength of a welded joint. Heating causes rapid expansion of the metal around the weld whereas rapid cooling results in contraction. If large areas of unaffected metal surround the weld area or if the section is rigidly clamped the solidifying weld metal readily develops internal or locked-up stresses. These stresses are not serious in welding ductile metal because the flow of metal will usually relieve the stresses. However, they may become very serious in metals of great rigidity unless provisions are made to compensate for expansion and contraction.

Residual stresses form cracks in the weld metal and in the base metal surrounding the weld. Control of residual stresses can be achieved by following a few simple procedures. Specifically these controls are:

Proper Edge Preparation and Assembly. Edges of plates should be correctly beveled and sufficient space provided between them to permit the movement of the component parts during welding. See Chapter XIII. Joints of maximum fixity should be welded first. Thus a long longitudinal seam is welded before a short transverse seam.

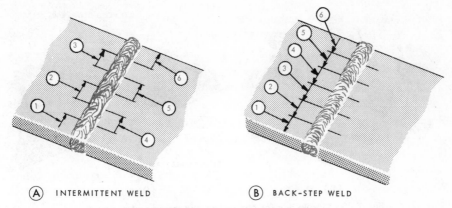

(A) INTERMITTENT WELD   (B) BACK-STEP WELD

Fig. 7-29. Method of heat input control.

Control of Heat Input. The faster a weld can be made the less the absorption of heat in the parent metal. A technique often used to minimize heat input is the intermittent, or skip, weld. Instead of making one continuous weld a short weld is made at the beginning of the joint; next a few inches are welded at the center of the seam, and then a short length is welded at the end of the joint. Finally the welding is returned to where the first weld ended and the cycle repeated. See Fig. 7–29A.

The back-step or step-back method involves depositing short sections of weld beads from right to left as shown in Fig. 7–29B instead of making one continuous weld.

Preheating. On many pieces, particularly alloy steels and cast iron, expansion and contraction forces can be controlled by preheating the entire structure before the welding is started. To be effective preheating must be kept uniform throughout the welding operation, and after the weld is completed the piece must be allowed to cool slowly.

Peening. To help the metal stretch as it cools a common practice is to peen it lightly with the round end of a ball peen hammer. However, peening should be done with care because too much hammering will impart stresses to the weld or cause the weld to work-harden and become brittle.

Heat Treatment. A common stress relieving method is heat treating. The welded component is placed in a furnace capable of uniform heating

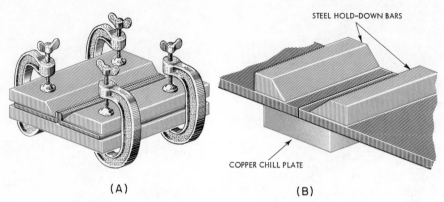

STEEL HOLD-DOWN BARS

COPPER CHILL PLATE

(A)   (B)

Fig. 7-30. Simple jigs for controlling expansion and contraction.

and temperature control. The metal must be kept in a soaking temperature until it is heated throughout. Correct temperatures are important to prevent injury to the metal being treated. For example, mild steels require temperatures of 1100° to 1200°F while other alloy steels must be heated to temperatures of 1600°F or more.

After the proper soaking period the heat must be reduced gradually to atmospheric temperature.

Jigs and Fixtures. The use of jigs and fixtures will help prevent distortion since holding the metal in a fixed position prevents excessive movements. A jig or a fixture is any device that holds the metal rigidly in position during the welding operation. Fig. 7–30A illustrates a simple way to hold pieces firmly in a flat position. These heavy plates not only prevent distortion but they also serve as *chill blocks* to avoid excessive heat building up in the work. Special chill plates made of

INCORRECT   CORRECT

Fig. 7-31. Use of few passes reduces distortion.

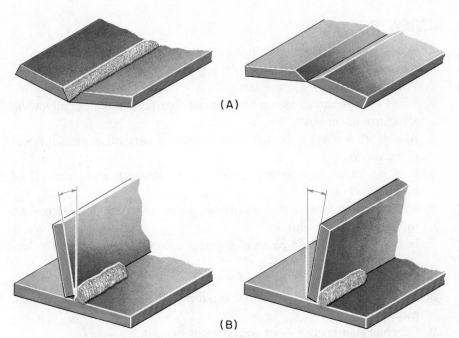

(A)

(B)

Fig. 7-32. Distortion is reduced by proper positioning of parts.

copper or other metal having good conductivity are particularly effective in dissipating heat away from the weld area. See Fig. 7–30B.

Number of Passes. Distortion can be kept to a minimum by using as few passes as possible over the seam. Two passes made with large electrodes are often better than three or four with smaller electrodes. See Fig. 7–31.

Parts Out of Position. When a single vee butt joint is welded the greater amount of hot metal at the top than at the root of the vee will cause more contraction across the top of the welded joint. The result is a distortion of the plate as shown in Fig. 7–32A.

In a T-joint the weld along the seam will bend both the upright and flat piece. See Fig. 7–32B.

To minimize these distortions the simplest procedure is to angle the pieces slightly in the opposite direction in which contraction is to take place. Upon cooling the contraction forces will pull the pieces back into position.

REVIEW QUESTIONS

1. What is the difference between an element, a compound, and a mixture?
2. What is meant by a space lattice?
3. What are the three basic space lattices that predominately affect the structure of metals?
4. Allotropic is a term used to describe what particular metallurgical phenomenon?
5. How do various space lattices affect the strength and ductility of metals?
6. In what way does a substitutional solid solution differ from an interstitial solid solution?
7. How does the atomic plane of slippage affect the general characteristics of solid solution alloys?
8. What is an eutectic?
9. How does an intermetallic compound differ from a mechanical mixture?
10. Of what significance is an equilibrium diagram?
11. The terms cementite, pearlite, austenite, and martensite are used to describe what structural characteristics?
12. Proper hardening of steel is contingent on what three factors?
13. Why are some metals often tempered after heat treating?
14. What three processes are used for case hardening metals?
15. What is cold working and what is its significance?
16. How does full annealing differ from spheroidizing and normalizing?
17. What is meant by solution heat treating?
18. What is the difference between age hardening and artificial hardening?
19. Why is metal sometimes subjected to a recrystallization process?
20. What are some of the conditions that generate corrosion?
21. What is the difference between a stress and strain?
22. Modulus of elasticity indicates what particular property of a metal?
23. How are the tensile and compressive strengths of a metal ascertained?
24. Of what significance are yield strength and yield point in a metal?
25. What is meant by elastic limit?
26. Why is the fatigue strength an important mechanical property?

27. Impact strength will indicate what particular characteristic of a metal?
28. In what way are cryogenic properties related to metals?
29. What is meant by creep and how does it affect the serviceability of a metal?
30. How does heat affect the grain growth of weld metal?
31. What are blowholes, inclusions, and segregations?
32. What are some of the ways residual stresses can be controlled?

Chapter 8

Weldability of Metals

Weldability refers to the ease with which a metal can be welded and how effectively a joint can be produced so it will have qualities comparable to the base metal. The ease with which a metal is welded is governed by:

1. *Melting point.* Metals with a low melting point such as aluminum are often difficult to control during welding especially from burning through.

2. *Thermal conductivity.* Metals with a high rate of heat transfer are hard to bring up to fusion heat. The weld often cools quickly because of the rapid withdrawal of heat from the weld into the area adjacent to the weld.

3. *Thermal expansion.* Rapid cooling from high welding heat usually results in warpage and excessive stresses.

4. *Electrical resistance.* Metals with low electrical conductivity may cause an overheating of electrodes.

5. *Surface condition.* Surfaces of metals to be welded must be free of dust, paint, oil, dirt, oxide, film, etc., otherwise proper fusion is hindered and porosity develops.

The quality of a welded joint is contingent on the control of such factors as:

1. *Cracking.* This results from improper heat control, and poor geometrical joint arrangement.

2. *Hardenability.* Welds made in steels with a high carbon content are likely to harden to a point of extreme brittleness. Poor heat and improper type of electrodes may induce excessive hardenability in the weld zone.

3. *Oxidation.* Welds that are not properly shielded from the atmosphere will have considerable porosity and their strength seriously weakened.

4. *Vaporization.* High weld heat destroys some of the elements in the metal and even causes large volumes of gas to become trapped in the weld.

5. *Structure change.* Welding heat if improperly controlled may induce grain growth or reduce corrosion resistance.

Effects of Elements in Steel

The various elements which are commonly found in metals have a definite effect on their weldability. Some of the more predominate elements and their resulting welding effects in steel are:

1. *Carbon* is the most important element in steel since it determines its degree of hardenability. The greater the carbon content the harder the steel. If carbon steels (over 0.30 percent) are welded and cooled suddenly from the welding heat a brittle zone develops adjacent to the weld. Furthermore, if additional carbon is picked up from the mixture of welding gases, the weld deposits may become so hard that they will readily crack. In general, for best weldability the carbon content in steel must be as low as possible.

2. *Manganese* in steel increases the hardenability and tensile strength. However, if the quantity of manganese is over 0.60 percent and especially when combined with a high degree of carbon, weldability is definitely impaired. This condition usually produces excessive cracking. If the manganese content is too low internal porosity and cracking may develop. Best welding results are obtained when the steel contains between 0.40 and 0.60 percent manganese.

3. *Silicon* is used in steel to improve its quality and tensile strength. Too high a silicon content especially with high carbon will contribute to excessive cracking.

4. *Sulphur* is often added to improve the machining properties of steels. However, the amount is kept low in other types of steel (0.035 percent with a maximum of 0.05 percent) because a high percentage of sulphur aggravates cracking conditions. High sulphur machining steels normally can be welded without difficulty by using low-hydrogen electrodes.

5. *Phosphorus* is considered an impurity in steel and therefore is kept

as low as possible. Insofar as welding is concerned a phosphorus content of over 0.04 percent will make welds brittle.

6. *Other elements* (nickel, chrome, vanadium, etc.) have varying effects on the weldability of metals. The welding of these alloys must be done with special care and usually require preheat and postheat to prevent hard and brittle weld zones.

STEEL CODE CLASSIFYING SYSTEMS

The two main systems used for classifying steels are known as the S.A.E. Code (Society of Automotive Engineers) and the A.I.S.I. Code (American Iron and Steel Institute).

The S.A.E. system is based on the chemical analysis of steel. The numbers of the code classifies the type of steel as follows: 1 represents carbon steel, 2 nickel, 3 nickel chromium, etc. In the case of alloy steels the second number of the series indicates the approximate amount of the predominating alloying element. The last two or three digits refer to the carbon content and are expressed in hundredths of 1 percent. For example, a 2335 S.A.E. steel indicates nickel steel of about 3 percent nickel and 0.35 percent carbon.

The following are the basic numerals for various S.A.E. steels:

TYPE OF STEEL	NUMERAL	TYPE OF STEEL	NUMERAL
Carbon Steels	1 xxx	Corrosion and Heat-resisting Steels .	30 xxx
Plain carbon	10 xx		
Free cutting (screw stock) ..	11 xx	Molybdenum Steels	4 xxx
Free cutting, manganese	x13 xx	Chromium	41 xx
High manganese	T13 xx	Chromium nickel	43 xx
		Nickel	46 xx and 48 xx
Nickel Steels	2 xxx		
.50% nickel	20 xx	Chromium Steel	5 xxx
1.50% nickel	21 xx	Low chromium	51 xx
3.50% nickel	23 xx	Medium chromium	52 xxx
5.00% nickel	25 xx	Corrosion and heat resisting .	51 xxx
Nickel Chromium Steel	3 xxx	Chromium Vanadium Steel	6 xxx
1.25% nickel; .60% chromium	31 xx		
1.75% nickel; 1.00% chrom.	32 xx	Tungsten Steel	7 xxx and 7 xxxx
3.50% nickel; 1.50% chrom.	33 xx		
3.00% nickel; .80% chromium	34 xx	Silicon Manganese Steel	9 xxx

The A.I.S.I. code is a refinement of the S.A.E. code. This code consists of letter prefixes and number designations. The prefix letters indicate the basic process used in making the steel. They are:

A—Open-hearth alloy steel.

B—Acid Bessemer carbon steel.

C—Basic open-hearth carbon steel.

D—Acid open-hearth carbon steel.

E—Electric furnace steel of both carbon and alloy steels.

The number designation classifies the kind of steel, the alloying elements, and range of carbon content. The first number indicates the type of steel and the second number in the series shows the amount of the predominating alloying element. The last two or three digits refer to the approximate permissible range of carbon content.

The following are the basic number designations:

TYPE OF STEEL	NUMERAL	TYPE OF STEEL	NUMERAL
Basic and acid open-hearth and acid Bessemer carbon steel grades, non-sulfurized and nonphosphorized	10 xx	Nickel 5.00%	25 xx
		Nickel 1.25%—chromium 0.60%	31 xx
		Nickel 1.75%—chromium 1.00%	32 xx
Basic open-hearth and acid Bessemer carbon steel grades, sulfurized but not phosphorized	11 xx	Nickel 3.50%—chromium 1.50%	33 xx
		Molybdenum	40 xx
		Chromium-molybdenum	41 xx
Basic open-hearth carbon steel grades, phosphorized	12 xx	Nickel-chromium-molybdenum	43 xx
		Nickel 1.65%—molybdenum 0.25%	46 xx
		Nickel 3.25%—molybdenum 0.25%	48 xx
Manganese 1.60 to 1.90%	13 xx	Low-chromium	50 xx
Nickel 3.50%	23 xx	Medium-chromium	51 xx

Examples of Steels with Codes:

C1078—Basic open-hearth carbon steel; carbon 0.72–0.85%.

E50100—Electric furnace chromium steel 0.40–0.60%; chromium, 0.95–1.10% carbon.

Preheating and Postheating

In welding high carbon or alloy steels there is always a danger that the weld deposit and heat-affected zone will form into a hard brittle metal known as martensite. When this happens the metal loses its ductility and may even crack while cooling. By resorting to some form of preheating and postheating the martensite content of the weld can be kept to a minimum. Preheating and postheating can be done in a furnace or by using a gas torch.

Preheating. Preheating raises the temperature of the metal surrounding the weld to keep the temperature differential between the weld and surrounding metal as small as possible. Low-carbon steel seldom requires preheating. Since the hardenability of a steel is directly related to its carbon content and alloying elements, the preheating temperatures will vary. In general preheating temperatures should be:

Equivalent Carbon (%)	Preheat Temperature
Up to 0.45	Preheat Optional
0.45 to 0.60	200° to 400°F
Above 0.60	400° to 700°F

Postheating. The purpose of postheating is the same as preheating. Actually postheating is used in conjunction with preheating. By heating the weldment as soon as the weld is completed the temperature of the work can be held at a sufficiently high level to permit the weld to cool slowly. As with preheating any postheating treatment keeps the weld zone more ductile.

Postheating temperatures and length of time will depend on the kind and thickness of the steel to be welded. The temperature may range from 600°F for 10XX type steels to 1200°F for 43XX steels, while postheating time may be from 5 minutes to several hours.

CARBON STEELS

Carbon steels are those which contain iron, carbon, manganese, and silicon. These steels are classified according to their carbon content and are called low-carbon, medium-carbon, and high-carbon steels.

Plain carbon steels are made in three grades: killed, semikilled, and rimmed. A killed steel is one that is deoxidized by adding silicon or aluminum (in the furnace ladle or mold) to cause it to solidify quietly without evolving gases. Killed steel is homogeneous, has a smooth surface and contains no blowholes. A semikilled steel is only partially deoxidized while a rimmed steel receives no deoxidizing treatment. Killed steels have greater toughness at low temperatures. Rimmed steels are usually considered more suitable for drawing and forming.

Properties

Low-carbon Steels. These have a carbon range between 0.05 and 0.30 percent. They are sometimes known as mild steels and represent the greatest bulk of steel for commercial fabrication of metal products and structures. Low carbon steels are tough, ductile, easily machined, and formed. As a rule, they do not respond to heat treatment but are readily case-hardened.

Medium-carbon Steels. Those with a carbon range of 0.30 to 0.60 percent are classed as medium-carbon steel. Because of their higher carbon content they can be heat treated. Steels in this category are strong, hard but not nearly as ductile as the low carbon types.

High-carbon Steels. A carbon content ranging from 0.60 to 1.7 percent is classed as high-carbon. Those with a carbon range of 0.75 to 1.7 percent are often called very-high carbon steels. Both groups respond very well to any form of heat treatment.

Some of the properties of basic carbon steels are shown in Table I.

Table I.   Properties of Carbon Steels.

AISI NO*	CARBON RANGE	TENSILE STRENGTH PSI	YIELD POINT PSI	% ELONG. IN 2"
C 1010	0.08-0.13	63,000	46,000	38
C 1015	0.13-0.18	61,000	45,000	39
C 1020	0.18-0.23	65,000	43,000	36
C 1025	0.22-0.28	67,000	45,000	36
C 1030	0.28-0.34	75,000	46,000	30
C 1040	0.37-0.44	93,000	58,000	27
C 1050	0.48-0.55	99,000	60,000	24
C 1060	0.55-0.65	95,000	59,000	25
C 1070	0.65-0.75	100,000	60,000	22
C 1080	0.75-0.88	103,000	62,000	22
C 1090	0.85-0.98	106,000	60,000	23
C 1120	0.18-0.23	69,000	36,000	32
C 1132	0.27-0.34	90,000	56,000	22

Weldability

Most carbon steels can readily be welded with standard welding processes such as oxyacetylene, shielded metal-arc, gas tungsten-arc, gas metal-arc, submerged arc, and various forms of resistance welding.

Low-carbon Steels. Particularly those in the lower carbon range generally present no difficulty in welding. As these steels approach the upper limit of their carbon range some degree of martensite may result when the weld is cooled rapidly. Low-carbon steels normally require no preheat or postheat for welding. On heavy sections, especially over one inch in thickness, some precaution must be taken for postwelding stress relief to avoid cracking in the weld deposits and in the heat-affected zone of the base metal.

Medium-carbon Steels. These can be welded but greater care must be exercised. Usually these steels when heated to welding temperatures and allowed to cool rapidly will harden and become brittle. Brittleness can be minimized by preheating to approximately 400°F to 500°F. By and large, very few problems are encountered in welding low and medium carbon steels when mild steel electrodes of the E-60XX or E-70XX classification are used. Moreover, preheating and postheating often can be eliminated by welding with low-hydrogen electrodes.

Gas tungsten-arc and gas metal-arc are being used more extensively in welding both low and medium carbon steels because of the ease with which the welding can be accomplished and the greater protection from atmospheric contamination. For economic reasons gas tungsten-arc welding is limited to materials under $\frac{1}{4}''$ in thickness. When the gas tungsten-arc process is used to weld some types of low and medium carbon steels without a filler rod there may be evidence of some pitting in the weld. This porosity can be eliminated by lightly brushing the seam with a mixture of aluminum powder and methyl alcohol. Ordinarily when filler rods, coated electrodes, or solid wire for gas metal-arc welding are used they contain sufficient deoxidizers to prevent porosity.

With gas metal-arc welding CO_2 is often considered the most economical shielding gas although argon or a combination of argon and oxygen is also used. Submerged arc welding is a very common automatic machine process for joining low and medium carbon steels. Resistance welding has wide application in production machine welding especially for steels with a low carbon content. Sound welds can be made with spot, seam, projection, flash, upset, and percussion welding processes.

High-carbon Steels. Weldable, but preheating, special welding techniques, and post-welding stress relief are required. Steels with a high amount of carbon are subject to rapid grain growth and unless precautions are taken the welded area loses its toughness, strength, and ductility. Quite often heat treating after welding brings about a grain refinement of the steel but this process adds to the problem and economy of fabrication. Very high-carbon steels are rarely recommended for welding since the required welding temperature tends to destroy their mechanical properties. The usual practice in repairing broken parts made with such steels is to use a brazing process where the heat is not sufficient to affect their metallurgical structure.

ALLOY STEELS

An alloy steel is a steel which contains one or more of other elements such as nickel, chromium, manganese, molybdenum, titanium, cobalt, tungsten, or vanadium. The addition of these elements gives steel greater toughness, strength, resistance to wear, and resistance to corrosion.

Alloy steels are usually called by the predominating element which has been added. The more common elements are:

Chromium (Cr). When quantities of chromium are added to steel the resulting product is a metal having extreme hardness and resistance to wear without making it brittle. Chromium also tends to refine the grain structure of steel thereby increasing its toughness. It is used either alone in carbon steel or in combination with other elements such as nickel, vanadium, molybdenum, or tungsten.

Manganese (Mn). The addition of manganese to steel produces a fine grain structure which has greater toughness and ductility.

Molybdenum (Mo). This element produces the greatest hardening effect of any element except carbon and at the same time it reduces the enlargement of the grain structure. The result is a strong, tough steel. Although molybdenum is used alone in some alloys it is often supplemented by other elements particularly nickel or chromium or both.

Nickel (Ni). The addition of nickel to steel tends to increase its strength without decreasing its toughness or ductility. When large quantities of nickel are added (25 to 35 percent) the steels not only become tough but develop high resistance to corrosion and shock.

Vanadium (V). Addition of this element to steel inhibits grain growth when the steel is heated above its critical range for heat treatment. It also imparts toughness and strength to the metal.

Tungsten (W). This element is used mostly in steels designed for metal cutting tools. Tungsten steels are tough, hard, and very resistant to wear.

Cobalt (Co). The chief function of cobalt is to strengthen the ferrite. It is used in combination with tungsten to develop red hardness; that is, the ability to remain hard when red hot.

The most common steel alloys are those which are classified as low-alloys, high strength low-alloys, and stainless steels.

Properties of Low-Alloy Steels

Low-alloy steels are steels that ordinarily are more ductile and easier to form than the higher carbon steels. The most common low-alloy steels are those in the 20XX series (nickel), 30XX series (nickel-chromium), and the 40XX series (molybdenum).

Nickel steels with a nickel range of 3 to 5 percent have higher elastic properties than mild steel of comparable strength. For example, a 2.7 percent nickel steel with a carbon content of 0.24 percent has a tensile strength of approximately 85,000 psi. An unalloyed steel to have the same strength would require a carbon content of over 0.45 percent. With the higher carbon it naturally follows that the steel will have less ductility. See Table II.

Table II.   Properties of Nickel Steels.

AISI NO*	CARBON RANGE	TENSILE STRENGTH PSI	YIELD POINT PSI	% ELONG. IN 2"
2317	0.15-0.20	85,000	56,000	29
2515	0.12-0.17	92,000	69,000	27

*NATURAL HOT ROLLED

The *nickel-chromium steels* have a higher responding range to heat treatment than for a given carbon content of straight nickel steels. See Table III. Chromium is the outstanding element that contributes to the hardenability of these steels.

Table III.   Properties of Nickel-Chromium Steels.

AISI NO*	CARBON RANGE	TENSILE STRENGTH PSI	YIELD POINT PSI	% ELONG. IN 2"
3120	0.17-0.22	75,000	60,000	30
3130	0.28-0.33	100,000	72,000	24
3140	0.38-0.43	147,000**	123,000	18

*NATURAL HOT ROLLED
**OIL QUENCHED AND DRAWN

Molybdenum steels are classified into three major groups: Carbon-Moly., Chrome-Moly., and Nickel-Moly. These steels are used where high strength is required at high temperatures. The addition of molybdenum to steel not only increases the hardenability but also tends to reduce temper brittleness. Low-carbon moly steels are not heat treatable but are often case-hardened. Chrome-moly steels are widely used for highly stressed parts. Some of the basic properties of molybdenum steels are included in Table IV.

Table IV.   Properties of Molybdenum Steels.

AISI NO*	CARBON RANGE	NICKEL RANGE	CHROMIUM RANGE	MOLYBDENUM RANGE	TENSILE STRENGTH PSI	YIELD POINT PSI	% ELONG. IN 2"
4023	0.20-0.25			0.20-0.30			
4027	0.25-0.30			0.20-0.30	68,000	55,000	40
4032	0.30-0.35			0.20-0.30			
4037	0.35-0.40			0.20-0.30			
4042	0.40-0.45			0.20-0.30			
4118	0.18-0.23		0.40-0.60	0.08-0.15	91,000	52,000	28
4130	0.28-0.33		0.80-1.10	0.15-0.25	89,000	60,000	32
4140	0.38-0.43		0.80-1.10	0.15-0.25	90,000	63,000	27
4150	0.48-0.53		0.80-1.10	0.15-0.25	105,000	71,000	21
4320	0.17-0.22	1.65-2.00	0.40-0.60	0.20-0.30	87,000	59,000	30
4615	0.13-0.18	1.65-2.00		0.20-0.30	82,000	55,000	30
4640	0.38-0.43	1.65-2.00		0.20-0.30	100,000	87,000	21
4815	0.13-0.18	3.25-3.75		0.20-0.30	105,000	73,000	24
5120	0.17-0.22		0.70-0.90		73,000	55,000	32
5150	0.48-0.53		0.70-0.90		103,000	68,000	22

* NATURAL HOT ROLLED

Weldability

Low alloy steels can be welded with comparative ease with standard welding processes. Best results are obtained by using steels with a carbon content below 0.18 percent. Standard E60XX electrodes are used in welding with the shielded metal-arc process, although E70XX electrodes are recommended where greater tensile strength is required. Special low alloy electrodes of the E80XX or E100XX series have wide application when the ultimate in strength is required and where there is a greater tendency to underbead cracking. The elimination of cracking can also be achieved by using low-hydrogen electrodes.

Special low-alloy electrodes are definitely needed when the carbon content of the steel exceeds 0.18 percent and where heat treatment of the weld is required to secure properties approximating those of the parent metal. Cracking and weld brittleness in most low-alloy steels of the higher carbon range can be minimized by preheating and slow postweld cooling.

Electrode wire for gas metal-arc welding is available for a variety of low alloy steels. Many of these wires are specially designed to eliminate the usual problems encountered in other forms of welding. The composition of the wire electrodes is balanced to produce weld strength similar to the base metals.

Properties of High Strength Low-Alloy Steels

High-strength low-alloy steels are categorized on the basis of their mechanical properties especially their higher yield point as compared with structural carbon steels. They are intended for applications where savings in weight can be achieved because of their greater strength and durability. For example, in certain thickness ranges high-strength low-alloy steels will have a minimum yield point of 45,000 to 65,000 psi as compared with 33,000 to 36,000 psi for carbon steel. The higher mechanical properties of these steels are obtained by a combination of several elements. The blending of these elements produces a steel that is ductile, corrosion resistant, has high strength, good forming, and is readily welded. Many high-strength low-alloy steels are often known by their trade names. See Table V.

Table V. Composition Ranges of Representative High-Strength Low-Alloy Steels.

Brand	Chemical Composition (Per Cent)								
	C	Mn	P	S (max.)	Si	Cu	Ni	Cr	Other
COR-TEN*	0.12 max.	0.20–0.50	0.07–0.15	0.05	0.25–0.75	0.25–0.55	0.65 max.	0.30–1.25	—
COR-TEN**	0.10–0.19	0.90–1.25	0.04 max.	0.05	0.15–0.30	0.25–0.40	—	0.40–0.65	V: 0.02–0.10
TRI-TEN	0.22 max.	1.25 max.	0.04 max.	0.05	0.30 max.	0.20 min.	—	—	V: 0.02 min.
MAN-TEN	0.28 max.	1.10–1.60	0.04 max.	0.05	0.30 max.	0.20 min.	—	—	
EX-TEN 50***	0.22 max.	1.25 max.	0.04 max.	0.05	—	—	—	—	Cb or V: 0.01 min.
GLX-50-W	0.20 max.	1.00 max.	0.04 max.	0.05	0.10 max.	—	—	—	Cb: 0.20 min.
Dynalloy	0.15 max.	0.60–1.00	0.05–0.10	0.05	0.30 max.	0.30–0.60	0.40–0.70	—	Mo: 0.05–0.15
Mayari "R"	0.12 max.	0.50–1.00	0.12 max.	0.05	0.20–0.90	0.50 max.	1.00 max.	0.40–1.00	Zr: 0.10 max.
NAX High Tensile	0.18 max.	0.50–0.90	0.04 max.	0.05	0.60–0.90	—	—	0.40–0.70	Mo: 0.20 max., Zr: 0.03–0.13
Yoloy "E" HSX	0.18 max.	0.90 max.	0.08 max.	0.05	—	0.20–0.50	0.40–1.00	0.20–0.35	
Jalten #1	0.15 max.	1.30 max.	0.04 max.	0.05	0.10 max.	0.30 min.	—	—	V: 0.035–0.065
Yoloy "E" ACR	0.10 max.	0.60 max.	0.05 max.	0.05	—	0.25–0.50	0.60 max.	0.35 max.	—

*½-inch and under in thickness.
**Over ½-inch in thickness.
***In addition to EX-TEN 50 steel (50,000 lb. per sq. in. minimum yield point), EX-TEN steels are also available with minimum yield points of 45,000, 55,000, 60,000, and 65,000 lb. per sq. in.

U.S. Steel Corp.

Weldability

High-strength low-alloy steels are weldable by the shielded metal-arc, gas metal-arc, and resistance-welding processes. For regular shielded metal-arc welding the ordinary E60 or E70 group electrodes produce satisfactory results. Welding heavier sections and steels with a higher carbon or manganese content, low-hydrogen electrodes are generally recommended. Preheating of the higher carbon alloys is usually advisable. Where maximum strength and ductility are required the electrodes used are often those in the E-80XX and E-100XX classes.

Properties of Stainless Steels

Stainless steel is classified into two general A.I.S.I. groups: 300 and 400. See Table VI. Each series includes several different kinds of steel each of which has some special characteristic. See Table VII.

The 400 series. The 400 series is further divided into two groups according to the crystalline structure. One group is known as *ferritic* and the other as *martensitic*.

The ferritic steels contain from 12 to 27 percent chromium and 0.08 to 0.35 percent carbon. They are sometimes grouped into two types: medium chromium with low carbon, and high chromium with somewhat higher carbon. Ferritic steels are about 50 percent stronger than plain carbon steels. Just as with austenitic steels the ferritic steels cannot be hardened by heat treatment.

Ferritic stainless steels differ from austenitics in their microstructure. In general, for steels of comparable strength the ferritics will contain a higher carbon content and slightly more chromium but no nickel. Ferritic steels are frequently used for decorative trim, and equipment subjected to high pressures and temperatures.

The martensitic steels are readily hardened by heat treating. Their carbon content will range from 0.15 to 1.20 percent with a chromium content of 12 to 18 percent. Martensitic stainless steels are designed for machine parts where creep properties are very critical and high strength, corrosion resistance, and ductility also are required.

The 300 series. These steels are known as austenitic steels and their strength cannot be increased by heat treating. However, surface hardness can be improved by cold working. Austenitic stainless steels have a carbon range of 0.08 to 0.25 percent, a chromium content of 16 to

Table VI. Chromium-Nickel Types.

CHROMIUM-NICKEL TYPES
Austenitic Non-Magnetic Non-Hardenable by Heat Treatment

AISI Type Number	Carbon %	Man- ganese Max. %	Phos- phorus Max. %	Sulfur Max. %	Silicon Max. %	Chromium %	Nickel %	Other Elements %
301	Over 0.08 to 0.15	2.00	0.04	0.03	1.00	16.0-18.0	6.0- 8.0	
302	Over 0.08 to 0.15	2.00	0.04	0.03	1.00	17.0-19.0	8.0-10.0	
302B	Over 0.08 to 0.15	2.00	0.04	0.03	2.0-3.0	17.0-19.0	8.0-10.0	
303	0.15 max.	2.00	X	X	1.00	17.0-19.0	8.0-10.0	
304	0.08 max.	2.00	0.04	0.03	1.00	18.0-20.0	8.0-11.0	
304L	0.03 max.	2.00	0.04	0.03	1.00	18.0-20.0	8.0-11.0	
305	0.12 max.	2.00	0.04	0.03	1.00	17.0-19.0	10.0-13.0	
308	0.08 max.	2.00	0.04	0.03	1.00	19.0-21.0	10.0-12.0	
309	0.20 max.	2.00	0.04	0.03	1.00	22.0-24.0	12.0-15.0	
309S	0.08 max.	2.00	0.04	0.03	1.00	22.0-24.0	12.0-15.0	
310	0.25 max.	2.00	0.04	0.03	1.50	24.0-26.0	19.0-22.0	
310S	0.08 max.	2.00	0.04	0.03	1.50	24.0-26.0	19.0-22.0	
314	0.25 max.	2.00	0.04	0.03	1.5-3.0	23.0-26.0	19.0-22.0	
316	0.10 max.	2.00	0.04	0.03	1.00	16.0-18.0	10.0-14.0	Mo-1.75-2.50
316L	0.03 max.	2.00	0.04	0.03	1.00	16.0-18.0	10.0-14.0	Mo-1.75-2.50
317	0.10 max.	2.00	0.04	0.03	1.00	18.0-20.0	11.0-14.0	Mo-3.00-4.00
317L	0.03 max.	2.00	0.04	0.03	1.00	18.0-20.0	11.0-14.0	Mo-3.00-4.00
330	0.25 max.	2.00	0.04	0.04	1.00	14.0-16.0	33.0-36.0	
321	0.08 max.	2.00	0.04	0.03	1.00	17.0-19.0	8.0-11.0	Ti-5 X C min.
347	0.08 max.	2.00	0.04	0.03	1.00	17.0-19.0	9.0-12.0	Cb-Ta-10XC min.
347F	0.08 max.	2.00	X	X	1.00	17.0-19.0	9.0-12.0	Cb-Ta-10XC min.

CHROMIUM TYPES
Martensitic Magnetic Hardenable by Heat Treatment

AISI Type Number	Carbon %	Man- ganese Max. %	Phos- phorus Max. %	Sulfur Max. %	Silicon Max. %	Chromium %	Nickel %	Other Elements %
403	0.15 max.	1.00	0.04	0.03	0.50	11.5-13.0		
410	0.15 max.	1.00	0.04	0.03	1.00	11.5-13.5		
414	0.15 max.	1.00	0.04	0.03	1.00	11.5-13.5	1.25-2.50	
416	0.15 max.	1.25	X	X	1.00	12.0-14.0		X
420	Over 0.15	1.00	0.04	0.03	1.00	12.0-14.0		
420F	Over 0.15	1.00	X	X	1.00	12.0-14.0		X
431	0.20 max.	1.00	0.04	0.03	1.00	15.0-17.0	1.25-2.50	
440A	0.60-0.75	1.00	0.04	0.03	1.00	16.0-18.0		Mo-0.75 max.
440B	0.75-0.95	1.00	0.04	0.03	1.00	16.0-18.0		Mo-0.75 max.
440C	0.95-1.20	1.00	0.04	0.03	1.00	16.0-18.0		Mo-0.75 max.
440F	0.95-1.20	1.00	X	X	1.00	16.0-18.0		X

CHROMIUM TYPES
Ferritic Magnetic Non-Hardenable by Heat Treatment

AISI Type Number	Carbon %	Man- ganese Max. %	Phos- phorus Max. %	Sulfur Max. %	Silicon Max. %	Chromium %	Nickel %	Other Elements %
405	0.08 max.	1.00	0.04	0.03	1.00	11.5-13.5		Al-0.10-0.30
430	0.12 max.	1.00	0.04	0.03	1.00	14.0-18.0		
430 Ti	0.12 max.	1.00	0.04	0.03	1.00	14.0-18.0		Ti-6 X C min.
430F	0.12 max.	1.25	X	X	1.00	14.0-18.0		X
442	0.20 max.	1.00	0.04	0.03	1.00	18.0-23.0		
446	0.35 max.	1.50	0.04	0.03	1.00	23.0-27.0		

X Free-Machining Stainless Steels—Phosphorus, Sulfur or Selenium—0.07% min., Zr or Mo—0.60% max.

26 percent, and a nickel content of 6 to 22 percent. They combine high tensile strength with exceptionally high ductility and possess the highest corrosion resistance of all stainless steels. The most prominent steel in this series is the 302 or 304 which are commonly referred to as 18-8, that is, a steel with 18 percent chromium and 8 percent nickel.

Table VII.   Mechanical Properties of Some Grades of Stainless Steel.

AISI NO*	TENSILE STRENGTH PSI*	YIELD STRENGTH PSI*	% ELONG. IN 2"
AUSTENITIC			
302, 303, 304	75,000-95,000	30,000-45,000	28
316	80,000-95,000	30,000-45,000	28
321	75,000-95,000	30,000-45,000	28
347	80,000-100,000	35,000-50,000	28
309	80,000-100,000	35,000-50,000	29
310	90,000-105,000	40,000-60,000	29
FERRITIC			
405	60,000-85,000	35,000-45,000	29
430	70,000-90,000	35,000-55,000	29
446	75,000-105,000	45,000-60,000	29
MARTENSITIC			
403, 410, 416	65,000-85,000	35,000-45,000	29

*ANNEALED CONDITION

Weldability

Stainless steels in the 300 series are generally conceded to have better welding qualities than those in the 400 series. However, this does not mean that the 400 series stainless steels are not weldable. Greater precautions have to be taken to avoid the destruction of the properties which are characteristic of them.

When stainless steels are heated to temperatures in the range of 800° to 1600°F or cooled slowly through that range carbon is precipitated from a solid solution at the grain boundaries. The result is an impairment of the corrosion resistance qualities. Normally the lowering of corrosion resistance is not unduly critical in the austenitic and ferritic steels, especially where the carbon content is low. Unless the metal is subsequently exposed to severe corrosive agents or environments the presence of a slight amount of precipitated carbides will not affect the serviceability of the steel. The danger of carbon precipitation is greater in martensitic stainless steels where carbon ranges are normally higher. Carbon precipitation can be controlled within certain limits by using electrodes that have been stabilized with columbium.

The primary consideration in the successful welding of any stainless steel is complete protection from atmospheric contamination. Without proper protection some of the alloying elements may become oxidized resulting in welds that are deficient in corrosion-resistance properties. Most standard welding processes (oxyacetylene, shielded metal-arc, gas tungsten-arc, gas metal-arc, submerged-arc, resistance) can be used to weld stainless steel. Best results are obtained with gas tungsten-arc and gas metal-arc because they provide more effective shielding.

Oxyacetylene Welding. Welding with an oxyacetylene flame requires a tip one or two sizes smaller than normally employed for welding plain carbon steel because of the lower heat conductivity of stainless steel (about 40 to 50 percent less than carbon steel). Although a neutral flame is required a very slight excess of acetylene is better since it is difficult to maintain exactly a neutral condition at all times. Too much acetylene will develop carburization thus making the weld hard, brittle, and lacking corrosion resistance. With an excess of oxygen the weld metal becomes oxidized and porous. The use of a flux is generally recommended to remove the highly infusible oxide and produce a smoother and better looking weld bead. Several commercial fluxes are available and they are applied by brushing on the seam of the weld as well as on the filler rod.

Special treated columbium 18-8 filler rods are essential for welding austenitic and ferritic steels. By and large filler rods should contain 1 to 1½ percent more chromium than the base metal. Ordinarily, martensitic stainless steels are not recommended to be welded with the oxyacetylene process.

Shielded Metal-Arc Welding. In welding with the metal arc direct current reverse polarity is recommended although AC current will also produce effective welds if proper electrodes are used. The low heat conductivity of stainless steel results in very little heat being dissipated to the surrounding metal and consequently sufficient penetration is usually obtained with reverse polarity.

Metal-arc welding will generally prove more satisfactory for welding sheets 20 gage and heavier. Special flux coated electrodes of the E308-15, 16, E309-15, 16, E310-15, 16, E316-15, 16 and E347-15, 16 are recommended for welding most types of stainless steels.

Although the austenitic steels are readily welded with the shielded metal-arc process greater care must be taken with the ferritic steels. Sound welds can be produced insofar as fusion and porosity are concerned but ferritic steels have a tendency to develop a coarse crystalline structure at welding temperatures. The result is that the weld metal and the heat affected zones of the base metal will lose some of their ductility. To some extent brittleness can be reduced by subsequently reheating to a red heat and cooling rapidly.

The procedure for shielded metal-arc welding of martensitic steels is the same as for welding austenitic stainless steel. Electrodes with a slightly higher chromium content are desirable. Danger of cracking is greatly reduced if the metal is preheated to a temperature of 300° to 500°F and annealed immediately after welding.

Gas Shielded-Arc. Both the TIG and MIG processes are highly adaptable for welding all stainless steels. Flux is not required because the welding zone is shielded from the contaminants of the atmosphere by a blanket of inert gas. The elimination of fumes, spatter, and flux residue tends to produce a smoother and cleaner weld, reduce clean-up operations, and minimize weld-metal porosity.

With gas tungsten-arc welding the arc will not pick up carbon and the loss of chromium and nickel is negligible. If a filler rod is needed it should be of the same composition as the parent metal. Either direct current straight polarity or alternating current with high-frequency stabilization can be used. The gas tungsten-arc is particularly adaptable for welding light gage stainless steels.

The gas metal-arc process can be used for welding both light gage and heavy stainless steels. The short-circuiting arc with a mixture of helium and argon as a shielding gas is better for thin sheets. Special stainless steel wire is available for welding various types of stainless steels. DCRP is recommended for most stainless steel welding with a 1 or 2 percent argon-oxygen mixture. For welding plate ¼" and thicker the torch should be moved back and forth in the direction of the joint and at the same time moved slightly from side to side. On thinner metal only back and forth motion along the joint is necessary.

Resistance Welding. All resistance welding methods are suitable for welding austenitic stainless steels. Electrical resistance of these steels is about 6 times that of mild steels. Since austenitic stainless steels heat rapidly by the current the heat is confined to a limited area. This minimizes warping and buckling. The short heating period reduces the tendency of carbon precipitation and leaves welds that are solid and ductile.

Ferritic stainless steels also can be resistance welded but only with due regard to the tendency of developing coarse grain structure at welding temperatures. Welds as a rule are not as tough or ductile as those in the austenitic group.

Martensitic stainless steels are weldable by various electrical resistance methods. However, these steels air harden readily and welds become hard and brittle. Although annealing is possible it cannot be confined exclusively to the weld region. Consequently, resistance welding of martensitic stainless steels is not recommended.

ALUMINUM

Properties

Aluminum is a very lightweight nonferrous metal with several very important characteristics. This metal is only about one-third as heavy as steel but has an extremely high strength-to-weight ratio. It has excellent corrosion resistance qualities, good electrical and thermal conductivity, and high reflectivity to heat and light. Aluminum has the added property of being easy to fabricate by casting, spinning, drawing, rolling, stamping, forging, machining, and extruding.

Commercial aluminum is divided into three groups: commercially pure aluminum, wrought alloys, and casting alloys. Commercially pure aluminum has a purity of at least 99 percent. Since it lacks alloying ingredients pure aluminum does not have a very high tensile strength. One of its chief qualities is ductility which makes the metal especially adaptable for drawing and other forming operations.

Wrought alloys are those which contain one or more elements such as copper, manganese, magnesium, silicon, chromium, zinc, and nickel. The wrought aluminums are either nonheat treatable or heat treatable.

The nonheat treatable types are those which cannot be hardened by any form of heat treatment. Their varying degrees of hardness are controlled only by cold working. The heat-treatable alloys are those in which hardness and strength are further improved by a heat-treating process.

Casting alloys are used to produce aluminum castings. The molten metal is poured into sand or permanent metal molds. A great many of these castings are weldable but extreme care must be exercised when welding the heat-treatable types to prevent any loss of the characteristics achieved by the heat-treating process.

Designation of Aluminum

Pure aluminum and wrought aluminum alloys are designated by a four digit index system. See Table VIII. The first digit of the desig-

Table VIII.   Designations for Aluminum.

		AA Number
Aluminum — 99.00% minimum and greater		1xxx
	Major Alloying Element	
	Copper	2xxx
Aluminum Alloys grouped by major Alloying Elements	Manganese	3xxx
	Silicon	4xxx
	Magnesium	5xxx
	Magnesium and Silicon	6xxx
	Zinc	7xxx
	Other Element	8xxx

nation serves to indicate aluminum groups. Thus 1XXX specifies an aluminum of at least 99.00 percent purity, 2XXX indicates an aluminum alloy in which copper is the major alloying element, 3XXX an aluminum alloy with manganese as the major alloying element, etc.

Pure Aluminum Group. In the 1XXX group for aluminum of at least 99.00 percent purity the last two of the four digits in the designation indicate the degree of aluminum purity which is expressed in hundredths of 1 percent.

The second digit in the designation indicates modifications in impurity limits. If the second digit in the desgnation is zero, it indicates that there is no special control on individual impurities; while numbers 1 through 9, which are assigned consecutively as needed, indicate special control of one or more individual impurities. Thus 1030 indicates 99.30 percent minimum aluminum without special control of individual impurities while 1130, 1230, 1330, etc. show the same purity with special control of one or more impurities. Likewise 1075, 1175, etc. indicate 99.75 percent minimum aluminum; and 1097, 1197, etc., indicate 99.97 percent aluminum.

Aluminum Alloy Group. In the 2XXX alloy groups the last two of the four digits in the designation have no special significance but serve only to identify the different alloys in the group. Generally these digits are the same as those used to designate the same alloys before the development of this system. Thus 2014 was formerly 14S, 3003 was 3S, and 7075 was 75S. See Table IX.

The second digit in the alloy designation indicates alloy modifications. If the second digit in the designation is zero, it indicates the orginal alloy; while numbers 1 through 9, which are assigned consecutively, indicate alloy modifications. In the former system letters were used to designate alloy modifications. These were assigned consecutively beginning with A. Thus 17S is now 2017 and A17S is 2117, 18S is 2018 and B18S is 2218.

Temper Designation A letter following the alloy designation and separated from it by a hyphen indicates the basic temper designation. The addition of a subsequent digit refers to the specific treatment used to attain this temper condition.

Alloys which are hardenable only by cold working are assigned "H" designations whereas alloys hardenable by heat treatment or by a combination of heat treatment and cold work are assigned "T" designations. The designations carry the following meanings:

--F	As Fabricated:	assigned to those products which have acquired some tempering properties during the shaping processes, but were not subsequently heat-treated or strain hardened.
--O	Annealed, Recrystallized:	assigned to those products which have the softest temper.

Table IX.   Aluminum Alloy Designation Conversions.

OLD	NEW	OLD	NEW	OLD	NEW
Commercial Designation	AA Number	Commercial Designation	AA Number	Commercial Designation	AA Number
		17S	2017	56S	5056
99.6, CD1S	1160	A17S	2117	XC56S	X5356
99.75	1175	18S	2018	C57S, K157	5357
99.87, EB1S	1187	B18S	2218	61S	6061
EC	EC	F18S, RR58	2618	62S	6062
AA1S	1095	24S	2024	63S	6063
BA1S	1099	25S	2025	66S	6066
CA1S	1197	B25S	2225	70S	7070
AB1S	1085	32S	4032	72S	7072
EB1S, 99.87	1187	43S, K145	4043	75S	7075
FB1S	1090	C43S, 44S, K143	4343	B77S	7277
AC1S	1070	XE43S	X4543	XA78S	X7178
BC1S	1080	44S, C43S, K143	4343	XB80S	X8280
CC1S, R998, 99.8	1180	45S	4045	K112	8112
JC1S	1075	50S	5050	K143, C43S, 44S	4343
AD1S	1050	A50S, K155, R305	5005	K145, 43S	4043
BD1S	1060	XD50S	X5405	K155, A50S, R305	5005
CD1S, 99.6	1160	A51S	6151	K157, C57S	5357
ED1S	1150	XB51S	X6251	K160, J51S	6951
AE1S	1030	J51S, K160	6951	X162, R306[4]	6003
BE1S	1145	52S	5052	LK183	5083
2S	1100	F52S	5652	K186	5086
3S	3003	53S	6053	R301 Core, 14S	2014
4S	3004	B53S	6253	R305, K155, A50S	5005
XA5S	X3005	XD53S	X6453	R306, K162	6003
11S	2011	E53S	6553	R308	1130
14S, R301 Core	2014	A54S	5154	R399	8099
XB14S	X2214	B54S	5254	R995, 99.35	1235
XC16S	X2316	X55S	X5055	R998, CC1S	1180

--H   Strain Hardened: assigned to those products which have been
strain hardened (for increased strength) with
or without subsequent heat treatment to attain partial softening. The --H is always fol-

lowed by two or more digits. The first digit defines the combination of basic temper operations:

--H1x Strain Hardened Only
--H2x Strain Hardened, Partially Annealed
--H3x Strain Hardened and Stabilized

The following digit or digits define the final degree of hardness. The numeral "8" designates the temper generally regarded as "full hard," whereas the numeral "4" defines a temper about midway between the "fully annealed" (--O temper) and the "full hard" (8) temper. Formerly, various tempers were designated by ¼H, ½H, ¾H, and H. Special applications utilize a "9" temper which, of course, is slightly harder than "8" temper.

--T Heat Treated
To Produce   with or without subsequent strain hardening.
Stable Tempers   The -T is always succeeded by one or more
Other Than   digits which are used to denote specific varia-
-F, -H, or -O: tions of the basic heat treatments listed:

-T3 Solution heat treated and then cold worked
-T4 Solution heat treated and naturally aged
-T5 Artificially aged
-T6 Solution heat treated and then artificially aged
-T7 Solution heat treated and then stabilized
-T8 Solution heat treated, cold worked, then artificially aged
-T9 Solution heat treated, artificially aged, then cold worked
-T10 Artificially aged and then cold worked.

Casting Alloys. Aluminum casting alloys are generally designated by a number and sometimes by a letter immediately followed by a number. However, this class of alloys does not always follow a specific alloy grouping system.

A letter following the alloy designation and separated from it by a hyphen indicates the basic temper designation. Heat treated castings are designated by the alloy number followed by the letter "T" and one or more numbers. Basic temper designations of castings are:

-T2 Annealed
-T4 Solution heat treated
-T5 Artificially aged
-T6 Solution heat treated and then artificially aged
-T7 Solution heat treated and then stabilized.

Weldability

Nonheat treatable wrought aluminum alloys in the 1000, 3000, and 5000 series are weldable. The heat of the welding may remove some of the material's strength developed by cold working but the strength will never be below that of the material in its fully annealed condition. Nevertheless, welding procedure should confine the concentration of heat in the narrowest zone possible.

The heat-treatable alloys in the 2000, 6000, and 7000 series can be

Table X.   Weldability of Aluminum.

TYPE	WELDING PROCESS			
	GAS	METAL–ARC	GAS SHIELDED–ARC	RESISTANCE
1060	A	A	A	B
1100	A	A	A	A
3003	A	A	A	A
3004	A	A	A	A
5005	A	A	A	A
5050	A	A	A	A
5052	A	A	A	A
2014	X	C	C	A
2017	X	C	C	A
2024	X	C	C	A
6061	A	A	A	A
6063	A	A	A	A
6070	C	B	A	A
6071	A	A	A	A
7070	X	X	A	A
7072	X	X	A	A
7075	X	X	C	A

A – READILY WELDABLE
B – WELDABLE IN MOST APPLICATION–MAY REQUIRE SPECIAL TECHNIQUE
C – LIMITED WELDABILITY
X – NOT RECOMMENDED

welded except that the 2000 and 7000 are not generally recommended for oxyacetylene welding. See Table X. Higher welding temperatures and speeds are needed to penetrate these alloys which is impossible with an oxyacetylene flame. In most instances resistance welding is preferred for high strength alloys.

Welding with the shielded metal-arc requires a special heavy coated aluminum electrode. An important factor is the choice of an electrode that will minimize weld cracking. Elimination of weld cracking can often be achieved with a rod having a higher alloy content than that of the base metal. The higher alloy will also add strength and ductility to the weld. While welding can be performed in any position the task is simplified and the quality of the completed joint is more satisfactory if the welding is done in a flat position. Metallic-arc welding is usually done with reverse polarity. Copper back-up blocks should be used wherever possible especially on plates ⅛" or less in thickness. In most cases, the electrode should be moved in a straight line without a weaving motion as in welding mild steel.

Welding with an oxyacetylene flame requires a flux to prevent rapid oxidation of the aluminum. After welding is completed all traces of the flux must be washed away otherwise the remaining flux will subsequently cause corrosion. Since aluminum has a high thermal conductivity it is necessary to use a welding tip slightly larger than one ordinarily used for steel of the same thickness. A properly balanced low gas velocity neutral flame is essential.

A high degree of skill is required to achieve sound welds with the oxyacetylene flame because aluminum has a relatively low melting point. Hot aluminum is flimsy and weak and must be supported during the welding operation. Due to its light color, there is practically no indication when the melting point is reached for when the metal begins to melt it collapses suddenly.

Best welding results are obtained with the gas tungsten-arc and gas metal-arc processes using ACHF current where the weld puddle is contained in a protective atmosphere of inert gases. Argon is recommended as the shielding gas because it produces better metal transfer and arc stability. For welding heavy aluminum plates between 1" and 2" in thickness with the Mig process a mixture of argon and helium is sometimes used. In this range of thickness the high heat input associated with helium is desirable.

Various electrical resistance processes have wide applications for welding both the nonheat-treatable and heat-treatable aluminum. Resistance welding is especially adaptable for joining high strength heat-treatable alloys which ordinarily are difficult to join by other fusion methods.

MAGNESIUM

Properties

Magnesium is a silvery white metal known for its extreme lightness. It weighs approximately one-fourth as much as steel and about two-thirds as much as aluminum. In its pure state it melts at 1202°F and boils at 2030°F. Its resistance to corrosion is about equal with some of the aluminum alloys in normal atmosphere.

Pure magnesium has insufficient strength for any structural purposes. Some other elements must be alloyed with the magnesium to give it strength. The common alloying elements are aluminum, manganese, zinc, and zirconium.

The two main groups of magnesium are the wrought alloys and the casting alloys. The wrought alloys are in the form of sheet, plate, and extrusions. Magnesium is often classified according to its use at room or elevated temperatures. Alloy designations for magnesium consists of one or two letters representing the alloying elements, followed by the percentages of the alloy content rounded out to whole numbers. A serial letter often is used after the coding number. Thus AZ61A indicates 6 percent aluminum and one percent zinc. The letter representation are as follows:

A aluminum   K zirconium
C copper   M manganese
E rare earths Z zinc
H thorium

The method of designating the temper of the magnesium is similar to that used for aluminum.

Wrought alloys of the nonheat-treatable types (A231B, A210, A251, AZ61A, M1A, etc.) are widely used because they combine high strength, good ductility, toughness, and formability. Some of the mechanical properties of magnesium alloys are listed in Table XI.

Table XI. Commercial Magnesium Alloys for Room Temperature Use.

ASTM Designation	Nominal Composition, % (Mg-Remainder)						Typical Mechanical Properties, 70°F			
	Al	Zn	Mn	RE*	Zr	Others	Tensile Strength, (x 10³ psi)	Tensile Yield Strength, (x 10³ psi)	Compressive Yield Strength, (x 10³ psi)	% Elongation in 2 In.
Sheet and Plate										
AZ31B-O	3.0	1.0	0.5	...	...	...	37	22	16	21
AZ31B-H24	3.0	1.0	0.5	...	...	...	42	32	26	15
LA141A-T7	1.3	...	...	...	...	14 Li	22	19	22	18
M1A-O	...	...	1.5	...	...	...	34	19	..	18
M1A-H24	...	...	1.5	...	...	...	39	29	..	10
ZE10A-O	...	1.2	...	0.17	...	...	33	20	16	23
ZE10A-H24	...	1.2	...	0.17	...	...	38	28	26	12
Extruded Shapes and Structural Sections										
AZ10A-F	1.2	0.4	0.5	...	...	...	35	22	11	10
AZ31B-F	3.0	1.0	0.5	...	...	...	38	29	14	15
AZ61A-F	6.5	1.0	0.2	...	...	...	45	33	19	16
AZ80A-F	8.5	0.5	0.2	...	...	...	49	36	22	11
AZ80A-T5	8.5	0.5	0.2	...	...	...	55	40	35	7
M1A-F	...	...	1.5	...	...	...	37	26	12	11
ZK21A-F	...	2.3	...	...	0.6	...	42	33	25	10
ZK60A-F	...	5.5	...	...	0.6	...	49	37	28	14
ZK60A-T5	...	5.7	...	...	0.6	...	52	41	34	9
Sand, Permanent Mold and /or Investment Castings										
AM100A-T6	10.0	...	0.2	...	...	...	40	22	..	1
AZ63A-F	6.0	3.0	0.2	...	...	...	29	14	..	6
AZ63A-T4	6.0	3.0	0.2	...	...	...	40	13	..	12
AZ63A-T6	6.0	3.0	0.2	...	...	...	40	19	..	5
AZ81A-T4	7.6	0.7	0.2	...	...	...	40	12	..	15
AZ91C-F	8.7	0.7	0.2	...	...	...	24	14	..	2
AZ91C-T4	8.7	0.7	0.2	...	...	...	40	12	..	14
AZ91C-T6	8.7	0.7	0.2	...	...	...	40	19	..	5
AZ92A-F	9.0	2.0	0.2	...	...	...	24	14	..	2
AZ92A-T4	9.0	2.0	0.2	...	...	...	40	14	..	9
AZ92A-T6	9.0	2.0	0.2	...	...	...	40	21	..	2
K1A-F	...	...	...	...	0.6	...	25	7	..	19
ZE41A-T5	...	4.2	...	1.2	0.7	...	30	20	..	4
ZH62A-T5	...	5.7	...	...	0.7	1.8 Th	40	25	..	6
ZK51A-T5	...	4.6	...	...	0.7	...	40	24	..	8
ZK61A-T6	...	6.0	...	...	0.8	...	45	28	..	10

American Welding Society

Weldability

The welding characteristics of magnesium are somewhat comparable to those of aluminum. Both, for example, have high heat conductivity, low melting point, high thermal expansion, and oxidize rapidly.

Although magnesium alloys can be welded with oxyacetylene and shielded-metal arc the most widely used processes are gas tungsten-arc and gas metal-arc. With gas tungsten-arc several current variations are possible. DC reverse polarity with helium gas produces wider weld deposits, higher heat, larger heat-affected zone, and shallower penetration. AC current with superimposed high frequency and helium, argon, or a mixture of these gases will join material ranging in thickness from 0.20″ to over 0.25″. Both DCRP and AC current provide excellent cleaning action of the base metal surface. Direct current straight polarity with helium as a shielding gas produces a deep penetrating arc but no surface cleaning. This technique is used for mechanized butt welding of sheets up to ¼″ in thickness without beveling.

The gas metal-arc with its continuous wire electrode and high current density arc is characterized by high welding speed and fast metal deposition rates.

Mechanized resistance welding also is used in fabricating magnesium alloy products. This includes spot, seam, and to a lesser extent, flash welding. Spot welding has wider applications especially for low-stressed assemblies or high-stressed parts that are not subjected to excessive vibration. Most spot and seam welding is done on sheets and extrusions up to ³⁄₁₆″ in thickness on materials such as AZ31B, AZ61A, HM21A, HM31A, and ZE10A.

COPPER AND COPPER ALLOYS

Properties

Pure copper melts at approximately 1980°F. It has a coefficient of expansion of about 1½ times that of steel and a thermal conductivity rate ten times greater than steel. Copper is a soft, tough, and ductile metal. It cannot be heat treated but will harden when cold worked. Commercially available coppers are divided into two groups: oxygen-bearing copper and oxygen-free copper.

Oxygen-bearing Copper. This electrolytic tough pitch copper is practically 99.9 percent pure and is considered to be the best conductor of heat and electricity. A small amount of oxygen in the form of copper oxide is uniformly distributed throughout the metal but it is insufficient to affect the ductility of the copper. However, if heated above 1680°F for prolonged periods the copper oxide tends to migrate to the grain boundaries, causing a reduction in strength and ductility. Also when exposed to this temperature the copper will absorb carbon monoxide and hydrogen which react with the copper oxide and release carbon dioxide or water vapor. Since these gases are not soluble in copper they exert pressure between the grains and produce internal cracking and embrittlement.

Oxygen-free Copper. This group of copper contains a small percentage of phosphorus or some other deoxidizer, thereby leaving the metal free of oxygen and consequently no copper oxide. The absence of copper oxide gives the metal superior fatigue resistance qualities and better cold working properties over the oxygen-bearing copper.

Weldability

Oxygen-bearing copper is not recommended for gas welding because of the formation of embrittlement. Some welds can be made with the shielded metal-arc process in situations where the tensile strength requirements are extremely low (19,000 psi or less), providing a high welding current and high travel speeds are used. The high current and speed will not give the embrittlement a chance to develop.

Deoxidized copper is the most widely used type for fabrication by welding. A properly made weld will have a tensile strength of about 30,000 psi. This copper can be welded with all standard welding processes including oxyacetylene, shielded metal-arc, gas tungsten-arc, gas metal-arc, and resistance. Since copper has a very high coefficient of expansion and contraction precautions must be taken to provide suitable jigging and clamping to prevent movement while cooling. Contraction forces will often cause cracking during the cooling temperature range.

Copper Alloys

There are many different kinds of copper alloys. Some of the more common include the copper-lead and copper-zinc (brasses), the copper-tins (bronzes), copper-aluminum, copper-silicon, copper-nickel, and copper-beryllium.

Copper-zinc Alloys (brasses). These are divided into three classifications: low brasses, high brasses, and alloy brasses. Low brasses contain 80-90 percent copper and 5-20 percent zinc. As the zinc content increases the strength, hardness, and ductility also increase. Their colors range from red to gold to green-yellows. Low brasses can be cold worked by any commercial process such as deep drawing, rolling, spinning, stamping, etc.

High brasses have a copper range of 55-80 percent and a zinc content of 20-45 percent. In general, these brasses are stronger but are not as ductile as the low brasses.

Alloy brasses are those which contain other elements such as tin, manganese, aluminum, phosphorus, antimony, and iron. These additions have a marked effect on the tensile strength, ductility, hardness, and corrosion resistance. Lead is sometimes added to improve machinability.

Copper-tin Alloys (bronze). Alloys containing a small amount of phosphorus are often known as phosphor bronzes. The phosphorus is added as a deoxidizing agent. The amount of tin in these alloys will range from 1.5 to 10 percent. The wrought forms of copper-tin alloys are tough, hard, and highly fatigue-resistant. They have a higher corrosion resistance than copper especially from sea water and acid reagents.

Copper-silicon Alloys. These are often called silicon bronze. They are known for high strength and corrosion resistance. These alloys can be readily rolled, forged, and extruded.

Copper-nickel Alloys. These alloys have a nickel content ranging from 5 to 30 percent. In addition they often contain a small amount of iron, manganese, or zinc. These alloys are relatively tough and have good ductility making them suitable for various deep drawing and stamping operations.

Copper-aluminum Alloys. Classified as aluminum bronzes they possess high tensile strength, toughness, excellent ductility, and moderate hardness. Aluminum bronzes resist oxidation and maintain their physical properties at high temperatures. All aluminum bronzes are highly resistant to corrosion.

Copper-beryllium Alloys. This group contains 1½ to 2½ percent beryllium with a small amount of iron, nickel, and silver. Their outstanding feature is high endurance under fatigue stresses.

Most copper alloys are weldable although some types are more easily brazed than welded. Thus the copper-lead alloys have poor welding properties because the lead has a low melting temperature which oxidizes or volatilizes before the copper melts leaving a weak and brittle weld. Copper-zinc alloys give off obnoxious zinc fumes that makes welding an unpleasant task. In general best welding results are obtained with the gas tungsten-arc and gas metal-arc processes.

NICKEL ALLOYS

Properties

Nickel-base alloys were developed to obtain a high strength, high corrosion resistant nonferrous metal. They are often referred to by trade names as Monel and Inconel. These metals have a high nickel content with varying percentages of copper, iron, manganese, and carbon. In general they are as ductile as copper, brass, and aluminum with corrosion resistant qualities compared with those of stainless steels. Their coefficient of expansion is practically the same as steel.

Weldability

The same processes used in welding steel will perform as well in welding nickel alloys. These alloys are usually sensitive to oxidation and consequently must be fully protected from atmospheric contamination. Both the shielded metal-arc and gas shielded-arc processes are very effective for welding nickel alloys. Special electrodes are available for welding Monel and Inconel. In any welding operation the removal of the oxide film from the surface of these metals is very essential.

TITANIUM

Properties

Titanium, a silvery colored metal, has become exceedingly important in the aircraft and missile industries because of its high strength-to-weight ratio. It has a melting point of approximately 3035°F with a density of about 60 percent that of carbon or stainless steels, and one and one-half that of aluminum. Another of its outstanding characteristics is its high resistance to corrosion and pitting in salt or oxidizing acid solutions. Like aluminum and magnesium titanium depends on the formation of a passive surface film for protection from corrosion.

Titanium is available in a pure or alloyed form. Pure titanium has a purity of around 99.5 percent. The remaining .5 percent consists of iron, oxygen, nitrogen, carbon, and tungsten. In its pure state it is nonheat-treatable and very ductile. The addition of alloying elements to titanium substantially increases its strength and toughness but at some sacrifice in ductility. Alloying elements frequently used are aluminum, zirconium, tin, molybdenum, columbium, vanadium, tantalum, manganese, copper, iron, and chromium. Alloyed titanium readily responds to heat treatment. As noted in Table XII, alloyed titaniums have much higher tensile and yield strengths than aluminum, stainless steel, and nickel alloys.

Table XII. Properties of Titanium.

PROPERTIES	COMMERCIALLY PURE TITANIUM*	ALLOYED TITANIUM
TENSILE STRENGTH PSI	80,000-120,000	145,000-200,000
YIELD STRENGTH PSI	70,000-100,000	140,000-185,000
PERCENT ELONGATION	20-15	15-5

*COLD WORKED

Weldability

One of the limitations of titanium is that it has an extremely high affinity for nitrogen, oxygen, carbon, and hydrogen especially in a molten state. Consequently the success of welding titanium depends on complete shielding from these atmospheric gases. For this reason the use of inert gas welding processes are almost mandatory. Not only is a protective shield required during welding but an equally protective cover is necessary during cooling. While the welding is underway a strong pressure of inert gas must also be applied on the root side of the weld. Shielding gas should be argon or helium or a mixture of the two gases. Carbon dioxide and nitrogen cannot be used. Both the oxyacetylene and the shielded metal-arc welding processes are not suitable for titanium.

One of the problems encountered in welding titanium is porosity. The elimination of porosity can be achieved by slightly preheating the metal to remove all traces of moisture and having a clean joint area. Foreign matter such as grinding particles, finger-prints, and other forms of dirt are detrimental to securing a sound weld. The rate of cooling also affects weld porosity. The slower the cooling rate the less porosity will likely develop in the weld.

As a rule the commercially pure titanium can be welded with greater ease than the alloys. Some of the alloys are weldable provided proper precautions are taken. Other alloys present serious problems when welding is attempted.

Special methods and chambers are often used because of the need to keep both sides of the weld covered with inert gas while welding and cooling.

ZIRCONIUM

Properties

Zirconium has an atomic structure very much like titanium. Its hardness is slightly above aluminum alloys but below that of low-alloy steels. Like titanium, zirconium undergoes a physical transformation at approximately 1589°F. The room temperature phase, called alpha, has a close-packed, hexagonal, crystal structure. Above 1589°F the crystal structure is body-centered cubic, called beta zirconium. The hexagonal close-packed structure in metals is generally considered to impart poor ductility because of the presence of only one plane of easy slip, but zirconium, like titanium, is an exception to this rule and can be worked with ease at room temperature.

Zirconium has a tensile strength ranging from 76,000 to 113,000 psi with a hardness varying from 175 to 275 Brinell, which is higher than aluminum alloys but less than steel. The tensile strength usually will decrease with rising temperature to 50 percent at temperatures from 600° to 700°F.

Zirconium is available in both the pure and alloy forms and is produced as sheet, strip, tube, and wire. The pure zirconium is hardenable by cold working only whereas the alloys can be heat treated and cold worked as well. One of the principal uses of zirconium is in nuclear reactor applications and aerospace structures.

Weldability

Zirconium like titanium and beryllium readily combines with oxygen, hydrogen, and nitrogen and becomes severely embrittled. Therefore success in welding zirconium requires adequate shielding by an inert gas. Complete shielding is necessary not only over the weld puddle, but at the underside of the weld bead plus a trailing stream of gas to cover the weld while it is hot and susceptible to atmospheric contamination.

The most successful welding process is tungsten-inert gas with argon as the shielding gas. Where the welding of zirconium is exceptionally critical the welding is done in a dry, closed chamber where the atmosphere has been evacuated and then refilled with argon at atmospheric pressure. The work, leads, and torch are all positioned in the chamber.

Proper torch adjustment is extremely important for suitable welds. For example, it has been found that with a number 5 cup, a 0.040 inch diameter tungsten electrode should project not more than $\frac{1}{8}''$ to $\frac{3}{16}''$ beyond the cup. The electrode must be clean with a minimum of rounding. Welds made under these conditions are usually sound and ductile with the weld metal as strong as the parent metal.

Resistance welding techniques where the molten nugget is shielded from the atmosphere by the parent metals can also be used for welding zirconium. Both spot and seam welding as used for welding stainless steels will produce effective welds.

BERYLLIUM

Properties

Beryllium probably has the highest strength-to-weight ratio of any stable metal. It has a density approximating that of magnesium, a high modulus of elasticity, a melting point far superior to the melting point of aluminum and magnesium, and extreme resistance to corrosion. See Tables XIII, XIV, and XV.

Beryllium is available either in a pure form or as an alloy. Pure beryllium is used principally in X-ray tube windows and in the atomic-energy field as a moderator and reflector of neutrons. The greatest usage of beryllium is in the form of copper alloys. All commercial beryllium-copper alloys are capable of precipitation-hardening. In the heat-treatable condition beryllium-copper alloy is ductile and can be severely formed. The alloy is non-magnetic and offers good resistance to corro-

Table XIII. Typical Physical Properties of Beryllium Copper in Heat Treated Condition.

	25	165	10	50	20C	275C
Specific Gravity	8.26	8.26	8.75	8.75	8.09	8.09
Density, Lb. per Cu. In.	0.298	0.298	0.316	0.316	0.292	0.292
Magnetic Properties	Non-magnetic	Non-magnetic	Non-magnetic	Non-magnetic	Non-magnetic	Non-magnetic
Melting Range, F	1600-1800	1600-1800	1885-1955	1850-1930	1600-1800	1570-1660
Elastic Modulus in Tension, Psi.	19,000,000	18,500,000	18,000,000	18,000,000	18,500,000	19,000,000
Thermal Expansion Coefficient						
Per Deg. C: 20-100 C	0.0000167	0.0000167	—	—	0.0000166	0.0000166
20-200 C	0.0000170	0.0000170	0.0000176	0.0000176	0.0000170	0.0000170
20-300 C	0.0000178	0.0000178	—	—	0.0000176	0.0000176
Per Deg. F: 68-212 F	0.0000093	0.0000093	—	—	0.0000092	0.0000092
68-392 F	0.0000094	0.0000094	0.0000098	0.0000098	0.0000094	0.0000094
68-572 F	0.0000099	0.0000099	—	—	0.0000098	0.0000098

The Brush Beryllium Co.

sion, wear, and fatigue at moderately elevated temperatures. It is furnished in wrought or cast forms. The wrought alloys are fabricated into strip, rod, wire, sheet, and tube. These materials are furnished in a fully annealed condition or in varying degree of hardness.

Beryllium-copper alloys are grouped as high strength or high conductivity. High strength alloys with over 1 percent beryllium offer

Table XIV. Typical Mechanical Properties of Beryllium Copper.

Alloy	Condition and Temper	Tensile Strength, Psi.	Elong. % in 2 In.
25	Sol.-Annealed	60-80,000	35-50
	¼ Hard	73-88,000	10-30
	½ Hard	80-100,000	5-10
	Hard	95-120,000	2-4
	Heat Treated from Sol.-Annealed	165-180,000	5-8
	Heat Treated from ¼ Hard	175-190,000	3.5-6
	Heat Treated from ½ Hard	185-200,000	2-4
	Heat Treated from Hard	190-205,000	1-2
165	Sol.-Annealed	60-80,000	35-50
	¼ Hard	73-88,000	10-30
	½ Hard	80-100,000	5-10
	Hard	95-120,000	2-4
	Heat Treated from Sol.-Annealed	150-165,000	5-8
	Heat Treated from ¼ Hard	160-175,000	3.5-6
	Heat Treated from ½ Hard	170-185,000	2-4
	Heat Treated from Hard	180-195,000	1-2
	Mill Hardened from Sol.-Annealed	100-110,000	21-30
	Mill Hardened from ¼ Hard	110-120,000	19-25
	Mill Hardened from ½ Hard	120-130,000	13-18
	Mill Hardened from Hard	135-145,000	9-12
10	Sol.-Annealed	40-55,000	25-45
	Hard	70-85,000	5-10
	Heat Treated from Sol.-Annealed	100-110,000	8-12
	Heat Treated from Hard	105-130,000	5-10

The Brush Beryllium Co.

maximum strength and hardness. High conductivity alloys with less than 1 percent beryllium are used where electrical and thermal conductivity are more important than maximum strength. The high strength-high hardness beryllium-copper alloys find wide applications in fabricating parts such as springs, diaphragms, bearings, cams, aircraft engine parts, gears, bushings, etc. The high electrical-high conductivity

Table XV.   Properties of Some Wrought Beryllium Alloys.

	ANNEALED AND HEAT TREATED	HARD AND HEAT TREATED
25		
TENSILE STRENGTH PSI	165-180,000	185-210,000
ELASTIC LIMIT PSI	75-85,000	10Q-135,000
YIELD STRENGTH PSI	85-95,000	120-150,000
165		
TENSILE STRENGTH PSI	150-175,000	170-195,000
ELASTIC LIMIT PSI	65-80,000	90-115,000
YIELD STRENGTH PSI	75-85,000	110-140,000
10		
TENSILE STRENGTH PSI	100-120,000	110-140,000
ELASTIC LIMIT PSI	55-65,000	65-85,000
YIELD STRENGTH PSI	60-70,000	75-90,000

alloys are used for welding dies, switches, circuit breakers, and other electrical current parts.

Weldability

The welding properties of beryllium are about the same as those of titanium and zirconium. Brazing has been found to be an effective joining process. Fusion welding has to be confined to methods where shielding is almost 100 percent. Excellent welds can be made with the electron beam where the metal can be enclosed and welded in a fully controlled atmosphere.

Beryllium is highly toxic and precautions must be taken to remove all dust and fumes by powerful ventilation.

CAST IRON

Properties

Cast iron is a term used to describe iron-base materials containing a high percentage of carbon (1.7 to 4.5 percent). There are four principal kinds of cast iron—gray, white, malleable, and nodular.

Gray Cast Iron. Whenever the silicon content is high and the metal is allowed to cool slowly gray cast iron is the result. The carbon separates in the form of graphite. It is this separation of the carbon from the iron that makes gray cast iron brittle. Gray cast iron is used a great deal in making castings for many kinds of machine parts. It can be easily identified by its dark gray, porous structure when the piece is broken. The tensile strength of common grades of gray cast iron run about 30,000 to 40,000 psi. Some gray cast irons are alloyed with nickel, copper, and chromium to give them higher corrosion resistances and greater strength.

White Cast Iron. Iron with a low silicon content in which the carbon has united with the iron instead of existing in a free state, as in gray cast iron. This condition is brought about through a process of rapidly cooling the metal leaving it very hard and brittle. In fact it is so hard that it is exceedingly difficult to machine and special cutting tools or grinders must be used to cut the metal. White cast iron is often used for castings having outer surfaces that must resist a great deal of wear. The structure of a piece of white cast iron will disclose a fine, silvery white, silky crystalline structure.

Malleable Cast Iron. Actually white cast iron which has been subjected to a long annealing process. The annealing treatment draws out the brittleness from the casting and leaves the metal soft but possessing considerable toughness and strength. The fracture of a piece of malleable cast iron will indicate a white rim and a dark center. Malleable irons have a tensile strength ranging from 40,000 to 100,000 psi.

Nodular Iron. This has the ductility of malleable iron, the corrosion resistance of gray iron and a greater tensile strength than gray iron. The tensile strength will range between 60,000 to 120,000 psi, depending on the type of iron. These special qualities in nodular iron are obtained by adding a small amount of magnesium to the iron at the time of melting and by using special annealing techniques. The addition of magnesium and control of the cooling rate causes the graphite to change from a stringer structure to rounded masses in the form of spheroids or nodules. This structural change is principally the reason for the improved properties of nodular iron.

Weldability

The important factor in welding cast iron is to keep the fusion penetration to a minimum to prevent the transformation of the metal into some other undesirable structure. In addition, preheating is necessary and followed by a slow rate of cooling. For example, if an excessive amount of heat develops because of the prolonged period of welding, followed by rapid cooling, it is possible that the malleable iron may turn into white cast iron. At the same time without proper preheating and cooling gray cast iron is likely to crack.

Preheating should be held to around 900°-1200°F. The critical temperature where structural changes will occur is about 1450°. Hence the preheat should never be near this temperature. The preheating should be applied as uniformly as possible and cooling should be controlled by postheating until the metal gradually reaches room temperature.

Prior to welding the surface layer of the cast iron should be removed. Elimination of the surface skin is important because it is full of impurities. Unless these impurities are removed they will interfere with the fusing of the weld metal.

Cast iron can be welded with the oxyacetylene or shielded metal-arc process. See Table XVI. Gas shielded-arc welding of cast iron is possible but ordinarily it is not considered as economical as the oxyacetylene and shielded metal-arc. The most economical process insofar as welding time and deposition rate is the shielded metal-arc process.

· Welding with an oxyacetylene flame requires a special cast iron filler rod with the correct amount of silicon. During the welding operation the silicon in the weld area has a tendency to burn away. Therefore, if the rod has sufficient silicon there will be a proper amount of this element in the weld area after the welding is completed. A flux also is essential with the oxyacetylene process to keep the molten puddle fluid. Otherwise, infusible slag mixes with the iron oxide leaving a weld with inclusions and blowholes. As a rule, it is more difficult to weld nodular iron than gray iron with the oyacetylene process because of the formation of gas pockets. Gas pockets are the result of the vaporization of magnesium which has a lower melting point than the nodular iron.

Broken castings are frequently repaired by bronze welding. See Chapter IX. The advantage of bronze welding is that the metal does not have to be heated to a molten condition and consequently there is less danger of destroying the characteristics of the base metal or developing stresses. The only restriction to this welding process is that a

Table XVI. Cast-Iron Welding Procedures Summary.

CAST-IRON TYPE	PROCEDURE	TREATMENT	PROPERTIES
Gray iron	Weld with cast iron	Preheat and cool slowly	Same as original
Gray iron	Braze weld	Preheat and cool slowly	Weld better; heat-affected zone as good as original
Gray iron	Braze weld	No preheat	Weld better; parent metal hardened
Gray iron	Weld with steel	Preheat if at all possible	Weld better; parent metal may be too hard to machine; if not preheated, needs to be welded intermittently to avoid cracking
Gray iron	Weld with steel around studs in joint	No preheat	Joint as strong as original
Gray iron	Weld with nickel	Preheat preferred	Joint as strong as original; thin hardened zone; machinable
Malleable iron	Weld with cast iron	Preheat, and postheat to repeat malleableizing treatment	Good weld, but slow and costly
Malleable iron	Weld with bronze	Preheat	As strong, but heat-affected zone not as ductile as original
White cast iron	Welding not recommended		
Nodular iron	Weld with nickel	Preheat preferred; postheat preferred	Joint strong and ductile, but some loss of original properties; machinable; all qualities lower in absence of preheat and/or postheat

American Welding Society

bronze weld cannot be used on metals that will later be subjected to high temperatures since bronze loses its strength when heated to 500°F or more.

Two types of electrodes are used when welding cast iron with the shielded metal-arc process—machinable and non-machinable. The non-machinable electrodes have a mild steel core covered with a special flux. They leave a very hard deposit and are used only when the welded section is not to be machined afterwards. These electrodes produce a tight and waterproof weld making them ideal for repairing motor blocks, water jackets, transmission cases, compressor blocks, pulley wheels and similar structures.

Machinable type electrodes have a copper-nickel or pure nickel core. They are used to repair all kinds of broken castings, correcting for machining errors, filling up defects, or to weld cast iron to steel. The deposited metal is soft enough so it can be readily machined.

Welding Galvanized Metal

One of the problems encountered in welding galvanized metal is that the zinc coating is burned away. Since the zinc is intended to protect the surface from corrosion the exposed weld area becomes readily contaminated unless measures are taken to recoat the area with some suitable protective material.

The development of Gal-Weld has simplified the process of replacing the zinc. Gal-Weld is a zinc alloy bar containing flux. After a seam is welded the surface is first brushed with a steel wire brush. A small amount of alloy bar is then melted over the weld using the residual heat from the welding or a gas torch. While the Gal-Weld is in a liquid state it is brushed lightly to spread it evenly over the heated area. The bar melts at approximately 450°F and it acts as its own pyrometer. If the surface is not hot enough the zinc will not spread. If the surface is too hot the zinc will gasify and melt away. The spreading of the galvanizing material should be done with a brass brush because the alloy will not readily adhere to brass.

REVIEW QUESTIONS

1. How does the thermal conductivity of a metal affect its weldability?

2. Improper control of welding heat will generally produce what results in a weld?
3. What is the effect of carbon in its relationship to the weldability of steel?
4. In a steel coding system what do the four digits represent?
5. How is the tensile strength of steel affected by the carbon content?
6. What type electrodes are recommended for welding low- and medium-carbon steel with the shielded metal-arc process?
7. Why are very high carbon steels rarely recommended for welding?
8. What are some of the elements added to steel to make alloy steels?
9. What are some of the more common types of low-alloy steels?
10. How can cracking and brittleness be minimized in welding low-alloy steels?
11. What is the difference between a low-alloy steel and a high strength low-alloy steel?
12. How do martensitic stainless steels differ from the ferritic stain-steel?
13. What is the difference between the austenitic and ferritic stainless steels?
14. What stainless steels generally respond better to welding? Why?
15. In welding stainless steel why will the gas shielded-arc process produce better results?
16. What are some of the specific properties of aluminum?
17. In the aluminum coding system what is the significance of the first digit?
18. How is the purity of aluminum designated in the coding system?
19. In what ways are the tempers designated for different classifications of aluminum?
20. When welding aluminum with an oxyacetylene flame why is a flux required?
21. Why is the Tig and Mig processes preferred in welding aluminum?
22. How does magnesium compare with aluminum in terms of strength, weight, and weldability?
23. What is the difference between oxygen-bearing copper and oxygen-free copper?
24. Why is deoxidized copper the most widely used where welding is required?
25. How do brasses differ from bronzes?

26. Nickel alloys are sometimes known by what trade names?
27. What are some of the outstanding mechanical properties of titanium?
28. How do the properties of zirconium and beryllium compare with those of titanium and beryllium?
29. What do beryllium, titanium, and zirconium have in common insofar as weldability is concerned?
30. What is the difference between gray, white, malleable, and nodular cast iron?

| Chapter 9 | # Brazing and Soldering |

In addition to fusion welding other means are often used to join metals. The two more common processes are brazing and soldering. These processes have wide application in joining most commercial metals, especially in situations where it is impractical or uneconomical to use regular fusion welding methods.

BRAZING

According to the American Welding Society brazing is defined as a group of welding processes where coalescence is produced by heating to suitable temperatures above 800°F and by using a nonferrous filler metal having a melting point below that of the base metals. The filler metal is distributed between the closely fitted surfaces of the joint by capillary action. Most commercial metals can be brazed. Although a brazed joint has a relatively high tensile strength this method of joining is not recommended when the full strength properties of a joint are required. An important characteristic of brazing is that there is less danger of destroying the mechanical properties of the base metal since lower bonding temperatures are used than normally required for regular fusion welding. This process is especially adaptable for joining dissimilar metals.

Brazing Requirements

The success of any brazing operation depends on joints having relatively small clearances and surfaces that are free of oxide and other contaminants. Cleaning is accomplished by coating the surfaces with a special flux which when heated is capable of dissolving all foreign matter. Once the surfaces are properly fluxed a brazing filler metal is melted at some point along the seam. Capillary action then draws the molten brazing metal between the surfaces of the joint. Upon cooling to room temperature the solidified brazing metal forms a solid bond at the interface of the workpiece. During heating and cooling precautions are taken to prevent any movement of the surfaces. Ordinarily, for most production work some type of fixture is used to hold parts in alignment during the brazing process.

Types of Joints for Brazing

The two basic joints for brazing are the lap and butt. (Tee and corner joints are considered as butt joints.) Fig. 9–1. The lap joint offers the

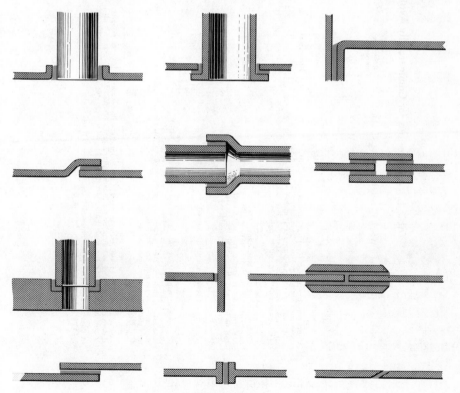

Fig. 9-1. Basic types of joints for brazing.

Table I. Applications for Commercially Available Brazing Fluxes.

AWS Brazing Flux Type No.	Metal Combinations for Which Various Fluxes Are Suitable — Base Metals	Filler Metals	Effective Temperature Range of Flux, °F	Major Constituents of Flux	Physical Form	Methods of Application†
1	Aluminum and aluminum alloys	BAlSi	700–1190	Fluorides; Chlorides	Powder	1,2,3,4.
2	Magnesium alloys	BMg	900–1200	Fluorides; Chlorides	Powder	3,4.
3A	Copper a: d copper-base alloys (except those with aluminum) iron base alloys; cast iron; carbon and alloy steel; nickel and nickel base alloys; stainless steels; precious metals (gold, silver, palladium. etc.).‡	BCuP BAg	1050–1600	Boric Acid, Borates, Fluorides, Fluoborate Wetting Agent	Powder Paste Liquid	1,2,3,
3B	Copper and copper-base alloys (except those with aluminum); iron base alloys; cast iron and alloy steel; nickel and nickel base alloys; stainless steels; precious metals (gold, silver, palladium. etc.)	BCu BCuP BAg BAu R BCuZn BNi	1350–2100	Boric Acid Borates Fluorides Fluoborate Wetting Agent	Powder Paste Liquid	1,2,3.
4	Aluminum-bronze; aluminum-brass §	BAg, BCuZn, BCuP	1050–1600	Borates Fluorides Chlorides	Powder Paste	1,2,3.
5	Copper and copper-base alloys (except those with aluminum) nickel and nickel-base alloys; stainless steels; carbon and alloy steels; cast iron and miscellaneous iron-base alloys; precious metals (except gold and silver)	BCu, BCuP BAg-(8-19) BAu, BCuZn BNi	1400–2200	Borax Boric Acid Borates	Powder Paste Liquid	1,2,3.

* This table provides a guide for classification of most of the proprietary fluxes available commercially.
For additional data consult AWS specification for brazing filler metal A5.8 ASTM B200; consult also AWS Brazing Manual. 1963 Ed.
† 1–Sprinkle dry powder on joint; 2–dip heated filler metal rod in powder or paste; 3–mix to paste consistency with water, alcohol, monochlorobenzene, etc.; 4–molten flux bath.
‡ Some Type 3A fluxes are specifically recommended for base metals listed under Type 4.
§ In some cases Type 1 flux may be used on base metals listed under Type 4.

American Welding Society

maximum strength. The overlap should be at least three times the thickness of the thinnest section. See Fig. 9–1. Since the cross sectional area of a butt joint is limited to the cross sectional area of the thinnest section maximum joint efficiency is impossible. Where a lap joint is objectionable the weakness of the butt joint can be minimized by using a scarf joint. See Fig. 9–1. However, the scarf joint is more difficult to prepare and often special care is required to keep the pieces in alignment. The strength of a butt joint can also be improved by increasing the cross sectional area as shown in Fig. 9–1 or by using a sleeve over the joint.

Joint clearance, which is the distance between the interface of the pieces, is an important factor in the performance of a brazed joint whether the joint is subjected to loadings of fatigue, impact, or static. Too tight a joint may hinder the plastic flow of the filler metal while too great a clearance may prevent the full effects of the capillary action leaving voids and uneven distribution of filler metal. In general joint clearance should range between 0.001″ and 0.010″.

Brazing Filler Metals

A brazing filler metal should meet the following requirements:
1. Sufficient fluidity so the metal will flow evenly by capillary attraction.
2. Good melting action to form a sound metallurgical bond.
3. Melting point consistent with the type of metal to be joined.

Brazing filler metals fall into seven groups: silver, aluminum-silicon, copper-phosphorus, gold, copper and copper-zinc, magnesium, and nickel. Table I lists the principal metals which are joined with these filler metals.

Fluxes

Any form of oxide on the surface of a metal will inhibit a uniform flow of the brazing metals. Accordingly a flux of some kind is necessary to eliminate the oxide. The common commercial fluxes are in paste, liquid, or powder form. Fluxes have as their main ingredients borates, fused borax, boric acid, fluorides, chlorides, and fluborates. There is no single flux which is applicable for all brazing operations. See Table I. All traces of flux residues must be removed after brazing to prevent corrosion.

In some situations where mass production is involved the application of fluxes is a time consuming task. Consequently, controlled atmospheres are used to remove oxide and prevent the formation of oxide during brazing. This method of oxide removal and control is often associated with induction brazing of titanium, zirconium, and other refractory metals. In a controlled atmosphere a gas is continuously supplied to a furnace and circulated within it at slightly higher than atmospheric pressure. Gas may consist of high-purity hydrogen, carbon dioxide, carbon monoxide, nitrogen, argon, ammonia or some form of combusted fuel gas.

Heating Methods

The application of heat for brazing purposes is accomplished by a variety of methods depending on the kind of material to be brazed, quantity of production, and sizes of parts to be joined. The following techniques are used:

Torch Heating. Torch heating is probably the most common for brazing purposes. See Fig. 9–2. The gas mixtures may be oxyacetylene,

Fig. 9-2. Torch heating is often used for brazing.

All-State Welding Alloys Co.

air-gas, gas-oxygen, or oxyhydrogen. To a large extent, the type of gas mixture depends on the thermal conductivity, type, and thickness of the material to be joined.

Oxyacetylene is often more versatile for torch brazing because of its wide range of heat control. A slightly reducing flame is required and care must be taken to prevent the core of the flame from coming in contact with the metal. Close flame contact may cause the base metal to melt and restrict the flow of the brazing metal.

The air-gas torch provides the lowest heat and is far more adaptable for brazing thin sections. The air-gas mixture may consist of air at atmospheric pressure and city gas or air and acetylene.

The gas-oxygen process uses oxygen with city gas, bottled gas, propane, or butane. This mixture produces a high flame temperature and is applicable where greater brazing heat is required.

The oxyhydrogen torch is very adaptable for brazing aluminum and other nonferrous metals because of its low heat producing temperatures. The low temperatures prevent possible overheating the metal. Hydrogen also provides additional cleaning action and shielding during the brazing process.

Furnace Heating. Furnace heating is a production process for brazing parts that can be assembled and positioned on trays. The trays are loaded in a furnace either manually or automatically. Some automatic loading consists of a conveyor belt on which the parts are placed. The conveyor belt then moves into the furnace at regulated speeds. See Fig. 9–3. The brazing filler metal, which can be in the form of wire, foil, powder or paste, is placed in the right position near the joint and the

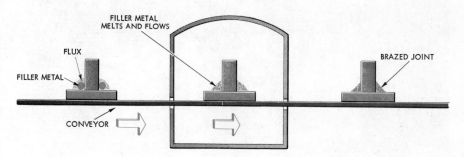

Fig. 9-3. Production brazing is frequently done in a furnace.

heat of the furnace melts the filler metal. Fluxing is used except when the heating is done in a controlled atmosphere.

Most of the high production brazing in a controlled atmospheric furnace is done with hydrogen and either endothermic or combusted gas. Inert gases such as helium or argon are also used for controlled atmospheric furnace brazing.

Some furnace brazing is done in a vacuum where continuous pumping prevents oxidation and removes volatile constituents that are liberated during the brazing process. Vacuum brazing has wider applications in aerospace and nuclear industries where reactive metals are joined and any form of entrapped fluxes must be avoided.

Induction Heating. In this process heat is generated by an inductor coil which is not in contact with the parts being brazed. See Fig. 9–4. A power supply unit converts regular line 60 hertz (cycle) current into a high frequency low voltage current. As the current flows through the inductor coil, which surrounds the object to be brazed, a magnetic field is created. When an electrically conductive object is placed in a magnetic field, an electro-motive force is induced in the conductive material. This sets up a current and the resistance of the object to the flow of current causes instant heating to occur. The heat is relatively near the surface of the metal and any interior heating results from thermal conduction from the hot surface.

The power supply for the high-frequency current is either a motor

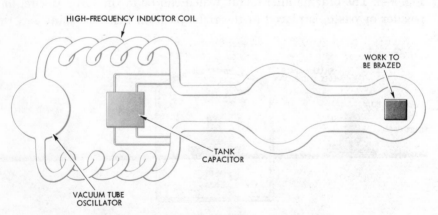

Fig. 9-4. Schematic of an induction heating coil.

generator, spark gap unit, or a vacuum tube oscillator. Motor generators are classified as low frequency units (up to 10,000 hertz (cycles)), spark gap as medium frequency (20,000 to 30,000 hertz (cycles)), and vacuum-tube oscillators as medium to high frequency (200,000 to 5,000,000 hertz (cycles)). The size of the power supply depends on the parts to be brazed and the production rate. Low frequency power units produce deeper heating and are designed for brazing heavy sections whereas high frequency units produce light or shallow heating.

The work or inductor coil usually is of tubing but can be a copper block with an internal passage for water. The work coil has to be kept cool with circulating water. The size of the coil, that is the number of turns of copper tubing or the thickness of the copper block, is governed by the required heat zone to be developed.

Heat input is also regulated to some extent by varying the gap (couple) between the induction coil and the parts to be brazed. The fixture should be such that it does not come too near the induction work coil particularly if the fixture is made of steel. A steel fixture is highly susceptible to induction heating and readily becomes overheated. The overheating of the fixture usually affects the proper heating of the area to be brazed. The most suitable material for fixtures that must come close to the work coil is ceramic or some kind of nonmagnetic material.

Induction heating is used in brazing parts which can be aligned in a fixture and require rapid heating. It is a very economical brazing technique for quantity production when parts can be adapted to the induction furnace.

Dip Brazing

There are two dip brazing methods, molten metal bath and molten flux bath.

Molten Metal Bath. This technique consists of immersing parts in a bath of molten brazing metal. The brazing material is melted in a crucible with a cover of flux maintained over the molten filler metal. The parts to be brazed are first cleaned and fluxed and then dipped into the bath. The process is limited to brazing small assemblies such as wire connections or metal strips when they can be easily held in fixtures.

Molten Flux Bath. Flux in the form of a chemical salt is melted in a container or flux pot by a gas flame or electrical resistance. The most common method is passing an electric current through the bath. With resistance heating the initial charge of flux is first melted by some other heating source and the molten salt poured into the fluxing container. Once the flux is in a molten state the resistance of the bath to the electrical current provides sufficient heat to keep the flux at the proper temperature. A thermocouple immersed in the molten bath and a temperature control unit maintains the molten flux at the required brazing heat.

The parts to be brazed are cleaned and assembled in a suitable fixture. Brazing filler metal in the form of rings, washers, slugs or paste mixture is preplaced on the base metal. In production brazing with molten salts the parts and holding fixtures are usually preheated in a furnace to a temperature near the melting point of the molten flux. The parts are then dipped in the molten flux where the heat is sufficient to melt the brazing material. Molten flux bath dip brazing is often used to fabricate radiators or other heat cooling units.

Braze Welding

Braze welding is slightly different from regular brazing. Whereas brazing involves the joining of two surfaces by a thin bond of brazing metal, braze welding is carried out much as fusion welding except that the base metal is not melted. The base metal is brought up to its tinning temperature and then a bead is deposited over the seam with a filler rod. Although the base metal is never actually melted the unique characteristics of the bond formed by the brazing metals are such that the results are often comparable to those secured through fusion welding.

Bronze welding is a typical braze welding operation. This technique is frequently used for joining or repairing such metals as cast iron, malleable castings, copper, brass, and various dissimilar metals. See Fig. 9–5. The bronze filler rod which is applied to the seam consists of copper and zinc with small quantities of tin, iron, manganese, and silicon.

The welding procedure involves cleaning the surface and then applying a coating of flux to diffuse the oxide. The flux is applied by dipping the heated rod into the powdered flux. The flux can also be mixed with water and spread over the seam.

Any type joint is suitable for bronze welding. On thick sections, the edges should be beveled to form a 90° vee-groove. The work is posi-

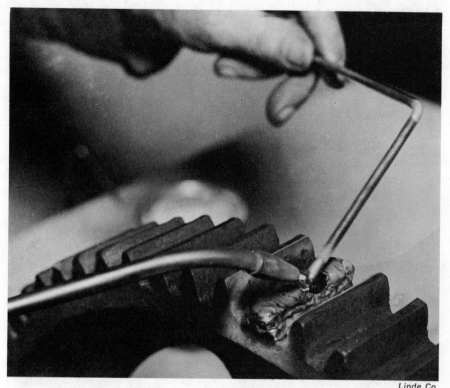

Fig. 9-5. Building up a missing gear tooth with a bronze weld.

tioned so the weld travels upward on an incline. See Fig. 9–6. In this position the molten bronze cannot flow ahead of the heated welding area and the surface in front of the weld is left open to heating.

Generally, bronze welding is performed with an oxyacetylene torch with a flame that is slightly oxidizing. The flame is concentrated on the starting end until the metal begins to turn to a dull red color. A small amount of bronze is melted on the surface and allowed to spread along the entire seam. The flow of this thin film of bronze is known as the tinning operation. Unless the surfaces are tinned properly the remaining bronzing procedure cannot be executed successfully. If the surface of the metal is heated to the proper temperature, the bronze should spread out evenly over the metal. A surface that is too hot will cause the bronze to bubble or run around like drops of water on a warm stove.

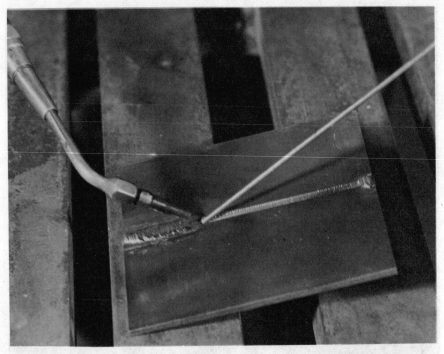

Fig. 9-6. In bronze welding, better results are obtained if the weld travels upward at an incline.

When the bronze forms into balls which tend to roll off the base metal is not hot enough.

Once the base metal is tinned sufficiently a bead is deposited over the seam using a slight circular torch motion. The rod must be constantly dipped into the flux as the weld progresses.

Brazing with Carbon Electrodes

Some brazing operations are performed with a carbon arc. The process is very similar to the metallic arc except that the electrode does not melt or provide filler metal. The carbon electrode in this instance serves only as the medium of generating the arc. The electrode is either of pure graphite or in the form of a copper coated carbon and held in a special holder. See Fig. 9-7. The regular metallic arc holder is not suitable since the carbon electrode becomes very hot and the intense heat

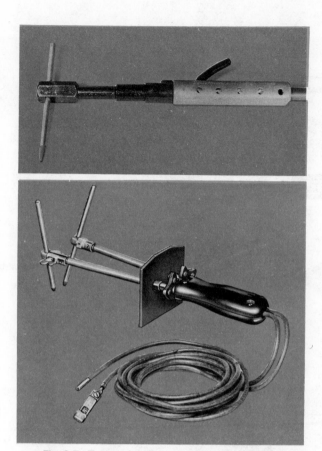

Fig. 9-7. Types of carbon arc electrode holders.

soon ruins the ordinary holder. A shield is often located near the handle to protect the operator's hand from the heat. The handle of the holder is made so air can circulate around it and keep it cool. When the carbon arc is used for continuous operations the holder is often water-cooled.

The electrode for brazing is shaped by grinding it to a long tapering point. See Fig. 9–8. Either a DC or AC power source can be used for brazing. With an AC unit the arc is formed between two carbon electrodes held in a holder as shown in Fig. 9-7. In the single carbon electrode with DC current the machine must be set for straight polarity. Reverse polarity produces too much of an unstable arc and causes

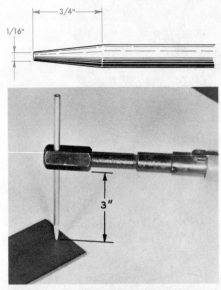

Fig. 9-8. Brazing with the carbon arc.

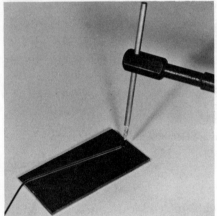

greater quantities of vaporized carbon to enter the molten brazing material.

The brazing operation is carried out by striking an arc and feeding a brazing rod into the arc. Most carbon arc brazing is done with a bare silicon bronze rod. The filler rod is held as illustrated in Fig. 9–8 and moved along the surface with the arc over the rod.

With the AC holder the formed arc between the two electrodes remains intact even though the electrodes are withdrawing from the work. The arc is established by moving the electrodes with a push button located on the holder until the electrodes touch. Pressure on the button then is released to permit the points of the electrode to part and establish the arc. When the distance between the two electrodes is correct there will be a quiet, soft flame. Striking the arc with a DC current is done in the same way as in metallic arc welding.

Brazing with a Liquifluxer

Liquifluxer is a product produced by the Rexarc Corporation which eliminates the need of applying flux to metal surfaces before beginning the actual brazing operation. A liquid fluxing material is placed in a separate tank that is connected to the regular gas line. The flow of gas into the tank vaporizes the flux and induces it into the gas stream where it passes through the welding torch and into the flame. As the flame contacts the metal the flux in the flame cleans and fluxes the surface for brazing. No pre-cleaning or post-cleaning of the parts is necessary. As a rule, less heat is required with Liquiflux than when a powder or wet flux is used. The weld metal is not oxidized and the strength of the brazed joint is greater.

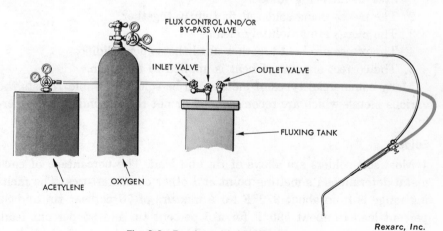

Fig. 9-9. Brazing with Liquifluxer.

Rexarc, Inc.

With the exception of the fluxing tank no other special equipment is needed as a standard welding torch is utilized. The tank is equipped with a three-valve control unit. The inlet valve regulates the entry of gas into the tank, the flux or bypass valve controls the amount of flux used or eliminates the flux entirely, and the outlet valve supplies the flux gas through the line to the torch. See Fig. 9–9.

Liquifluxer is applicable for brazing brass, bronze, copper, cast aluminum, cast and malleable iron, nickel and steel.

SOLDERING

Soldering is a process of joining two pieces of metal by using a nonferrous filler metal having melting temperatures below 800°F or bellow those of the base metal. The filler metal is called solder and is distributed between the surfaces by capillary attraction.

Soldering is used to join surfaces where they are not subjected to high strength forces since solder has a relatively low tensile strength. Soldered joints are also not suitable where temperatures approach the melting point of solder.

Soldering is a simple operation providing the following five basic requirements are observed:

1. The right type and amount of solder and flux are used for the base metal being joined.
2. The pieces being soldered fit tightly together.
3. The pieces are absolutely clean.
4. The pieces are held together until the solder solidifies.
5. The correct amount of heat is applied to the seam.

Many commercial types of metals can be soldered. Table II lists the various metals which are recommended or not recommended for soldering.

Solders

Most soft solders are alloys of tin and lead. The percentage of each metal determines its melting point and other characteristics. The melting range is from about 370°F for a mixture of 70 percent tin and 30 percent lead to about 590°F for a 5 percent tin and 95 percent lead. The most common general-purpose solder is known as half-and-half or

Table II. Solderability of Metals.

Base Metal, Alloy or Applied Finish	Flux Requirements			Soldering Not Recommended
	Non-Corrosive	Corrosive	Special Flux and/or Solder	
Aluminum			X	
Aluminum-Bronze			X	
Beryllium				X
Beryllium Copper		X		
Brass	X	X		
Cadmium	X	X		
Cast Iron			X	
Chromium				X
Copper	X	X		
Copper-Chromium		X		
Copper-Nickel		X		
Copper-Silicon		X		
Gold	X			
Inconel			X	
Lead	X	X		
Magnesium			X	
Manganese-Bronze (High Tensile)				X
Monel		X		
Nickel		X		
Nichrome			X	
Palladium	X			
Platinum	X			
Rhodium		X		
Silver	X	X		
Stainless Steel			X	
Steel		X		
Tin	X	X		
Tin-Bronze	X	X		
Tin-Lead	X	X		
Tin-Nickel	X	X		
Tin-Zinc	X	X		
Titanium				X
Zinc		X		
Zinc Die Castings			X	

American Welding Society

Brazing and Soldering   251

50-50 solder. It contains 50 percent tin and 50 percent lead and melts at about 471°F.

A variety of other alloys are used for soldering purposes. See Table III. In general alloys with a low tin content have higher melting points and do not flow as readily as the high tin alloys. The low-tin alloys are less expensive and find application where large volumes of soldering is done. Solders with a high amount of tin have better wetting properties and produce less cracking. High-tin solders are used considerably in electrical work. The solders with a tin content of 60 percent or more are classified as fine solders and are employed in soldering instruments where temperature requirements are critical.

Special solders are also available for specific purposes. Thus a tin-antimony solder is designed to solder food-handling vessels where lead contamination must be avoided. Tin-zinc solders are intended primarily for joining aluminum. Lead-silver solders are used where strength at elevated temperatures are required.

Solders are available in bar, cake, solid wire, flux-core wire, ribbon and paste forms. Flux-core wire solder has an acid or resin flux in the center of the wire. With these solders no additional flux is needed.

Table III.   Composition of ASTM Solders.

| ASTM NUMBER | COMPOSITION % | | | | MELTING RANGE |
	TIN	LEAD	ANTIMONY	COPPER	OF
5A	5	bal	0.12 max	0.08	572-596
5B	5	bal	0.50 max	0.08	572-596
10B	10	bal	0.50 max	0.08	514-573
20B	20	bal	0.50 max	0.08	364-535
20C	20	bal	1.0-1.2	0.08	364-535
30A	30	bal	0.25	0.08	361-477
30B	30	bal	0.50	0.08	361-477
40A	40	bal	0.12	0.08	361-455
50A	50	bal	0.12	0.08	361-421
60A	60	bal	0.12	0.08	361-374
63A	63	37	0.12	0.08	361-361

American Welding Society

Fluxes

Oxides and rust will form on surfaces of most metals when exposed to air. Solder will not adhere if such impurities are present. By applying a flux the oxide is removed and the formation of new oxide during the soldering process is prevented. Fluxes also increase the wetting action enabling the solder to flow more freely.

Fluxes come in paste, liquid, powder, and cake form. Some are general-purpose fluxes usable on most metals. Others are special fluxes such as those for aluminum soldering.

All fluxes are classified as corrosive or noncorrosive. Although the corrosive types are most effective they must be washed from the metal after soldering. They should never be used for electrical or electronic work. Rosin is the most common noncorrosive flux. Zinc chloride is the most frequently used corrosive flux.

Heating Devices

In any soldering operation both pieces of metal to be joined must be hot enough to melt the solder. A strong bond is achieved only if the molten solder spreads evenly over the surface. A number of devices are available for heating purposes. The type used depends on the size and configuration of the assembly.

Soldering Coppers. A soldering copper consists of a forged piece of copper fastened to an iron rod with a wooden handle on one end. These coppers vary in size with heads forged in several shapes. See Fig. 9–10. Generally a lightweight copper is used for soldering light-gage metal and a heavyweight copper for soldering heavy-gage metal. A lightweight copper on heavy metal does not hold enough heat to heat the metal or allow the solder to flow smoothly. Soldering coppers are heated in a furnace or with a blowtorch.

Electric Soldering Irons and Pencils. These devices are often more convenient than soldering coppers because they maintain a uniform heat. See Fig. 9–10. They vary in sizes from 25 watts to 550 watts. Lightweight, low-voltage irons with replaceable heating elements and tips are called soldering pencils and are preferred for electrical and electronic work. Electric soldering guns produce instant heat at the tip of a long small point when the trigger is pulled. On most guns the trigger also turns on a light which focuses at the point. For these reasons the soldering gun is very popular for electronic work.

Soldering copper

Electric soldering iron

Fig. 9-10. Types of heating coppers for soldering.

Soldering pencil

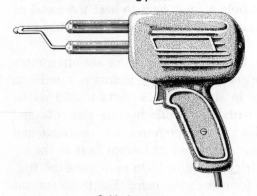

Soldering gun

Flame-burning Devices. Some soldering operations are impossible or very difficult to perform with a soldering copper or iron. For such tasks a flame is used as the source of heat. See Fig. 9–11. The flame is produced either with an ordinary Bunsen burner or a gas torch depending on the nature of the job. The most efficient, safe, and versatile gas torch is one that burns city gas and compressed air. These torches are often equipped with changeable tips which can produce a wide range of flame sizes. The gas-air torch has two needle valves, one valve controls the pressure of gas and the other valve the compressed air. To light this torch the gas-needle valve is opened slightly and ignited. Then the oxygen valve is turned on and adjusted until a blue flame results. The length of the flame is controlled by the amount of gas and air allowed to flow to the tip.

Bottled-gas torches are also used for soldering especially when the

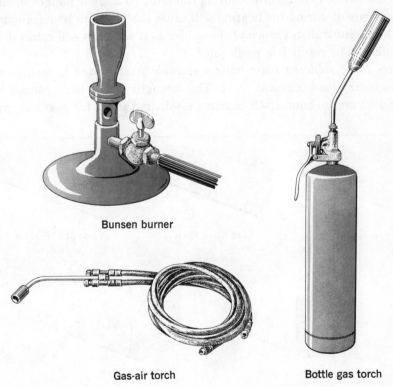

Bunsen burner

Gas-air torch   Bottle gas torch

Fig. 9-11. Flame-burning devices for soldering.

work is not at a fixed station where a gas-air torch is available. The bottled-gas torch must be operated with care, therefore the manufacturer's instructions should always be followed carefully.

Soldering Procedure

Parts to be soldered must fit perfectly so the solder can travel by capillary action between the two surfaces. Solder will cease to flow where there is a gap between the two workpieces.

Parts to be soldered must be absolutely clean because the solder will not stick to a dirty, oily or oxide-coated surface. Dirt and grease can be removed with a cleaning solvent. Steel wool or some form of abrasive cloth is used to eliminate the oxide. Application of a flux completes the cleaning process and keeps the metal free from oxide during the heating and soldering operation.

Parts must be held together during soldering so there is no movement. Any movement during the heating will cause the pieces to be misaligned and the slightest disturbance of the solder as it solidifies will cause it to crystallize. The result is a weak joint.

Parts to be soldered must have a suitable joint design to withstand the necessary load imposed on it. The strength of the joint cannot be dependent on the bond itself because a soldered joint, for example, will

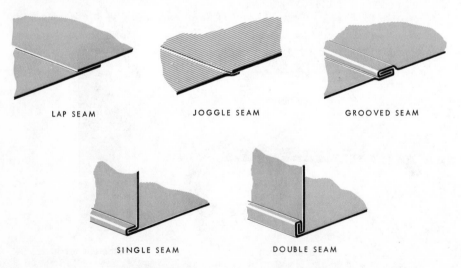

LAP SEAM   JOGGLE SEAM   GROOVED SEAM

SINGLE SEAM   DOUBLE SEAM

Fig. 9-12. Types of joints used for soldering.

seldom develop shear strength greater than 250 psi. If greater strength is needed, some type of mechanical joint should be made before soldering. Typical joints for soldering are shown in Fig. 9–12.

Tinning a Copper. The point of a soldering copper must be covered with a thin coat of solder to operate properly. A tinning operation is performed by first smoothing the surfaces of the point with a file and then heating the copper until it is hot enough to melt solder. The point is next rubbed on a block of sal ammoniac and a small amount of solder melted during the rubbing action. Tinning can also be accomplished by dipping the point in a liquid or paste flux and applying solder. See Fig. 9–13.

Sweat-soldering. This is a process whereby two surfaces are soldered without the solder being visible. This operation is performed by applying a coating of solder to each of the surfaces to be joined. The coated surfaces are placed together and a heated copper held over the seam. See Fig. 9–14. To avoid smearing the exposed surfaces of the metal with solder the excess solder on the copper is removed by quickly wiping the joint with a damp cloth.

When the solder between the two surfaces begin to melt and show evidence of flowing out from the edges, pressure is applied on the metal. As the copper is drawn slowly along the seam, pressure must be maintained until the solder solidifies.

Seam Soldering. Seam soldering involves running a layer of solder along the outside edge of the joint. See Fig. 9–15. As a rule the surfaces

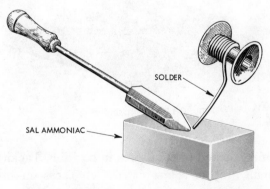

SOLDER

SAL AMMONIAC

Fig. 9-13. Tinning a copper over a block of sal ammoniac.

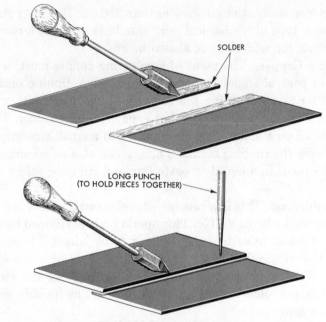

Fig. 9-14. Sweat soldering.

Fig. 9-15. Seam soldering.

of the joint are first tacked to hold them in position. Tacking is done by depositing a small drop of solder at several places on the seam. The point of the copper is then moved along the edge of the seam and solder applied directly in front of the point.

Flame Soldering. Many soldering jobs are difficult to execute with a soldering copper or electric-soldering iron. In such cases a torch flame is often more practical especially where fast soldering is required. The flame can be adjusted to the required size and directed to the exact spot where it is needed and moved to heat both pieces properly as well as the solder.

Induction Soldering. Used where large-scale production is involved and the pieces can be kept in alignment by some kind of fixture. Induction heating is produced by a motor generator, spark-gap unit, or vacuum tube oscillator and the rate of heating is governed by the amount of induced current flows. The pieces to be soldered are aligned near the induction work coil with the solder preplaced on the joint. Cleaning and fluxing must be completed before the soldering process is started.

Solders for induction heating should have rapid spreading and good capillary flow characteristics. Preforms combining solder and flux are the best means of supplying the correct amount of solder and flux for induction soldering.

Resistance Soldering. The electrical resistance of the metal provides the necessary heat for soldering. The work to be soldered is placed between a ground and a movable electrode or two movable electrodes. Solder is fed directly into the joint when the surfaces of the workpieces reach the proper temperature.

Dip Soldering. This consists of a pot with molten solder. The joint to be soldered is simply dipped into the molten solder. The bath supplies both the heat and solder to complete the soldering operation. This soldering technique is very economical since an assembly with several joints can be soldered simultaneously. Any dip soldering operation requires suitable jigs and fixtures to keep the unit in alignment until solidification of the solder is completed. These fixtures must be well made to prevent movement of the parts during the operation when they are being dipped and handled.

Oven Soldering. Gas or electric heated ovens are used for soldering when the entire assembly can be heated without damaging any of its components. Since the assembly must be moved before and during the solidification of the solder suitable jigs and fixtures are required to hold parts together. Without proper clamping the joint will fail because solder is easily disturbed when it is at its solidifying temperature.

REVIEW QUESTIONS

1. How does brazing differ from regular fusion welding?
2. In some instances why is brazing preferred over fusion welding?
3. Why is joint design an important factor in brazing?
4. What factors must be considered in selecting a brazing metal?
5. Why is a flux required for brazing?
6. What type of heating devices may be used for brazing?
7. How does the molten metal bath differ from the molten flux bath dip brazing?
8. In brazing with the carbon arc, how does the carbon electrode for DC current differ from the one used with AC power supply?
9. When brazing with a DC power supply why is straight polarity used?
10. Why is a special fluxing liquid induced into the gas stream from a special fluxing tank a convenient method of brazing?
11. What is the basic principle of induction heating for brazing purposes?
12. What is the difference between braze welding and regular brazing?
13. When is soldering not recommended for joining metals?
14. What are the basic ingredients of soft solder?
15. How does the tin content of solder affect its flowing properties?
16. What kind of fluxes are used for soldering purposes?
17. What is the advantage of an electric soldering iron over the soldering copper?
18. When are flame burning devices used in preference to soldering coppers or irons?
19. What are some of the basic requirements which contribute to effective soldering?
20. What is meant by tinning a copper?
21. How does sweat soldering differ from seam soldering?
22. How can the strength of a soldered joint be increased?
23. What is dip soldering?
24. Why must the parts be tightly clamped for dip or oven soldering?

Chapter 10 | Surfacing

Surfacing is a term used to describe processes of depositing metal on surfaces of tools and machine parts in order to increase their dimensional sizes or to obtain some specific mechanical property. Thus, the process may involve building up a worn shaft, extending the life of machine parts such as gears, improving the wear resistance of cutting tools, or replacing metal which has corroded away. See Fig. 10–1. As a matter of fact, surfacing is considered to be one of the most economical ways of conserving and extending the life of machines, tools, and construction equipment.

The two main categories of surfacing are known as *hardfacing* and *metallizing*. Hardfacing is a fusion technique and includes processes which are designed to produce hard, tough overlays to resist severe abrasion, corrosion, and impact loads. Metallizing is a spray coating procedure where finely divided particles of metal are deposited on surfaces where metal has worn away or where it is necessary to build up certain contours.

HARDFACING

Hardfacing is actually a fusion weld process in which a special hardfacing rod, wire, or electrode is intermixed with the base metal. The success of any hardfacing application depends on the use of the correct filler metal to meet the service requirements of the pieces which are to be conditioned. In general, parts which are to be hardfaced are subjected to three types of wear: abrasion, impact, or corrosion.

The Lincoln Electric Co.

Fig. 10-1. Examples of parts which have been repaired by surfacing.

Abrasion is associated with surfaces that are subjected to continuous grinding, rubbing, or gouging actions. For example, low stress scratching abrasion or erosion is characterized by sand sliding down a chute, high stress grinding action is typified by stone being crushed between metal faces, and gouging abrasion is exemplified by dipper teeth crushing into bed rock.

Impact refers to the absorption of kinetic energy and its release by elastic action. With impact wear metal is lost or deformed as a result of chipping, upsetting, cracking, or crushing forces. In abrasion the wearing forces move parallel to the surface of a component while impact forces are more or less perpendicular to the absorbing members.

Corrosion involves the destruction of a surface from atmospheric contamination, chemicals, and oxidation or scaling at elevated temperatures.

Without an understanding of the service condition of the parts to be repaired it becomes impossible to select the correct wear-resistant material. Thus a hardfacing metal for corrosion wear may not be suitable for impact resistance wear.

In most cases, hardfacing application requires the use of a single product. However, there are instances when the use of two alloys provide the best results. For example, if a part is extremely worn a less expensive material may be used followed by the appropriate surface deposit. The buildup metal may be a carbon steel of sufficient strength to resist deformation and still support the surface. On the other hand, if service conditions are subject to impact loading or high temperatures carbon steel will not be adequate for buildup and an alloy build-up material will have to be used.

Properties of Parts to be Hardfaced

An additional requirement of any hardfacing operation is a knowledge of the composition of the component to be serviced. By and large, metals of these parts can be grouped into two categories. In one group are the metals whose physical properties are not changed significantly or are subject to cracking when heated and cooled in a hardfacing operation. These metals include the low range carbon and medium carbon steels, the low alloy steels, and the stainless steels. The second group includes metal parts made of steels whose physical characteristics are changed with the application of hardfacing materials. These metals usually have been hardened by some heat treating process and any subsequent expo-

Table I.   Classification of Surfacing Alloys.

Classification by Basic Types	Important Features	Successful Applications
Tungsten carbide deposits	Maximum abrasion resistance	Oil well rock drill bits and tool joints
Granules or inserts		A wide range of severely abrasive conditions
Coarse granule tube rods	Worn surfaces become rough	
Fine granule tube rods	Best performance when gas welded	
High chromium irons	Excellent erosion resistance	Abrasion by hot coke
Multiple alloy type	Hot hardness from 800–1200° F with W & Mo	Erosion by (1000° F) catalysts in refineries
Martensitic type	Can be annealed and rehardened	Agricultural equipment in sandy soil
Austenitic type	Oxidation resistant	General abrasive conditions with light impact
Martensitic alloy irons	Excellent abrasion resistance	Machine parts subject to repetitive metal-to-metal
Chromium-tungsten type	High compressive strength	wear and impact
Chromium-molybdenum type	Good for light impact	General erosion conditions with light impact
Nickel-chromium type		
Austenitic alloy irons	More crack-resistant than martensitic irons	
Chromium-molybdenum type		
Nickel-chromium types		
Chromium-cobalt-tungsten alloys	Hot strength and creep resistance	Hot wear and abrasion above 1200° F
High carbon (2.5%) type	Brittle and abrasion-resistant	
Medium carbon (1.4%) type		Exhaust valves of gasoline engines. Valve trim of
Low carbon (1.0%) type	Tough and oxidation-resistant	steam turbines
	Good hot hardness and erosion resistance	Oil well slush pumps
Nickel base alloys		
Nickel-chromium-boron type		Exhaust valves of trucks, buses and aircraft
Nickel-chromium-molybdenum-tungsten type	Corrosion resistance	
Nickel-chromium-molybdenum type	Resistant to exhaust gas erosion	Bearing surfaces
Nickel-chromium type	Oxidation resistant	General abrasive conditions with medium impact
Copper base alloys	Anti-seizing; resistant to frictional wear	
Martensitic steels	Fair abrasion resistance	Hot working dies
High carbon (0.65–1.7%) type	Good resistance to medium impact	
Medium carbon (0.30–0.65%) type	Tough, economical	General low-cost hard facing
Low carbon (below 0.30%) type	Tough, crack resistant	Base for surfacing or a build-up to restore dimensions
Semi-austenitic steels	Crack resistant and low in cost	
Pearlitic steels	Suitable for build-up of worn areas	
Low alloy steel	A good base for hard facing	
Simple carbon steel	Tough; excellent for heavy impact	General metal-to-metal wear under heavy impact
Austenitic steels	Fair abrasion and erosion resistance	Railway trackwork
13% manganese—1% molybdenum type		
13% manganese—3% nickel type	Lower yield strength	
13% manganese-nickel-chromium type	High yield strength for austenitic types	Frictional wear at red heat; furnace parts
High carbon nickel-chromium stainless type	Oxidation and hot wear resistant	Corrosion resistant surfacing of large tanks
Low carbon nickel-chromium stainless type	Oxidation and corrosion resistance	

American Welding Society

sure to heat may jeopardize this hardness or produce cracks. Metals in such a group include the higher range medium-carbon steels, high-carbon steels, cast irons, and other alloy steels.

Metals in the first group can be hardfaced without any particular precautions since no harmful cracking on adjacent weld hardness will result. With metals in the second group special care must be taken to minimize the sudden shock of localized heat. This is done by reducing the hardness by annealing or through gradual and uniform preheating and post cooling. Preheating from 300° to 500°F will usually prevent weld hardening in medium and high-carbon steels. High-carbon alloy steels and wear-resistance alloy steels require preheating to the same temperature. After the surfacing operation is completed, reheating to a temperature of 800° to 1300°F and slow cooling should follow.

Hardfacing Materials

There are many different types of hardfacing materials. Most of them have a base of iron, nickel, copper, or cobalt. Auxiliary elements may be carbon, chromium, molybdenum, tungsten, silicon, manganese, nitrogen, vanadium, and titanium.

The alloying elements form hard carbides which contribute to the matrix properties of the hardfacing metals. Thus a high percentage of tungsten or chromium with a high carbon content will form high carbide crystals that are harder than quartz. Materials having a high chromium content provide excellent resistance to oxidation and scaling. Nickel, cobalt, and chromium are particularly effective for corrosion resistance.

The matrix of hardfacing metals are either martensitic, pearlitic, or austenitic. Martensite is the hardest and strongest. Pearlite is moderately tough and hard. Austenite is soft, strong, tough and has good impact applications.

Hardfacing metals are available as rods for oxyacetylene welding, electrodes for shielded metal-arc welding, and hard wire for automatic welding. Some hardfacing materials come in tubular rods which contain a mixture of powder metal, powder ferralloys and fluxing ingredients. The same material is available in powder form for hardfacing with the carbon arc.

Table I lists some of the more common hardfacing alloys, their features and applications according to metal types.

Although many hardfacing materials are designated by manufacturers' trade names some also carry AWS-ASTM codes. AWS-ASTM has classified surfacing materials into the following types: high speed steel (Fe5), austenitic manganese steel (FeMn), austenitic high chromium iron (FeCr), cobalt-base metals (CoCr-A and CoCr-C), copper-base alloy metals (CuZn, CuSi, and CuAl), and nickel chromium-boron metal (NiCr). The coding system identifies the important elements of the hardfacing material. The prefix R designates a welding rod and the prefix E an electrode. Suffixes are sometimes used to identify a classification within a basic group. Certain materials also carry a further subdivision in the form of digits after the suffix.

Hardfacing with Oxyacetylene

The oxyacetylene hardfacing technique is very useful in depositing overlays on small parts such as engine valves, plowshares, tools, and other similar items. With the oxyacetylene flame tiny areas can be surfaced and thin layers applied smoothly. Preheating and slow cooling are readily controlled minimizing cracking even with brittle wear-resistant surfacing materials. The principal limitation of the oxyacetylene technique is its low deposition speed.

Metals for hardfacing with oxyacetylene generally consist of low melting high carbon filler rods such as high chromium or a Cr-Co-W alloy. As a rule, a slightly reducing flame is recommended inasmuch as this will add carbon to the deposit.

The hardfacing operation is started by preheating the surface to produce a "sweating" condition. During the preheating cycle the tip of the hardfacing rod is held on the fringe of the flame. The rod is then moved into the center of the flame and melted. The actual deposition of the filler rod is carried out with a regular forehand welding technique using a slightly weaving motion.

Hardfacing with Shielded Metal-Arc

Hardfacing with the shielded metal arc is probably used much more extensively because of its high deposition rate. It also has wide application where large areas have to be surfaced or for heavy parts that normally would require excessive time to heat with the oxyacetylene flame. This method of hardfacing is especially suitable for depositing overlays on manganese steel and other steel alloys where heat buildup must be restricted.

Either AC or DC current produces satisfactory welds. The electrodes may be of the coated solid wire type or hollow tube containing alloy powder and flux.

Hardfacing electrodes are normally classified into five groups: resistance to severe impact, resistance to very severe abrasion, resistance to corrosion and abrasion at high temperatures, resistance to severe abrasion with moderate impact, and resistance to abrasion with moderate to heavy impact.

1. *Resistance to severe impact.* These electrodes are intended for repair and buildup of manganese steel, of which most impact resistance parts are made. The electrodes are either solid drawn, that is, they contain all alloying elements in the wire itself, or composite in which all of the alloying elements are in the flux coating.

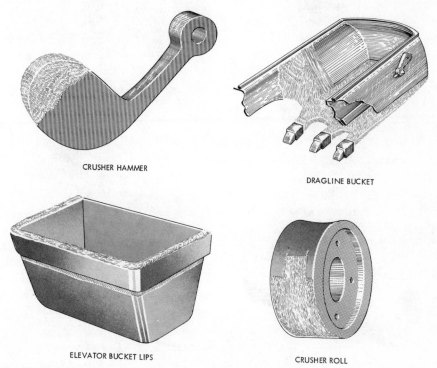

CRUSHER HAMMER

DRAGLINE BUCKET

ELEVATOR BUCKET LIPS

CRUSHER ROLL

Fig. 10-2. Typical parts that require hardfacing materials to produce severe impact resistance.

Deposits of these electrodes are not hard but very tough. They are often referred to as self-hardening because the deposited surface hardens as it is pounded. While the outside surface is hard the material underneath remains soft. Typical equipment requiring impact protection includes: crusher parts, bucket fronts, hammermill and rolling mill parts, jaws and teeth on dredges, ditchers and power shovels, and other similar heavy industrial and construction equipment. See Fig. 10–2.

Stainless steel electrodes are often used for hard surfacing parts that must resist impact forces without cracking. These electrodes offer the least resistance to abrasion in the "as deposited condition," however, they will work harden. Stainless steel and low hydrogen electrodes are frequently used as base layers for other hardfacing electrodes.

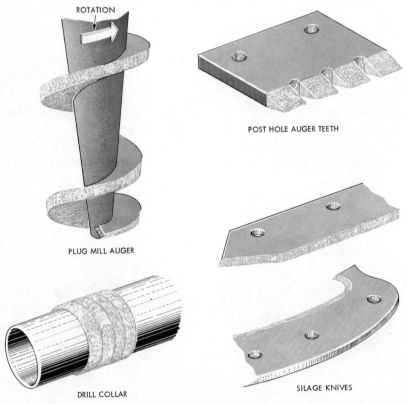

ROTATION

POST HOLE AUGER TEETH

PLUG MILL AUGER

DRILL COLLAR

SILAGE KNIVES

Fig. 10-3. Typical parts that require hardfacing materials to produce severe abrasion resistance.

2. *Resistance to very severe abrasion.* Electrodes in this group are of the tungsten carbide and chromium carbide types. These electrodes deposit a very hard abrasive-resistant material. They are not suitable for impact wear, since the material they deposit chips and cracks when subjected to shock.

Tungsten electrodes have tiny crystals of tungsten carbide embedded in the steel alloy. When applied on a surface, the steel wears away leaving toothlike particles of tungsten carbide exposed. Since tungsten carbide is very hard, the exposed particles make the edge of the part self-sharpening. The property is particularly desirable for earth digging equipment, scraping tools, plowshares, rotary digger blades, cultivator sweeps, and similar machinery. See Fig. 10–3.

Still another type of tungsten carbide electrode deposits fine particles of tungsten carbide that are so close that they form a smooth cutting edge. These electrodes are useful in repairing steel cutting edges such as lathe tool bits.

Chromium carbide electrodes are slightly less hard and less abrasion-resistant than the tungsten carbide type but are tougher. Most of them are not affected by heat treatment and are too hard to be machined. In addition to being hard chromium carbide electrodes produce surfaces that provide better protection against corrosion.

3. *Resistance to corrosion and abrasion at high temperature.* A great deal of wear on metal surfaces occurs at high temperatures, particularly where metal-to-metal contact is involved. In cases where friction is the cause of high operating temperature special cobalt base alloy electrodes are used for hardfacing. These electrodes are capable of withstanding temperatures of 1200°F and higher. Cobalt alloy electrodes are excellent for hardfacing valves and valve seat facings, chain-saw guide bars, pump shafts, rocker-arms, and hot trimming dies. See Fig. 10–4.

Nickel-base alloys also are effective for hardfacing surfaces that are subjected to corrosion and oxidation at high temperatures. For hardfacing corrosion-resistant parts which are not used at elevated temperatures, copper-base alloys are sometimes recommended. Copper-base alloy filler metals have wide application in surfacing areas that come in contact with corrosion generating materials such as acids, mild alkalies and salt water.

4. *Resistance to severe abrasion with moderate impact.* Industrial, transportation and construction equipment is frequently subject to a

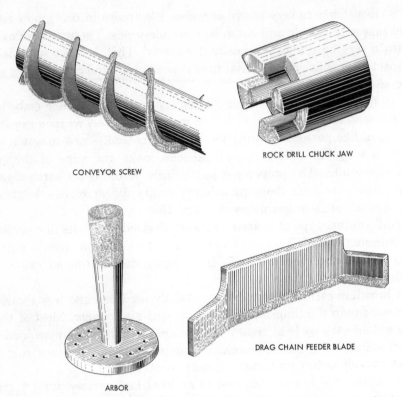

CONVEYOR SCREW

ROCK DRILL CHUCK JAW

ARBOR

DRAG CHAIN FEEDER BLADE

Fig. 10-4. Typical parts that require hardfacing materials to produce corrosion and abrasion resistance at high temperatures.

combination of wear factors—abrasion and impact. Alloys used for this combination of wear are carbon-chromium types ranking in hardness from 54 to 65 Rockwell C. In abrasion resistance they are second only to the tungsten carbide alloys. These carbon-chromium alloys lay down a smooth-surfaced deposit that is excellent for applications involving the moving, cutting, or handling of loose sliding materials such as earth, sand, gravel or cement. See Fig. 10–5.

5. *Resistance to abrasion with moderate to heavy impact.* A great deal of industrial equipment such as cable tools, mud pumps, bucket fronts, and churn drills requires hardfacing alloys that are resistant to abrasion as well as to moderate or heavy impact. These alloys have an iron base and can be heat treated and sharpened to a knife edge. See Fig. 10–6.

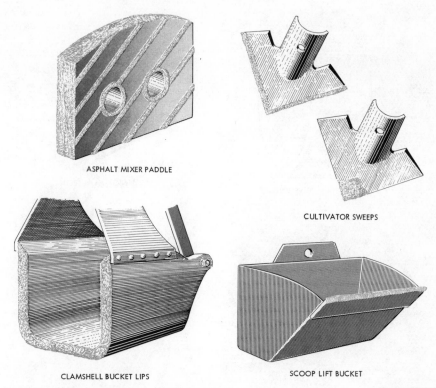

ASPHALT MIXER PADDLE

CULTIVATOR SWEEPS

CLAMSHELL BUCKET LIPS

SCOOP LIFT BUCKET

Fig. 10-5. Typical parts that require hardfacing materials to produce severe abrasion with moderate impact resistance.

Hardfacing Procedure. The surface to be hardfaced should be thoroughly cleaned of rust, scale, and all other foreign matter. The power supply should be set to provide only enough amperage to maintain the arc. This is important in order to prevent dilution of the deposit by the base metal. Best results are obtained if the work is arranged in a flat position. See Fig. 10–7.

A medium long arc is recommended and the torch moved with a straight or weaving motion. If several layers are to be deposited all slag must be removed before additional layers are made. Careful manipulation of the torch is necessary to secure adequate penetration into the adjoining beads. This can be done by holding the electrode a moment over the deposited bead to allow the heat to build up in the adjoining beads. Such a procedure will also minimize undercutting.

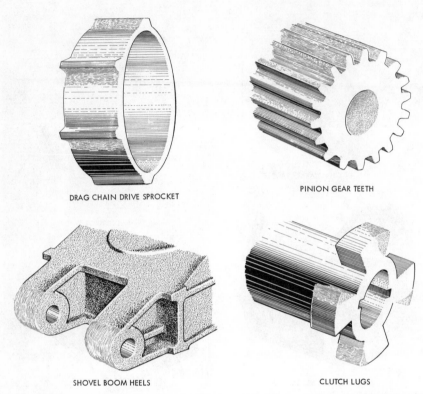

DRAG CHAIN DRIVE SPROCKET

PINION GEAR TEETH

SHOVEL BOOM HEELS

CLUTCH LUGS

Fig. 10-6. Typical parts that require hardfacing materials to produce resistance to abrasion with moderate to heavy impact.

A whipping action is often used when surfacing an area along a thin edge. The arc is held over the heavy portion and then whipped out to the thin edge. In this manner a shallow deposit is made before the heat builds up enough in the base metal to burn through.

Hardfacing with Gas-Shielded Arc

Both the gas tungsten-arc and gas metal-arc processes are ideal for hardfacing. In many cases, the gas shielded-arc processes are considered superior because of the ease with which an overlay can be made. The surfacing materials are readily deposited to form smooth, uniform, porosity-free surfaces.

Hardfacing with Tig is somewhat slower than with Mig but the overlays are of slightly higher quality. Tig is particularly effective in applying

Fig. 10-7. Hardfacing with the shielded metal-arc is best done in a flat position.

cobalt base alloys. The process ordinarily requires very little pre-heating. Since the heat buildup is minimal there is less distortion and very little of the base metal is affected by the heat.

The gas metal-arc process with its continuous wire is faster than Tig and produces excellent overlays. With both Tig and Mig the shielding gas provides an added feature when aluminum bronze surfacing materials are used. The shielding gas prevents oxidation and loss of alloying ingredients. A variety of special wires are available for practically every conceivable hardfacing operation.

Care must be taken in using Tig and Mig for surfacing to avoid dilution of the deposited weld metal. Helium and a mixture of helium-argon generally produces a higher arc voltage than pure argon and therefore increases the tendency for greater dilution, hence, argon or a mixture of argon and oxygen is recommended for hardfacing with the gas-shielded arc processes.

Hardfacing with Submerged Arc

The submerged arc process is considered the most economical for hardfacing parts where heavy deposits are required and extensive areas are to be surfaced. Since the submerged arc utilizes high welding current its deposition rate is high and its deposits are of high quality. Smooth overlays can be made with little or no welding experience required of the operator.

The filler metal may be either solid or tubular and is especially suitable for surfacing that requires high compression strength. However, the relatively deep penetration of the submerged arc plus its protective flux covering usually develops more intensive heat in the welded area. Consequently, greater precautions must be taken to provide suitable preheat and post heat treatment.

Very often the full strength of the hardfacing metal is attained only by depositing two or more layers. The initial layer frequently becomes diluted when fused into the base metal and therefore an additional layer is necessary to secure the required results.

Hardfacing with Plasma Arc

Plasma arc surfacing is a mechanized tungsten-arc process that uses

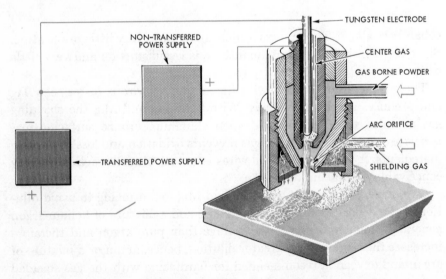

Fig. 10-8. Schematic view of hardfacing with plasma arc.

a metal powder as surfacing material. The metal powder is carried from a hopper to the electrode holder in an argon gas stream. See Fig. 10–8. From the torch the powder moves into the arc stream where it is melted and then fused to the base metal. The surfacing is an actual welding process and not a metal spray process. A wide variety of cobalt, nickel, and iron-base surfacing powders are available from manufacturers of welding supplies. These powders are fused materials and consequently are homogeneous in composition. They are classified as high-alloy materials having varying degrees of impact resistance qualities, abrasion resistance qualities, and corrosion resistance qualities. The application should be determined before selecting the powder to be used. See Table II.

The power source consists of a conventional DC power supply unit with straight polarity. A second DC unit is connected between the tungsten electrode and arc constricting orifice to support a nontransferred arc. The second power supply supplements the heat of the transferred arc and serves as a pilot arc to start the transferred arc. Argon gas is used to form the plasma as well as the shield.

Table II.   Hardfacing Powders.

POWDER ALLOY BASE	APPLICATION	ROCKWELL HARDNESS	FINISHING
NICKEL	BUILD UP FOR GLASS MOLDS	13-17	MACHINEABLE
NICKEL	CAN BE USED AS BRAZING ALLOY	28-32	MACHINEABLE
NICKEL	FOR USE ON CAST IRON OR STEEL PARTS. EXCELLENT CORROSION RESISTANCE AND HIGH RED HARDNESS. HIGH IMPACT SERVICE.	38-42	MACHINEABLE
NICKEL	HIGH CARBON AND CHROMIUM CONTENT FOR INCREASED RESISTANCE TO OXIDATION AND ABRASION	48-52	USE CARBIDE TOOLS
NICKEL	FOR SEVERE WEAR APPLICATIONS WITH HEAT AND/OR CORROSION RESISTANCE, ALSO FOR METAL WEAR	60-62	GRINDING
COBALT	FOR APPLICATIONS REQUIRING TOUGHNESS, METAL TO METAL WEAR, RESISTANCE AND EXCELLENT PERFORMANCE IN CORROSIVE CONDITIONS	46-51	USE CARBIDE TOOLS
NICKEL WITH TUNGSTEN CARBIDE	FOR EXTREMELY SEVERE ABRASION PROBLEMS. TOUGH NICKEL BASE MATRIX SUPPORTS TUNGSTEN CARBIDE PARTICLES.	60 MATRIX 86 CARBIDE	CANNOT BE MACHINED
PHOSPHORUS AND COPPER	COPPER TO COPPER BINDING		GRINDING AND MACHINING

Table III. Comparison of Surfacing Processes.

Process	Average Deposition Rate (lb/hr)	Minimum Weld Dilution (%)	Minimum Deposit Thickness (in.)	Surfacing Material Form	Type of Operation
Plasma Arc Weld Surfacing	7	5	0.010	Powder	Mechanized
Oxy-Acetylene	4	1	1/32	Rod	Mechanized & Manual
Tungsten-inert-Gas	5	10	3/32	Rod, Wire	Mechanized & Manual
Submerged-Arc Single Wire	15	20	1/8	Wire	Mechanized & semi-automatic
Submerged-Arc Series Circuit	30	15	3/16	Wire	Mechanized
Metal-Inert-Gas Single Wire	12	30	1/8	Wire	Mechanized & semi-automatic
Metal-Inert-Gas With Aux. Wire	25	20	3/16	Wire	Mechanized

Overlays made by the plasma arc possess the metallurgical characteristics of weld overlays produced by the gas tungsten-arc, gas metal-arc, or submerged-arc processes. Table III compares the deposition rates of various surfacing processes. Since plasma arc deposits are completely homogeneous and metallurgically bonded to the base metal they are very suitable for high stress applications and ideal for thin precision-controlled corrosive and severe wear parts.

METALLIZING

Metallizing, sometimes referred to as metal spraying, is a process of depositing fine semi-molten metal particles or metal powder onto the surface of a metal to form an adherent coating. The powder or metal particles are impelled through an intense heat and are impinged on the surface where they form thin layers of metallic lamellates. The powder or wire rod is fed into an oxygen-fuel gas flame and the small semi-molten droplets are driven onto the surface by a stream of high pressure air. The minute particles strike the surface at estimated speeds of 250 to 500 ft per second depending on the gun design. Cohesion is achieved by the mechanical interlocking and fusion of the tiny metallic particles

Fig. 10-9. Metallizing involves spraying tiny particles of metal onto the surface of a part to be rebuilt.

and the bonding of the thin oxide film which forms on the particles while in motion. See Fig. 10–9.

Metallizing has made wide application in the machine field where worn surfaces need to be restored to their original sizes but where low tensile strength and porosity are not objectionable. It is a very functional process for jobs where welding or brazing heat is impractical or for applying deposits of dissimilar metals which otherwise are not possible. The process is unique since there is no limit to the size of the object or structure which can be coated. The added feature of metallizing is that no preheating or postheating is required. There is little or no distortion resulting from the spray and consquently no rigid sequence of operations are necessary.

Metallizing Process

The success of any spraying process depends on having a clean surface that has been properly roughened. All traces of oil, dirt, scale, rust, etc., must be removed. The roughing of the surface provides mechanical anchorages for the sprayed metal particles. Grit blasting is probably the most common method used for surface roughing. Blasting abrasives may be steel grit, hard sand, aluminum oxide, or silicon carbide. Sometimes after a surface is prepared for spraying a thin layer of molybdenum is sprayed on. This produces a fusion bond which gives greater adherence to the subsequent spraying coats. Another method employed in roughing a surface is to run a chasing tool over the area. The chasing tool produces ragged type threads.

The sprayed metal coatings are somewhat porous but in machine elements this is an advantage. The porous coating absorbs oil which provides more complete lubrication. Porosity becomes more critical in applications that are subject to severe attacks of corrosive materials.

To some extent porosity can be controlled by the adjustment of gas and air as well as the distance of the gun from the workpiece. However, too much reduction of porosity will usually result in hard, brittle and highly oxidized coatings.

Oxidation normally occurs in the melting flame and during the flight of the metal particles to the surface. As a rule, little oxidation will take place as the metal is melted unless the gas-fuel mixture is oxidizing. The greatest causes of oxidation are overheating of coating, excessive use of oxygen, and spraying too great a distance from the workpiece.

Wire Spray Guns. Metallizing is done with a special spray gun which weighs from 3 to 6 pounds and handles 20 gage to $\frac{3}{16}''$ diameter wire. Usually these guns will spray 4 to 12 pounds of metal per hour. Larger type guns are often mounted on a fixture and are designed for spraying large machine components.

The gun consists of two major parts: the power unit and the gas head. See Fig. 10–10. The power unit feeds the wire into the nozzle of the gun. The gas head controls the flow of oxygen, fuel gas, and the compressed air. The nozzle has a center orifice through which the wire is fed. Around this orifice are a number of gas jets which provides the flame and high velocity air stream. As the wire comes through the orifice, it is melted and atomized by the flame. The fine molten particles are picked up by the air stream and projected against the work. The most common gas for the oxy-fuel flame is acetylene which produces a temperature exceeding 5600°F, although hydrogen or propane is sometimes

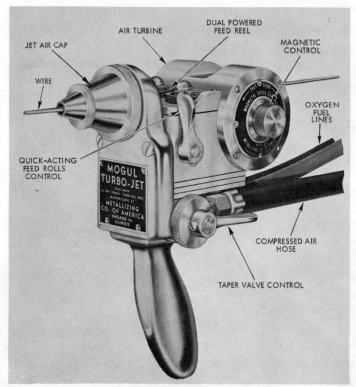

JET AIR CAP

AIR TURBINE

DUAL POWERED
FEED REEL

MAGNETIC
CONTROL

WIRE

OXYGEN
FUEL
LINES

QUICK-ACTING
FEED ROLLS
CONTROL

MOGUL
TURBO-JET

METALLIZING
CO. OF AMERICA
CHICAGO 14,
ILLINOIS

COMPRESSED AIR
HOSE

TAPER VALVE CONTROL

Metallizing Co. of America.

Fig. 10-10. Metallizing gun.

used for lower melting metals. A schematic sketch of a complete metallizing unit is shown in Fig. 10–11.

Powder-Type Guns. Guns used for spraying metal powder are somewhat similar to the wire type guns except that they do not have a power
unit. The powder is carried from a reservoir by compressed air to the
gun where it is forced through the central orifice of the gun. The gas
flame then melts the tiny particles which are carried to the work by
compressed air as in the wire-spray gun. See Fig. 10–12.

Spraying Operation. Any metal spraying operation should be carried
out in a well ventilated area. Adequate ventilation is necessary to remove dust particles and fumes which are extremely hazardous to health.
If positive ventilation is not possible the operator should wear an effective respirator.

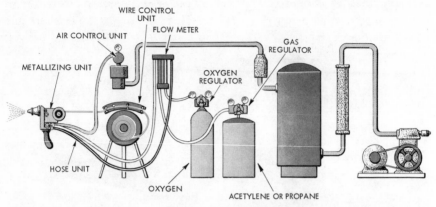

Fig. 10-11. A schematic of a typical metallizing unit.

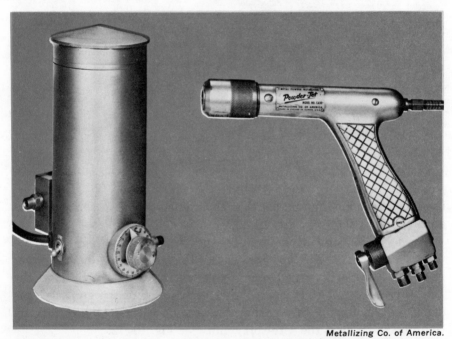

Metallizing Co. of America.

Fig. 10-12. Metallizing is also done with a special powder spray gun.

The wire speed, amount of spray, gas, and oxygen pressure must be regulated according to the recommendations established for the equipment to be used and the type of metallizing to be done. As a general rule air pressure is normally set for 60 psi. The use of a flowmeter will insure more accurate control of the gases. A slight increase in air pressure provides a finer coating and similarly, a decrease of air pressure produces a coarser coating.

The tip of the melting wire should project beyond the end of the air cap. This length depends to a large extent on the material being used. A recommended practice is to speed up the wire until chunks are ejected. Then the wire feed is reduced until the ejection of chunks discontinues.

Each coating should be kept as light as possible somewhere around 0.003" to 0.005" in thickness. Too heavy a coat will produce an irregular and stratified surface. The actual movement of the gun is very similar to paint spraying. The nozzle should be kept approximately 4" to 10" away from the surface and moved with a uniform motion. If the gun is held too close to the work minute cracks will form in the coating. Too great a distance will produce a soft spongy deposit with low physical properties. The rate of gun travel is also important. When the travel is too rapid the coating develops a high oxide content.

In spraying a flat surface the gun is moved back and forth to allow a full uniform deposit. Spraying should begin beyond the edge of the area to be covered and continued beyond the end of the area. After the first layer the work or gun is often rotated 90 degrees and this technique repeated for each subsequent coating until the required thickness is built

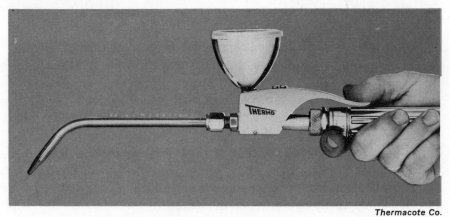

Thermacote Co.

Fig. 10-13. Spray metal torch.

up. On cylindrical pieces the work is generally fastened in a lathe with the gun mounted on the traveling carriage.

Torch Spraying

Metal spraying of small objects is simplified with the use of a spray metal torch as shown in Fig. 10–13. This consists of a hopper and a spray control mechanism which fits on any conventional oxyacetylene torch. Metal powder is simply placed in the hopper, the torch is lighted, and the gun moved over the area to receive an overlay. The powder is pulled into the oxygen-acetylene stream and becomes fluid when it hits the flame.

REVIEW QUESTIONS

1. What is the difference between hardfacing and metallizing?
2. Parts which are hardfaced are subject to what three basic types of wear?
3. Why is it impossible to use a single type of surfacing material for all types of wears?
4. In hardfacing metals that are very susceptible to cracking, what can be done to minimize the sudden shock of localized heat?
5. Hardfacing materials have as a base what kinds of alloying elements?
6. What do the following symbols represent: FeMn, FeCr, CoCr, NiCr, CuZn, CuSi, CuAl?
7. Why is the shielded metal-arc used more extensively than oxyacetylene in hardfacing?
8. Electrodes of what alloy base produce the hardest type of overlay?
9. Electrodes of what alloy base are best for corrosion resistant hardfacing?
10. When is the submerged-arc process used for hardfacing?
11. How does hardfacing with the plasma arc differ from the regular shielded metal-arc process?
12. In metallizing how do the tiny particles of metal become bonded to a surface?
13. How should a surface be prepared for metallizing?
14. How can porosity be controlled in metallizing?
15. What is the difference between metal spraying and powder spraying?

Chapter 11

Flame and Arc Cutting

Metal cutting involves the severing or removal of metal by a flame or arc process. The more common cutting processes are oxygen, arc, and plasma.

Cutting may be performed manually or with mechanized equipment. In manual cutting the operator manipulates a cutting torch over the cutting areas while in machine cutting the cutting torch is guided entirely by automatic controls. See Fig. 11–1.

Cutting processes have wide applications in many industries. They are used to cut plate or pipe to required shape or size, to remove risers from castings, and to scrap obsolete structures. Whether the oxygen, arc, or plasma cutting process is used depends often on the kind of cutting that is to be done, the type of metal to be severed, or the economy of the cutting process itself for the particular operation involved.

OXYGEN CUTTING

Severing metal by the oxygen process is possible because of the reaction of metal to oxidation. For example, when a piece of iron or steel is exposed to various atmospheric conditions a reaction known as rusting takes place. This rusting is simply the result of the oxygen in the air uniting with the metal causing it gradually to decompose and wear away. Naturally this action is very slow but if the metal is heated and permitted to cool, heavy scales form on the surfaces, showing that the iron oxidizes much faster when subjected to heat. If a piece of steel were to be heated red hot and dropped in a vessel containing oxygen a burning action would immediately take place.

Fig. 11-1. Mechanized cutting is used extensively in industry.

To make possible the rapid cutting of metal it is necessary to have a device that heats the metal to a certain temperature and then throw a blast of oxygen on the heated section. The cutting torch whether manually or machine operated functions in just such a manner.

Oxygen cutting can be used on plain carbon steels, low alloy steels, manganese steels, and low content chromium steels. Nonferrous metals and stainless steels or steels with a high chromium or tungsten content that form refractory oxides cannot be cut with the oxygen process.

The Cutting Process

The process of cutting is actually a progressive action especially in cutting thick sections. The cut is started by preheating a section of

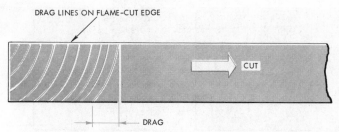

DRAG LINES ON FLAME-CUT EDGE

CUT

DRAG

Fig. 11-2. Efficient cutting is also related to drag.

metal to the ignition temperature. Oxygen is then turned on which ignites the iron or steel at the upper surface. This burning releases heat thereby raising the temperature of the metal below the surface and melting it. The cutting oxygen now ignites the molten metal which, in turn, releases heat to the metal below. The same action progresses downward until the metal is severed.

During cutting a slag is formed. The slag is principally iron oxide (Fe_3O_4), alloying elements in the iron, plus some of the pure iron itself. This slag is blown out of the cut by the force of the oxygen stream.

Effective cutting depends on the amount of oxygen supplied for combustion and correct tip size. Once the metal is ignited the type of fuel gas used for preheating has no influence on the cutting.

The volume of cutting oxygen has a direct relationship to the formation of what is known as *drag*. The drag of a cut is the distance between the point where the oxygen stream enters the top of the metal and the point where the slag emerges from the bottom of the cut. See Fig. 11–2. Conceivably the most efficient cut is where there is zero drag, that is, no lag between the top and bottom distance of the cutting reaction. However, to achieve zero drag a high volume of oxygen is necessary and this may not always be the most economical. On the other hand, if the oxygen volume is insufficient an excessive amount of drag may develop. Too long a drag will leave uncut corners and even cause the steel fire to extinguish before the kerf reaches the bottom of the workpiece. When the supply of oxygen is too high the top edges of the kerf usually burn over, producing rough surface edges, and considerable waste of oxygen. Fig. 11–3.

Most shops where a great deal of cutting is done attempt to ascertain what is judged to be the most economical drag for the type of cutting.

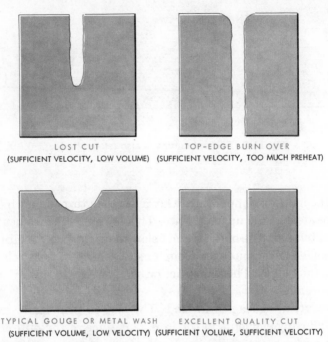

LOST CUT
(SUFFICIENT VELOCITY, LOW VOLUME)

TOP-EDGE BURN OVER
(SUFFICIENT VELOCITY, TOO MUCH PREHEAT)

TYPICAL GOUGE OR METAL WASH
(SUFFICIENT VOLUME, LOW VELOCITY)

EXCELLENT QUALITY CUT
(SUFFICIENT VOLUME, SUFFICIENT VELOCITY)

Fig. 11-3. Effective cutting depends on the volume of oxygen.

This so-called "standard drag" is generally the longest possible drag that will produce the required accuracy of cut with a minimum amount of uncut corners.

The use of a correct tip size is important in cutting because it affects the speed, accuracy, and economy of the cutting process. A tip that is too small will fail to generate sufficient kindling temperature to keep the cutting progressing forward. If the tip size is too large the waste of oxygen becomes extremely high.

Fuel Gases

The function of the fuel gas is to raise the temperature of the metal to start and continue the cutting process. Several types of gases are used for his purpose. The most common are acetylene, propane, Mapp, and acetogen. The type of fuel gas selected is governed by such factors as cost of gas, ease of handling gas containers, heat intensity of the flame,

Table I.   Flame Temperature of Fuel Gases for Cutting.

FUEL GASES	FLAME TEMPERATURE
ACETYLENE	5,420° F
PROPANE	5,190° F
HYDROGEN	4,600° F
MAPP	5,301° F
ACETOGEN	5,400° F
NATURAL GAS	5,000° F

type of cutting to be done, and kind of metal to be cut. Table I lists the flame temperatures of different fuels for cutting purposes. Special cutting tips must be used with each type of fuel gas.

Manual Cutting

Equipment. The cutting torch is equipped with a special lever for the control of the oxygen and a cutting tip which has an orifice in the center surrounded by several smaller ones. See Fig. 11–4. The center open-

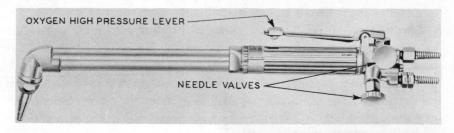

OXYGEN HIGH PRESSURE LEVER

NEEDLE VALVES

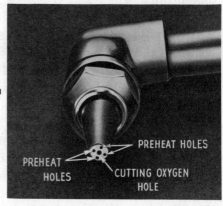

Fig. 11-4. Oxyacetylene cutting torch and tip.

PREHEAT HOLES

PREHEAT HOLES

CUTTING OXYGEN HOLE

Table II. Cutting Pressure.

Tip No.	Thickness of Metal (inches)	Acetylene Pressure (pounds)	Oxygen Pressure (pounds)
0	1/4	3	30
1	3/8	3	30
1	1/2	3	40
2	3/4	3	40
2	1	3	50
3	1 1/2	3	45
4	2	3	50
5	3	4	45
5	4	4	60
6	5	5	50
6	6	5	55
7	8	6	60
7	10	6	70

ing permits the flow of the cutting oxygen and the smaller holes are for the heating flame. Tip size will depend on the thickness of the metal to be cut. Table II includes the various tip sizes for different thicknesses of metal as it applies to a particular cutting torch.

Cutting Procedure. The torch is lighted by turning on the gas needle valve and igniting the gas in the same way as for welding with an oxyacetylene torch. The oxygen needle valve is then turned on and adjustment made for an oxidizing flame. Very often a neutral preheat flame is used but in most instances an oxidizing flame is more efficient. An oxidizing flame set at a 1.5 oxygen to 1 acetylene is hotter and more concentrated. The result usually is greater economy in oxygen use and reduced labor time.

The torch is placed over the workpiece with the tip vertical to the surface of the metal and the inner cone of the heating flame about $\frac{1}{16}''$ above the surface. When the metal is heated to a bright red heat the oxygen pressure lever is depressed and the torch moved slowly forward. If the cut is proceeding correctly a shower of sparks should be seen to fall from the underside. See Fig. 11–5. The torch should be advanced steadily at a proper speed without any waver motion. Too great a speed will cause the slag to emerge from the bottom and trail or lag behind at too great an angle. This may result in the stream failing to penetrate the metal completely. If a cut should become lost the cutting procedure must

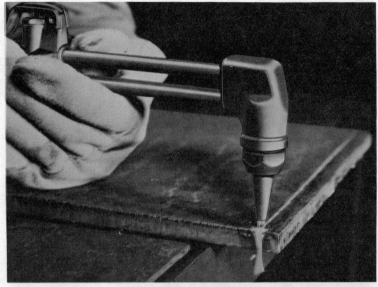

Fig. 11-5. Cutting with the oxyacetylene torch.

be restarted by preheating as it was originally begun. When the travel movement is too slow the edges of the cut appear to melt and have a very ragged appearance.

Fig. 11-6. Typical oxygen cutting machine.

Machine Cutting

Machine cutting is used for both straight line cutting and making irregular cuts. The machine is designed to maintain a preset cutting speed with the proper amount of oxygen flow. See Fig. 11–6.

Where irregular cuts are required the machine is equipped with a sensing and actuating device that guides the cutting along the path. The tracing device is combined with a driving wheel which steers a torch bar around a template or drawing. The tracing device may be manually

Fig. 11-7. An automatic flame cutting machine.

controlled, however, in most industrial practices the tracer is automatically guided. With the automatic guided operation the tracer is magnetized so it adheres to and follows the contour of the steel template. See Fig. 11–7. Another automatic technique is the use of an electronic tracer which utilizes a photocell. The photocell follows the outline of the drawing and steers the driving wheel. A more current innovation in automatic control is to punch the program of profile cutting on tape. This tape then controls the shape cutting by sending impulses to the cutting machine drive motors.

Auxiliary Oxygen Cutting Operations

Oxygen cutting is frequently used for piercing holes, stack cutting, scarfing, lancing, washing, gouging, and powder cutting.

Piercing Holes. Piercing involves the cutting of specific size holes in the base metal. For small holes the cutting torch is held over the spot where the hole is to be cut until the flame has heated a small, round

Fig. 11-8. Cutting holes.

spot. The oxygen lever then is pulled and at the same time the nozzle is raised slightly. In this manner a small, round hole can be pierced quickly through the metal.

When larger holes are required the size of the opening is traced with a piece of chalk. If the hole is located away from the edge of the plate a small hole is first pierced and then the cut is started from this point, gradually working to the chalk line and continuing around the outline. For holes near the outer edge of the plate the cut is made as shown in Fig. 11–8.

Stack Cutting. Stack cutting involves cutting several plates simultaneously while stacked together. See Fig. 11–9. The advantage of stack cutting is increased productivity at lower unit cost. The resulting pieces will usually have less stressed edges with squarer surfaces and fewer burrs and drag.

Stack cutting requires that the pieces form a solid slab. The surfaces must obviously be free of all foreign matter and be absolutely flat. Sufficient clamping devices are often required to eliminate all air gaps between the stacked plates.

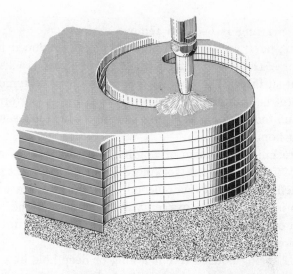

Fig. 11-9. Typical example of stack cutting.

Fig. 11-10. Scarfing is a process used to remove surface defects.

Scarfing. Scarfing is used to remove surface defects from blooms, billets, ingots, bars, and other unfinished pieces produced in the steel-making process. Defects may be in the form of scabs, cracks, breaks, or decarburized surfaces. The operation actually involves removing a thin layer of metal containing these defects. Scarfing is performed with a hand manipulated torch or a scarfing machine which has several oxygen acetylene burners.

Scarfing technique consists of holding the torch at an angle of approximately 75 degrees to the workpiece with the tip almost touching the surface. When the preheat has reached the right temperature the angle of the torch is reduced, the oxygen valve opened and the torch moved forward at a constant speed. Successive passes are made until the entire surface is covered. See Fig. 11–10.

Lancing. Lancing is used to cut a long or deep hole in a thick metal body. The lance is simply a piece of steel pipe $\frac{1}{8}''$ or $\frac{3}{4}''$ diameter equipped with a globe valve to control the flow of oxygen. See Fig. 11–11. Preheating is done with a welding or cutting torch or any other means for supplying heat. As soon as the preheating is completed the lance pipe is brought over the area where the hole is to be pierced and the oxygen turned on. The lance pipe is rotated slightly to produce a hole larger than the diameter of the lance, permitting the slag to be blown out of the hole.

Washing. Washing is an oxygen cutting process used to remove superfluous metals from castings such as risers, pads, gates, or fins. The operation is almost identical to scarfing.

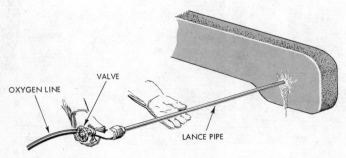

OXYGEN LINE

VALVE

LANCE PIPE

Fig. 11-11. Lancing is used to cut deep holes in thick metal sections.

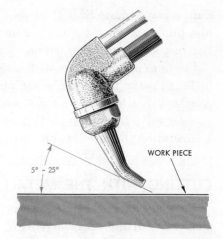

Fig. 11-12. Gouging is used to produce an even groove.

Gouging. Gouging is another form of scarfing used to produce a fully controlled groove. The torch is held stationary until the correct preheat is reached and the nozzle then lowered and moved forward as shown in Fig. 11–12.

Powder Cutting. The regular oxygen cutting process is ineffective for cutting metals that form refractory oxides such as aluminum, bronzes, and high-nickel alloys. To cut these metals an iron powder is fed into the oxygen stream. A mixture of iron powder and aluminum powder is sometimes used for cutting brass, copper, and high-nickel alloys. The alumi-

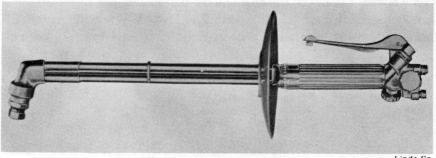

Linde Co.

Fig. 11-13. Powder-cutting torch.

num releases more heat than just the iron powder alone. Iron powder also produces a rapid cutting action on stainless steels and is very effective in getting smoother cuts in cast iron.

Powder cutting is done with a special powder-cutting torch as shown in Fig. 11–13. The torch is equipped with a powder tube, nozzle, and powder valve. The powder is stored in a dispenser and is carried to the powder valve by compressed air or nitrogen where it is fed to the flame. In operation the powder valve is opened first and then the oxygen valve.

CUTTING WITH THE METALLIC ARC

Coated mild steel electrodes, such as E-6010 or E-6011, are sometimes used for cutting. Special cutting electrodes are also available. The diameter of the electrode will depend on the thickness of the metal to be cut and the amperage capacity of the machine. For most general purpose cutting $\frac{3}{32}''$ electrodes are satisfactory for cutting metals up to $\frac{1}{8}''$ in thickness and $\frac{5}{32}''$ electrodes for materials which exceed $\frac{1}{4}''$ in thickness.

Cutting with the arc is possible because of the high temperature produced by the arc. Heat of the arc ranges from about 6500° to 10,000°F whereas steel, for example, melts around 2600°F. The limitation of the arc process for cutting is that it leaves a very rough, ragged edge. However, it is effective for cutting cast iron and steel for salvage purposes and areas which are hard to reach. Table III gives the approximate ampere setting for cutting.

In a cutting operation the metal is placed in a flat position and the cut started at the bottom edge of the plate. When the diameter of the electrode is larger than the thickness of the plate being cut the electrode is simply moved in a straight line as in Fig. 11–14A. When the material is heavier than the electrode a weaving motion is used to make the cut.

Table III. Suggested Amperage Setting for Cutting.

Metal Thickness (inch)	Electrode Diameter (inch)	Ampere Range
up to 1/8	3/32	75-100
up to 1/8	1/8	125-140
over 1/4	5/32	140-180

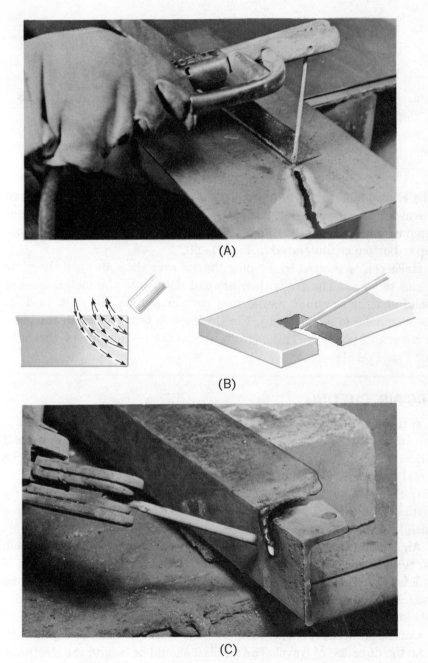

(A)

(B)

(C)

Fig. 11-14. Cutting with the metallic arc.

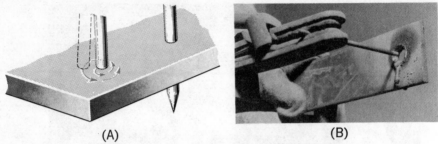

(A)   (B)

Fig. 11-15. Piercing holes with the metallic arc.

The electrode is moved upward with a quick motion and then pushed downward as shown in Fig. 11–14B. Flat stock over ⅛" in thickness is often easier to cut if placed in a vertical position. The cut is made from top to bottom as illustrated in Fig. 11–14C.

Holes can be pierced by keeping the arc over the spot until the plate begins to sweat. The arc is then brought down into the molten pool of metal and the electrode moved in a circular motion. See Fig. 11–15A. Piercing holes in metal over ¼" in thickness is best done with the plate in a vertical position. This permits the metal to run out of the hole. See Fig. 11–15B.

ARC-AIR CUTTING

A regular AC or DC welding machine and compressed air are used in the arc-air cutting process. A carbon-graphite electrode held in a special holder (Fig. 11–16) provides the electric arc to melt the metal. As the metal melts a jet of compressed air is directed at the point of arcing to blow the molten metal away. The compressed air line is fastened directly to the torch. The jet air stream is controlled by simply depressing a push-button valve on the holder.

Air is supplied by an ordinary compressor. In general, pressure will range from 80 to 100 psi. For light work a pressure as low as 60 psi is sufficient. Either plain or copperclad carbon-graphite electrodes can be used. Plain electrodes are less expensive, however, copperclad electrodes last longer, carry higher current, and produce more uniform cuts.

Cutting Procedure. The torch is held so the electrode slopes back from the direction of travel. The air blast should be behind the electrode. See Fig. 11–16. The arc should be kept as short as possible and moved fast enough to keep up with metal removal. The depth of the cut is controlled by the electrode angle and travel speed. For a narrow deep cut a steep electrode angle and slow speed is used. A flat electrode angle and fast speed is required for shallow cuts.

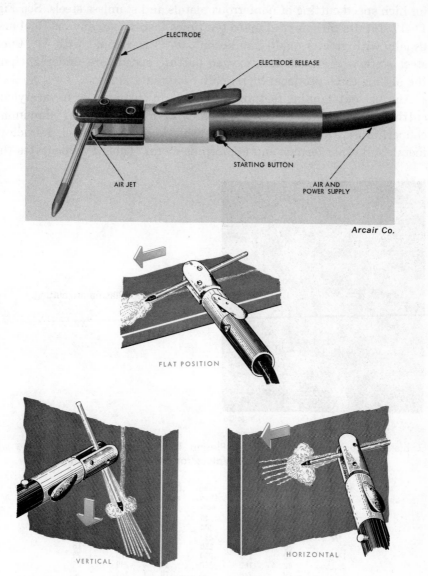

Arcair Co.

Fig. 11-16. Arc-air cutting.

PLASMA ARC CUTTING

Plasma arc cutting is regarded as one of the most effective processes for high speed cutting of nonferrous metals and stainless steels. See Fig. 11–17. It cuts carbon steel up to ten times faster than any oxy-fuel and usually with greater quality and economy. For example, with ½″ thick steel plate where maximum oxygen cutting speeds are under 25 ipm, the plasma torch produces high quality cuts at 200 ipm.

Another outstanding feature of plasma arc cutting over oxyacetylene cutting is its ability to sever all kinds of metals such as aluminum, copper, brass, cast iron, carbon steel, nickel, stainless steel, refractory metal, and other ferrous and nonferrous metal. Table IV illustrates the

Fig. 11-17. Plasma arc cutting.

Linde Co.

Table IV. Correlation of Maximum Plasma Torch Cutting Speeds Attained with Maximum Recommended Values for Oxyacetylene Cutting.

Plate thick-ness, in.	Arc Torch			Oxyacetylene Torch		
	Cutting speed, ipm	Kw	Gas flow, scfh	Cutting speed, ipm	Oxygen flow, scfh	Acetylene flow, scfh
¼	525	115	240	26	70	14
½	200	130	300	22	100	18
¾	100	130	230	20	120	18
1	80	130	230	18	130	18
1½	50	150	230	16	170	25
2	40	150	250	13	230	25
3	20	210	250	10	270	32

cutting speed of plasma cutting as compared with that of oxyacetylene cutting.

Plasma Cutting Principle[1]

In plasma cutting the tip of the electrode is located within the nozzle of the torch. The nozzle has a relatively small opening (orifice) which constricts the arc. The high-pressure gas must flow through the arc where it is heated to the plasma temperature range. Since the gas cannot

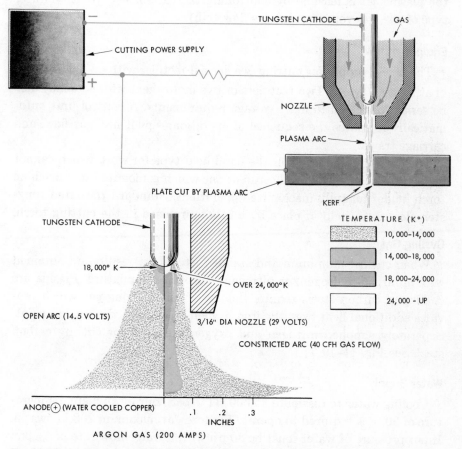

Fig. 11-18. In plasma-arc cutting, the gas flowing through the torch produces a supersonic jet which is hotter than a flame.

[1]Courtesy Linde Co.

expand due to the constriction of the nozzle it is forced through the opening and emerges in the form of a supersonic jet, hotter than any flame. This heat melts any known metal and its velocity blasts the molten metal through the kerf. See Fig. 11-18.

Because transfer of heat to work is essential in cutting Plasmarc torches use a transferred arc (the workpiece itself becomes an electrode in the electrical circuit). The work is thus subjected to both plasma heat and arc heat. Direct current straight polarity is used. Precise control of the plasma jet is feasible by controlling the variable—current, voltage, type of gas, gas velocity, and gas flow (cfh).

Equipment

The power supply for cutting is a special rectifier with an open circuit of at least 100 volts. Two rectifiers or two motor-generator sets may also be series connected to meet voltage requirements. A control unit automatically regulates the sequence of operations—pilot arc, gas flow, and carriage travel.

The cutting torch is either the hand held type for work which cannot be adapted to a mechanized setup or one which is mounted in a machine torch holder for fully mechanized operations. Standard thoriated tungsten electrodes, held in place by a collet, are used in the cutting torch.

Cutting Gases

When cutting aluminum and stainless steel best results are obtained with an argon-hydrogen, or nitrogen-hydrogen gas mixture. Plasma arc cutting of carbon steels require the use of an oxidizing gas which provides additional heat from the iron-oxygen reaction at the cutting point. Separately supplied nitrogen and oxygen are used for cutting carbon steel. See Fig. 11-19.

Water Supply

Cooling water to dissipate 35,000 BTU per hour at ambient temperature of 90°F is required to operate the torch at maximum 200 kilowatts. Input pressure of water must be 80 psi with a water flow rate of 3 gpm.

Cutting Procedure

The first steps in making a Plasmarc cut are to adjust the power supply and the gas flow to the appropriate settings. See Table V. When the operator pushes the start button on the remote control panel, the control unit performs all on-off and sequencing functions. The cooling water must also be turned on or the waterflow interlock will block the starting circuit.

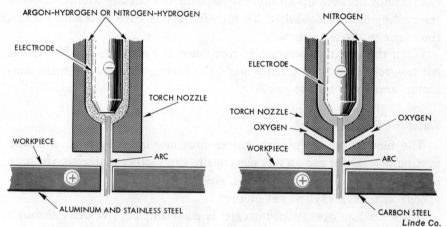

Fig. 11-19. Schematics of plasma cutting nozzles for different inert gases.

Linde Co.

Table V. Typical Plasmarc Cutting Conditions.

	THICKNESS (IN.)	SPEED (IPM)	ORIFICE INSERT TYPE	ORIFICE INSERT DIA. IN	POWER (KW)	GAS FLOW (CFH)
STAINLESS STEEL	1/2	25	4 x 8	1/8	45	130 N_2
	1/2	70	4 x 8	1/8	60	130 N_2
	1 1/2	25	5 x 10	5/32	85	130 N_2 10 H_2
	2 1/2	18	8 x 16	1/4	150	175 N_2 15 H_2
	4	8	8 x 16	1/4	160	175 N_2 15 H_2
ALUMINUM	1/2	25	4 x 8	1/8	50	100 ①
	1/2	200	4 x 8	1/8	55	100 ①
	1 1/2	30	5 x 10	3/32	75	100 ①
	2 1/2	20	5 x 10	5/32	80	150 ①
	4	12	6 x 12	3/16	90	200 ①
CARBON STEEL	1/4	200	4 x 12M ②	1/8	55	250
	1	50	5 x 14M ②	5/32	70	300
	1 1/2	35	6 x 16M ②	3/16	100	350
	2	25	6 x 16M ②	3/16	100	350

① 65% Argon, 35% Hydrogen Mixture ② Multiport Orifice

To make a mechanized cut the operator locates the center of the torch about ¼" above the surface of the plate to be cut and pushes the start button. Current flows from the high frequency generator to establish the pilot arc between the electrode (workpiece) and the cathode in the nozzle. Gas starts to flow and welding current flows from the power supply.

The pilot arc sets up an ionized path for the cutting arc. As soon as the cutting arc is established the high-frequency current is shut off and the carriage starts to move.

When the cutting operation is completed the arc goes out because it has no ground and the control stops the carriage, opens the main contactor, and shuts off the gas flow.

Water-Arc Plasma Cutter[2]

The new water-arc plasma cutter provides industry with a shape cutting process for use on any electrically conducting materials (ferrous or non-ferrous). With this process, steels can be cut at plasma cutting speeds with oxyacetylene cut quality.

In the cutting operation, nitrogen is passed at high velocity through a high-frequency arc. This creates a plasma which is maintained at a high temperature in a DC arc between the electrode and the work. A spray of filtered tap water is impinged upon this high-velocity, high-temperature plasma to constrict and accelerate it. This water-arc plasma system produces a highly concentrated and very stable plasma column, resulting in a very square, narrow kerf in the material being cut. The

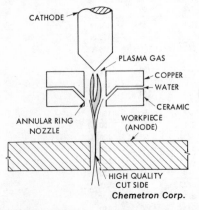

Fig. 11-20. Diagram of water-arc plasma cutting operation.

---

[2]Courtesy Chemetron Corp.

rounded top edge of the kerf characteristic of other plasma cutting systems is further eliminated by the cooling water spray on the work piece.

The use of the water jet constricts the arc as it exits from the torch. The resultant extremely intense, high concentration of the arc energy produces a clean, narrow cut. The water spray is directed to the arc between a ceramic outer nozzle and the copper plasma nozzle. This water spray maximizes the heat transfer from the nozzle to prolong nozzle life. The ceramic outer nozzle controls the water spray and prevents double arcing which is a major cause of nozzle failure on other torches.

REVIEW QUESTIONS

1. In oxygen cutting what is the reaction that makes possible the actual cutting process?
2. How does the oxygen supply affect the cutting action of steels?
3. What is meant by drag?
4. What is the effect of drag in a cutting operation?
5. What determines the most appropriate fuel gas to use for cutting?
6. How is it possible to determine if a cut is proceeding properly?
7. If too great a cutting speed is used what is likely to happen?
8. In machine cutting what devices are used to guide the torch for cutting intricate patterns?
9. What is meant by stack cutting?
10. What is the principal function of a scarfing operation?
11. When is a lancing operation used?
12. What is the difference between a washing and gouging operation?
13. What is meant by powder cutting?
14. What is the limitation of the metallic arc for cutting purposes?
15. How does the arc-air cutting process differ from the regular metallic arc process?
16. Why is it possible to cut types of metals with the plasma arc which normally cannot be cut effectively with the oxygen process?
17. What gases are used in plasma arc cutting?
18. What kind of power supply units are used for plasma arc cutting?

Chapter 12

Strength of Materials

Strength of materials is the branch of mechanics which involves the action of forces and their effects on structural members. In the design and manufacture of any product, size and strength of the materials must be considered. Correct material size and strength are significant because they are directly related to the serviceability of the product. Components and structural members usually are subjected to various kinds of loads. If the materials are to withstand these loads they must be of a specific size and strength. Similarly, weldments also are subjected to varying loads; hence weld strengths are equally important in any fabrication process. (See Chapter XIII.) Only by identifying these loads and determining their magnitude can these variables be resolved in any product planning.

Most structural members and weldments are subjected to one or more forces. These forces are classified as either simple or complex. Simple forces are referred to as tensile, compression, and shear. See Chapter VII. More complex forces generally are categorized as bending and torsion.

Simple Stresses

The internal stress produced on a straight bar of a constant cross section by a central load may be assumed to be uniformly distributed over the cross-sectional area. Since unit stress is the internal stress per unit of area it is constant due to the central load.

If the intensity of a tensile and compressive stress is equally distributed over an area then the following relationship exists:

$$P = \sigma A \qquad \text{\textit{Equation} 1}$$

Where

P = total external load (lbs)

σ = average stress (psi) (Greek *sigma*)
A = stressed area (sq. in.)

A corresponding equation involving shear stress is:

$$P = \tau A \qquad\qquad Equation\ 2$$

Where

P = total external load (lbs)
τ = average unit shear stress (psi) (Greek *tau*)
A = area being sheared (sq. in.)

A study of equations 1 and 2 indicates that given a load and area the stress may be determined; given a stress and area the load may be obtained; and given a stress and load the required area may be found. For example, if a 2 in. diameter shaft is subjected to a tensile load of 10,000 lbs, the average stress σ is 10,000 ÷ $\pi \times 1^2$ = 3,190 psi.

Allowable Stresses

A unit stress which safely may be used in design is designated as an *allowable stress*. This stress represents the maximum load which should be applied to a material according to the judgment of some competent authority. Table I includes the allowable stresses for some of the more common materials. They represent average values where the usage of the member being analyzed is not unduly severe.

In the design of any welded component it is important that the entire structure be analyzed for the optimum usage of materials and methods of fabrication. Likewise, it is well to recall the basic engineering principle that a machine or structure is no stronger than its weakest member. Since design is based on the premise that materials will not fail the allowable stress provides a margin of safety.

Table I.   Allowable Stresses.

MATERIAL	TENSION PSI	COMPRESSION PSI	SHEAR PSI
STRUCTURAL STEEL	20, 000	20, 000	12, 000
CAST IRON	3, 000	15, 000	3, 000
ALUMINUM ALLOY	15, 000	15, 000	10, 000
BRASS	12, 000	12, 000	8, 000

Factor of Safety

As shown in Chapter VII a stress greater than the elastic limit will produce a permanent deformation of the material. Repeated stresses of this magnitude usually cause failure.

In order to avoid exceeding certain values that might cause failure a safety margin is used. Safety margin is referred to as *factor of safety* (FS) and represents the ratio of ultimate stress to allowable stress. Items involved in determining the proper factor of safety are:

1. Uniformity of material—the greater the potentials of inclusions, blowholes, corrosion, etc. in a material, the greater should be the FS.

2. Danger to human life—the greater the possibility of personal injury, the greater should be the FS.

3. Type of load—the greater the unpredictability of the load, the greater should be the FS.

4. Permanency of design—the longer the life of a product or component, the greater should be the FS.

Table II lists Factors of Safety for several common materials where their usage is not too severe.

Table II. Factor of Safety.

MATERIAL	STEADY STRESS	REPEATED STRESS
STRUCTURAL STEEL	4	10
HARD STEEL	6	12
CAST IRON	6	18
TIMBER	10	15

Strains

The changes in size that occur due to stresses within a body are called *deformations* or *strains*. The total change of length is called deformation and is denoted by δ. The unit change of length is strain and is indicated by ϵ. Therefore, the equation relating deformation to strain is:

$$\epsilon = \frac{\delta}{L} \qquad\qquad Equation\ 3$$

Where

ϵ = strain (in. per in.)

δ = deformation (in.)

L = total length (in.)

For example, if a 10 in. long bar has a total change in length (δ) of .075 in. when a load is applied, then the strain (ε) is .075 ÷ 10 = .0075 in. per in.

The relationship between the stress applied to a member and the resulting strain is plotted to give a stress-strain diagram as shown in Chapter VII.

The slope of the stress-strain diagram up to its proportional limit was indicated to be the *modulus of elasticity*. The formula showing this relationship is:

$$E = \frac{\sigma}{\epsilon}$$

Equation 4

Where

E = modulus of elasticity (psi)

σ = stress (psi)

ε = strain (in. per in.) (Greek *epsilon*)

Table III shows the modulus of elasticity of some of the more commonly used materials.

Table III. Modulus of Elasticity.

MATERIAL	MODULUS OF ELASTICITY (PSI)
STEEL	30,000,000
CAST IRON	15,000,000
BRASS, BRONZE	14,000,000
ALUMINUM	10,000,000
MAGNESIUM	6,500,000

From equations **1, 3,** and **4,** the following relationship is established:

$$\delta = \frac{PL}{AE}$$

Equation 5

Where

δ = deformation (in.) (Greek *delta*)

P = load (lb)

L = length (in.)

A = stressed area (sq. in.)

E = modulus of elasticity (psi)

Example

A cylindrical steel bar having a length of 10 in. is subjected to a tensile force of 8,000 lbs. Determine the required diameter if the stress is not to exceed 18,000 psi or an elongation of .005 in.

Solution

Since this problem involves two conditions, both must be solved and the final diameter must satisfy both conditions. (Note from Table III that $E = 30,000,000$ psi for steel)

From Equation 1

$$A = \frac{P}{\sigma} = \frac{8,000}{18,000} = .445 \text{ sq. in.}$$

From Equation 5

$$A = \frac{PL}{\delta E} = \frac{8,000 \times 10}{.005 \times 30,000,000} = .533 \text{ sq. in.}$$

Using the larger of the two areas, the result is:

$$A = \frac{\pi D^2}{4}$$

$$D = \sqrt{\frac{4A}{\pi}} = \sqrt{\frac{4 \times .533}{\pi}} = .825 \text{ in.}$$

Therefore, by using a shaft with a diameter of .825 in., the elongation will be limited to .005 in. and the stress will be less than the limit of 18,000 psi.

Thermal Expansion

If a member is subjected to changes in temperature it will expand when the temperature increases and contract when it decreases. The

Table IV.   Thermal Expansion Coefficients.

MATERIAL	COEFFICIENT – IN/IN.(PER ° F)
STEEL	.0000067
CAST IRON	.0000056
BRASS, BRONZE	.0000102
ALUMINUM	.0000128
MAGNESIUM	.0000145

change of length per unit of length for each degree of temperature change have been measured through experiments. This change is called *thermal expansion coefficient.* Table IV lists the coefficients of some of the more commonly used materials.

The equation showing the total change of length in a member is:
$$\delta = \alpha \,(\Delta T)\,(L) \qquad\qquad Equation\ 6$$

Where
$\delta =$ total deformation (in.)
$\alpha =$ thermal expansion coefficient (in./in. Per °F) (Greek *alpha*)
$\Delta T =$ temperature change (°F) (Greek capital *delta*)
$L =$ length of member (in.)

Example

New steel rails 30 feet in length were positioned during a temperature of 50°F. In the summer the temperature rose to 100°F. Determine the total elongation of the rails. (Note from Table IV that $\alpha =$ .0000067 in./in. for steel)

Solution

Using equation 6
$$\delta = \alpha \,(\Delta T)\,(L) = (.0000067)\,(100\text{-}50)\,(30 \times 12) = .121 \text{ in.}$$

Beams

A beam is a structural member which is subjected to loads acting transversely to its longitudinal axis. Various kinds of beams are in use. Loads on beams are (1) concentrated loads, (2) distributed loads, or (3) a combination of both. A concentrated load is supported on an area so small that it is assumed to be at a point. A distributed load, as the term indicates, is distributed over larger areas.

Beams may be classified according to the type of support. Fig. 12–1A shows a simple beam with a concentrated load. Fig. 12–1B shows an overhanging beam with a uniformly distributed load. Fig. 12–1C shows a double overhanging beam with a combination of two concentrated loads and a uniformly distributed load. Fig. 12–1D shows a cantilever beam with a distributed load.

The calculation of the reactions, R_1 and R_2, shown in Fig. 12–1 is

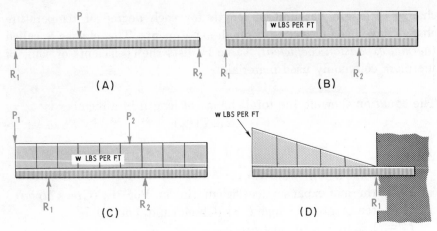

Fig. 12-1. Types of beams.

a problem in statics involving the equilibrium of forces. Equilibrium is assured in these cases since the member is not in motion. The equations involved in equilibrium are:

$$\Sigma P_x = 0, \ \Sigma P_y = 0, \ \Sigma M_o = 0 \qquad \textit{Equation 7}$$

Σ = sum of
P_x = forces in X direction (lbs)
P_y = forces in Y direction (lbs)
M_o = moments (lbs-ft about point 0)

A moment is defined as the product of a force and the perpendicular distance from a point to the line of action of that force. Therefore M_o means the moment with respect to point 0.

Example No. 1

Determine the reactions at the supports of a simple beam 15 feet long which carries a uniform load (including its own weight) of 100 lbs per linear foot and a concentrated load of 1000 lbs at a point 5 feet from the left support. (See Fig. 12–2)

Solution

1. The beam being 15 ft. long, its total weight is 100 $\times$ 15 = 1500 lb. Since the center of gravity of a uniform load is at the middle of the load, the distance to it is 7.5 feet from either support.

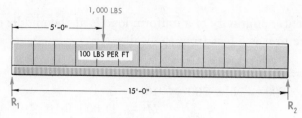

Fig. 12-2. Simple beam.

2. $\Sigma M_1 = 0$

   $15_{R2} - 1500 \times 7.5 - 1,000 \times 5 = 0$

                   $R_2 = 1083$ lb.

3. $\Sigma M_{R2} = 0$

   $15_{R1} - 1,000 \times 10 - 1500 \times 7.5 = 0$

                   $R_1 = 1417$ lb.

4. Check $\Sigma y = 0$ (To see if errors were made in steps 2 and 3).

   $1,000 + 1500 - 1083 - 1417 = 0$

                      $0 = 0$

Example No. 2.

Determine the reaction and moment at the wall of a cantilever beam 10 feet long which carries a uniform load (including its own weight) of 60 lb per linear foot and a concentrated load of 800 lb at the free end of the beam. (See Fig. 12–3)

Solution

1. With the beam 10 ft. long, its total weight is $60 \times 10 = 600$ lb.

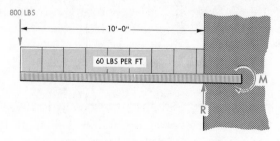

Fig. 12-3. Cantilever beam.

Since the center of gravity of a uniform load is at the middle of the load, the distance to it is 5 feet from the wall.

2. $\Sigma M_R = 0$
   $M \quad -600 \times 5 -800 \times 10 = 0$
   $$M = 11,000 \text{ lb ft}$$

3. $\Sigma Y = 0$
   $R \quad -800 -600 = 0$
   $$R = 1400 \text{ lb}$$

Vertical Shear and Bending Moment

When a beam is loaded the loads and the reactions cause it to bend. The bending is resisted by forces which are set up within the beam. In order to visualize this phenomenon the beam is cut as shown in Fig. 12–4 and the force and moment equivalent to the portion removed are applied to the remaining section.

Since the entire beam is in equilibrium the cut portion of the beam is also in equilibrium. By using equation 7,

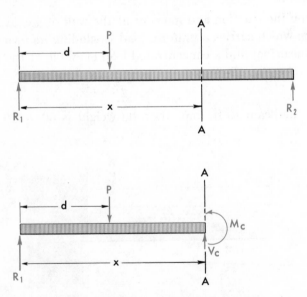

Fig. 12-4. Portion of a beam in equilibrium.

$$\Sigma y = 0$$
$$R_1 + V_c = P$$
$$V_c = P - R_1$$

$$\Sigma M_A = 0$$
$$R_1 (X) - P (d) - M_c = 0$$
$$M_c = R_1 (X) - P (d)$$

The force V_c at an arbitrary section is caused by the adjoining portion of the beam and acts as a shearing force on the section. The moment M_c is likewise caused by the adjoining portion and acts to bend the beam and so is called the bending moment.

Vertical-shear and bending-moment diagrams are graphical representations of the variation of the vertical-shear and bending-moment along the full length of the beam. In order to understand the construction of these diagrams check the examples shown in Figs. 12–5, 12–6, and 12–7.

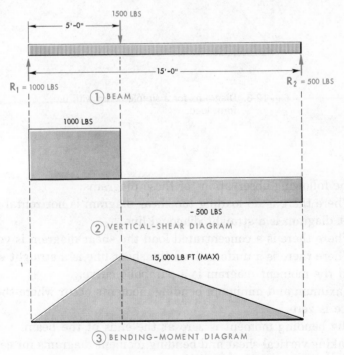

Fig. 12-5. Diagrams for a simple beam with concentrated load.

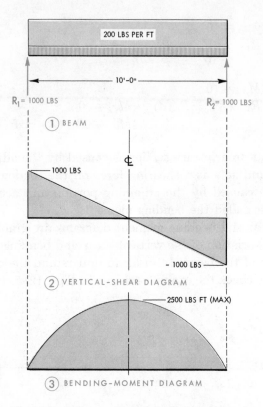

Fig. 12-6. Diagrams for a simple beam with uniform load.

Note the following observations of these diagrams:

1. Where there is no loading the shear diagram is horizontal and the moment diagram is a straight diagonal line.

2. Where there is a concentrated load the shear diagram is vertical.

3. Where there is a uniform load the shear line is a straight diagonal line and the moment diagram is a parabolic curve.

4. Maximum and minimum bending moments occur where the shearing force is zero.

5. The bending moment is zero at the ends of the beam.

In making vertical-shear and bending moment diagrams for cantilever beams it should be noted that there is only one reaction. The reactive

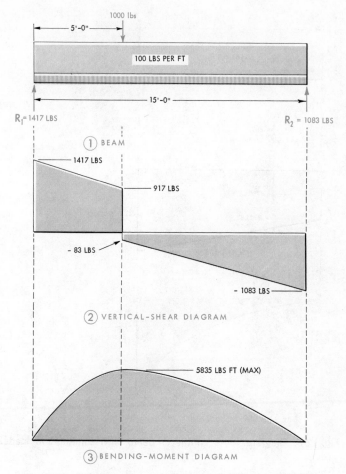

Fig. 12-7. Diagrams for a simple beam with combination loads.

force is the summation of the loads. The bending moment is zero at the free end and reaches its maximum at the support. See Fig. 12–8.

Stresses in Beams

A beam subjected to loads tends to bend. The resisting bending moment in the beam must be analyzed to determine the effect on the material. In Fig. 12–4 the moment was shown to be a twisting action.

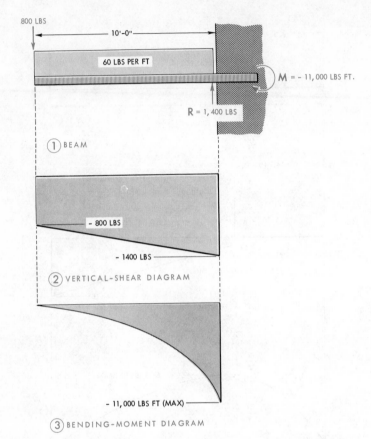

Fig. 12-8. Diagram for a cantilever beam, with combination loads.

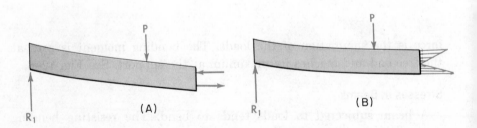

Fig. 12-9. Distribution of stress due to bending moment.

A moment can also be shown as two parallel forces opposite in direction and not along the same line as in Fig. 12–9A. Tests have indicated that a force is not concentrated as shown in Fig. 12–9A but is distributed as in Fig. 12–9B. Therefore, the stress is zero at the neutral axis and reaches its maximum at the outer fibers. In this case, the upper fiber is in compression and the lower in tension.

The formula that is used to determine this stress is:

$$\frac{M}{\sigma} = \frac{I}{c}$$
Equation 8

Where

M = bending moment (lbs in.)

σ = stress (psi)

I = moment of inertia of the section (in.^{4})

c = distance from neutral axis to outer fiber (in.)

Example

Determine the maximum tensile stress in a 10 foot long simple beam with a uniform load of 200 lbs per linear foot. The cross section of the beam is 2 in. thick by 4 in. high.

Solution

In Fig. 12–6 the maximum bending moment was determined to be 2500 lbs ft.

$$I = \frac{bh^3}{12} = \frac{2(4)^3}{12} = 10.67 \text{ in.}^4$$

The distance from the neutral axis to the lower fiber (to give the maximum tension) is 2 inches.

Using equation 8

$$= \frac{CM}{I}$$

$$= \frac{2(2500 \times 12)}{10.67}$$

$$= 5630 \text{ psi}$$

The maximum shearing stress on the cross section of a rectangular beam is given as

$$\tau = \frac{3}{2} \times \frac{V}{A} \qquad\qquad Equation\ 9$$

Where

τ = shear stress (psi)
V = vertical shearing force (lbs)
A = area of cross section (in.^{2})

The corresponding formula for a beam with a circular cross section is

$$\tau = \frac{4}{3} \times \frac{V}{A} \qquad\qquad Equation\ 10$$

Example

Refer to Fig. 12–7 and determine the maximum shear stress if the beam has: (a) cross section 2 in. × 2 in. (b) cross section 2 in. dia.

Solution

(a) using equation **9**

$$\tau = \frac{3}{2} \times \frac{V}{A}$$

$$= \frac{3}{2} \times \frac{1417}{2 \times 2}$$

$$= 531 \text{ psi}$$

(b) using equation **10**

$$\tau = \frac{4}{3} \times \frac{V}{A}$$

$$= \frac{4}{3} \times \frac{1417}{\pi\, 1^2}$$

$$= 603 \text{ psi}$$

Deflection of beams

When a load is placed on a beam the beam tends to sag or deflect. Although no damage to the structure may result its appearance is not necessarily pleasing. Also, a floor may be so out of level that its usefulness may be impaired.

Several methods are available for determining the deflection of a beam. However, for most beams where the loading is not too complex the use of developed formulas is recommended. Many of these formulas are shown in Fig. 12–10.

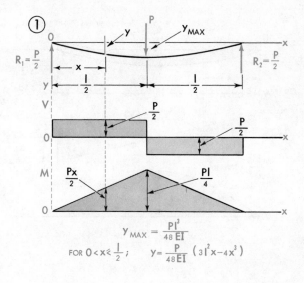

$$y_{MAX} = \frac{Pl^3}{48\,EI}$$

FOR $0 < x \leqslant \frac{l}{2}$; $\quad y = \frac{P}{48\,EI}\,(\,3\,l^2 x - 4\,x^3\,)$

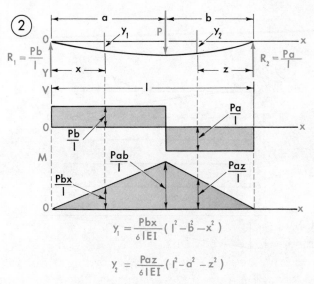

$$y_1 = \frac{Pbx}{6\,l\,EI}\,(\,l^2 - b^2 - x^2\,)$$

$$y_2 = \frac{Paz}{6\,l\,EI}\,(\,l^2 - a^2 - z^2\,)$$

Fig. 12-10. Standard formulas for determining beam deflection.

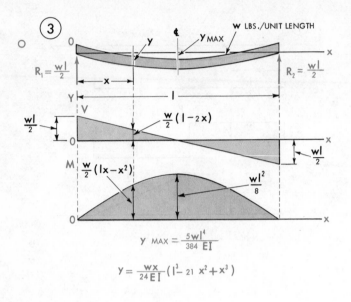

$$y_{MAX} = \frac{5wl^4}{384\,EI}$$

$$y = \frac{wx}{24\,EI}(l^3 - 2l\,x^2 + x^3)$$

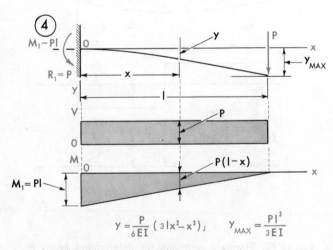

$$y = \frac{P}{6\,EI}(3l\,x^2 - x^3); \qquad y_{MAX} = \frac{Pl^3}{3\,EI}$$

Fig. 12-10. Continued.

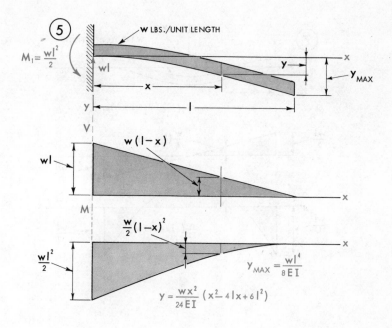

$M_1 = \dfrac{wl^2}{2}$

⑤

w LBS./UNIT LENGTH

y_{MAX}

V

wl

$w(l-x)$

M

$\dfrac{wl^2}{2}$

$\dfrac{w}{2}(l-x)^2$

$y_{MAX} = \dfrac{wl^4}{8EI}$

$y = \dfrac{wx^2}{24EI}(x^2 - 4lx + 6l^2)$

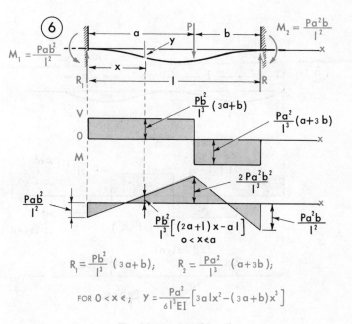

⑥

$M_1 = \dfrac{Pab^2}{l^2}$

$M_2 = \dfrac{Pa^2b}{l^2}$

V

0

$\dfrac{Pb^2}{l^3}(3a+b)$

$\dfrac{Pa^2}{l^3}(a+3b)$

M

$\dfrac{Pab^2}{l^2}$

$\dfrac{Pb^2}{l^3}\left[(2a+l)x - al\right]$
$0 < x \leqslant a$

$\dfrac{2Pa^2b^2}{l^3}$

$\dfrac{Pa^2b}{l^2}$

$R_1 = \dfrac{Pb^2}{l^3}(3a+b); \qquad R_2 = \dfrac{Pa^2}{l^3}(a+3b);$

FOR $0 < x \leqslant$; $\quad y = \dfrac{Pa^2}{6l^3EI}\left[3alx^2 - (3a+b)x^3\right]$

Fig. 12-10. Continued.

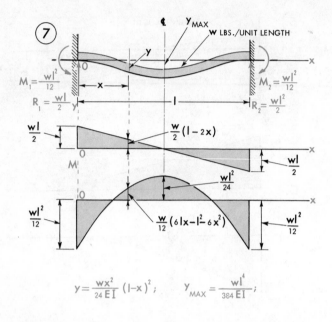

$$y = \frac{wx^2}{24\,EI}\,(l-x)^2\;;\qquad Y_{MAX} = \frac{wl^4}{384\,EI}\;;$$

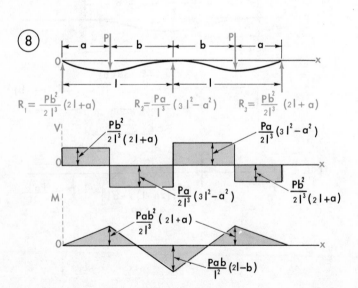

Fig. 12-10. *Continued.*

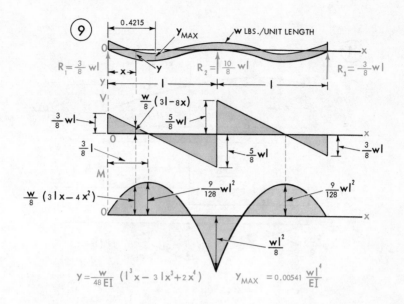

$$y = \frac{w}{48\,EI}\,(l^{3}x - 3\,l\,x^{3} + 2\,x^{4})$$ $$y_{MAX} = 0.00541\,\frac{w\,l^{4}}{EI}$$

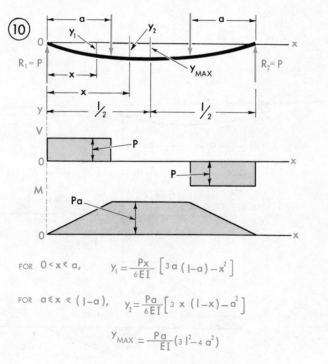

FOR $0 < x \leq a,$ $y_1 = \frac{Px}{6EI}\left[\,3a\,(l-a) - x^2\,\right]$

FOR $a \leq x \leq (l-a),$ $y_2 = \frac{Pa}{6EI}\left[\,3\,x\,(l-x) - a^2\,\right]$

$$y_{MAX} = \frac{Pa}{EI}\,(3\,l^2 - 4\,a^2)$$

Fig. 12-10. Continued.

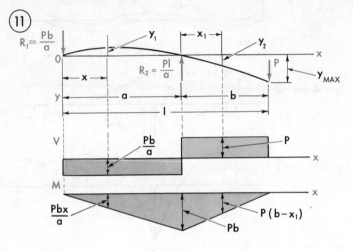

$$\text{FOR} \quad 0 \leqslant x \leqslant a, \qquad y_1 = \frac{Pbx}{6aEI}\ (x^2 - a^2)$$

$$\text{FOR} \quad 0 \leqslant x_1 \leqslant b_1 \qquad y_2 = \frac{Px_1}{EI}\ (2ab + 3bx_1 - x_1^2)$$

$$y_{MAX} = \frac{Pb^2l}{3EI}$$

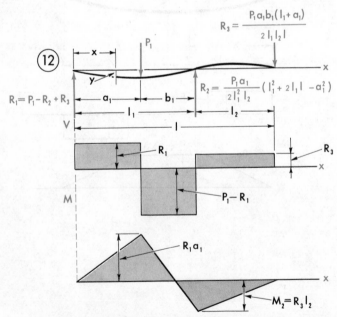

Fig. 12-10. Continued.

Design of Beams

The design of a beam involves making sure that it is within safe limits with regard to (a) tensile and compressive stresses, (b) shear stress, and (c) deflection. Companies which manufacture standard shapes such as I-beams and channels have prepared tables which give the properties of the shape such as area, moment of inertia, and dimensions. Many companies will furnish these tables to users of their products.

Many of the applications of beams involve simple loading so standard diagrams and formulas can be used to determine the distribution of shear, bending-moment, and deflection. Fig. 12–10 shows some of the more commonly used beams. Many more can be found in engineering handbooks.

Torsion

External forces which cause a member to twist are called torsional loads. The product of an external force and the distance to the axis of the member is known as twisting moment or torque. A torque is normally applied in a plane perpendicular to the longitudinal axis of the member.

When the shaft shown in Fig. 12–11 is held stationary at the left end and a torque applied to the right end, the right end is rotated until line OB assumes the position OB^1. The relative movement of the sections of the shaft is resisted by the material, thereby setting up internal shear stresses known as torsional stress.

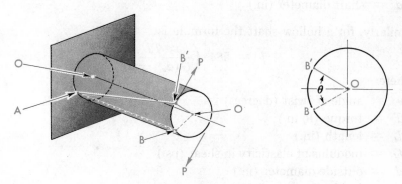

Fig. 12-11. Shaft in torsion.

The formula relating the torque and stress for a solid circular shaft is:

$$T = .196\ \tau d^3 \qquad Equation\ 11$$

Where

T = torque (lb in.)
τ = shear stress at outer fiber (psi)
d = shaft diameter (in.)

Likewise, for a hollow shaft, the formula is:

$$T = .196\ \tau \times \frac{(d_o{}^4 - d_i{}^4)}{d_i} \qquad Equation\ 12$$

Where

T = torque (lb in.)
τ = shear stress at outer fiber (psi)
d_o = outside diameter (in.)
d_i = inside diameter (in.)

It was noted in Fig. 12–11 that when the shaft was subjected to a torque, the line OB rotated through an angle θ. The formula relating torque and angle of twist for a solid circular shaft is given by:

$$\theta = 584 \left(\frac{TL}{Gd^4}\right) \qquad Equation\ 13$$

Where

θ = angle of twist, degrees
T = torque (lb/in.)
L = length (in.)
G = modulus of elasticity in shear (psi)
d = shaft diameter (in.)

Similarly, for a hollow shaft the formula is:

$$\theta = 584 \left(\frac{TL}{G(d_o{}^4 - d_i{}^4)}\right) \qquad Equation\ 14$$

Where

θ = angle of twist (degrees)
T = torque (lb in.)
L = length (in.)
G = modulus of elasticity in shear (psi)
d_o = outside diameter (in.)
d_i = inside diameter (in.)

Example

Calculate the minimum diameter of a steel shaft 4 feet long subjected to a torque of 100,000 lb in. if the maximum shearing stress is not to exceed 10,000 psi and the angle of twist is not to exceed 1°. (For steel, $G = 12,000,000$ psi)

Solution

1. Using equation **11**
$$T = .196 \ \tau \ d^3$$
$$100,000 = .196 \times 10,000 \times d^3$$
$$d = 3.71 \text{ in.}$$

2. Using equation **13**
$$\theta = 584 \left(\frac{TL}{Gd^4} \right)$$
$$L = 584 \left(\frac{100,000 \times 12 \times 4}{12,000,000 \times d^4} \right)$$
$$d = .391 \text{ in.}$$

Therefore to satisfy both conditions, the diameter of the shaft should be at least 3.91 inches in diameter.

Horse Power Relationships

Shafts are generally used to transmit power from a motor to a machine. The following formula shows the relationship between horsepower, torque, and RPM.

$$H.P. = \frac{TN}{63,000} \qquad \qquad \textit{Equation 15}$$

Where
H.P. = horsepower
T = torque (in. lb)
N = revolutions per minute (RPM)

Example

If a 5 horsepower motor is turning at 3,000 RPM determine the torque being transmitted.

Solution

Using equation **15**

$$T = \frac{63,000 \ (H.P.)}{N}$$

$$T = \frac{63,000 \times 5}{3,000} = 105 \text{ in. lb}$$

Thin-walled Pressure Vessels

A pressure vessel is described as thin walled when the ratio of the wall thickness to the radius of the vessel is less than 1/10.

In order to obtain the stress in the wall of a cylinder subjected to an internal pressure, the equation used is: (See Fig. 12–12A)

$$\sigma = \frac{Pr}{t} \qquad \qquad \textit{Equation 16}$$

Where

σ = stress (psi)
P = pressure (psi)
r = radius (in.)
t = wall thickness (in.)

The corresponding equation for obtaining the stress in the wall of a hemisphere is: (See Fig. 12–12B)

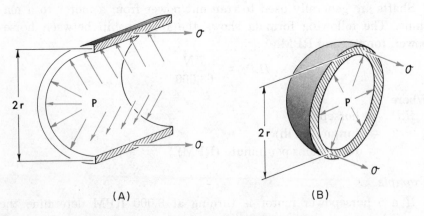

(A)   (B)

Fig. 12-12. Stresses in thin-walled vessels.

$$\sigma = \frac{Pr}{2t}$$ *Equation* **17**

Where

σ = stress (psi)
P = pressure (psi)
r = radius (in.)
t = wall thickness (in.)

Example

A steel pipe is made by wrapping a .063 in. plate around a mandrel and welding it parallel to the axis of the mandrel. If the pipe has a diameter of 2 in. and is subjected to a pressure of 100 psi, determine the stress at the weld.

Solution

Using equation **16**

$$\sigma = \frac{Pr}{t} \times \frac{100 \times 1}{.063} = 1590 \text{ psi}$$

REVIEW QUESTIONS

1. A short steel part with an 8″ OD and 6″ ID carries a compressive load of 30,000 lb. What is the unit stress developed?
2. What is the allowable tensile load on a standard steel bar 2″ in diameter?
3. A 30″ long brass bar elongates .03″ under a load of 10,000 lb. Determine its size if it has a square cross section.
4. What will be the increase in length of an aluminum bar 10′ long which was set in place at 32°F and then the temperature rises to 100F°?
5. A steel bar 1″ in dia. and 3′ long is placed between two supports when the temperature is 60°F. If the ends are rigidly held what is the stress developed when the temperature rises to 110°F?
6-9. Make shear and bending moment diagrams of beams shown in Problems 6, 7, 8, and 9.

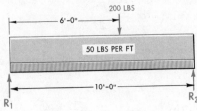

Problem 6. Simple beam.

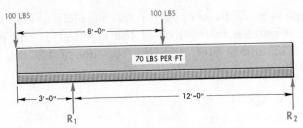

Problem 7. Overhanging beam.

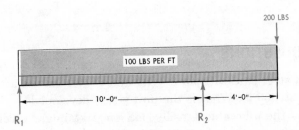

Problem 8. Overhanging beam.

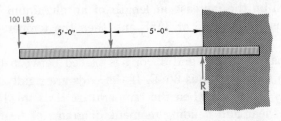

Problem 9. Cantilever beam.

10. If the beam in problem 6 is a 2″ dia steel rod, determine its maximum tensile stress and shear stress.
11. If the beam in problem 7 is a 4″ OD by 3″ ID steel tube, determine the maximum stress due to bending.
12. Determine whether the beam in problem 8 is safe if it is made of 1″ × 2″ aluminum. Does it make any difference whether the beam is installed in a vertical or horizontal position?
13. What minimum size square aluminum bar is needed for the beam shown in problem 9?
14. Determine the deflection in the middle of a simple 12′ long beam with a uniform load of 100 lbs per ft. including the weight of the beam. The beam is made of steel with a cross section 10″ wide by 2″ deep.
15. Determine the deflection of a cantilever beam 6′ long. It has a uniform load of 50 lbs per ft. including the weight of the beam and a concentrated load of 200 lbs at the free end. The beam is made of 2″ dia. aluminum.
16. A steel shaft 4″ in dia. is 10′ long. Compute the allowable torque and the angle of twist if the maximum shear stress is not to exceed 8,000 psi.
17. Determine the horsepower a 2″ diameter steel shaft will transmit at 600 rpm if the maximum shear stress is not to exceed 8,000 psi.
18. Determine the horsepower a steel tube with a 2″ OD and 1¾″ ID transmits at 3,000 rpm. Also determine the angle of twist in a length of 6′. Maximum shear stress is not to exceed 10,000 psi.
19. A certain pipe has an inside diameter of 4″ and a wall thickness of .250″. If the allowable stress of the material is 20,000 psi, determine the maximum safe pressure.
20. A sphere is made by welding two halves together. If the radius is 6″, the pressure 100 psi, and the wall thickness .125″, determine the stress on the welded joint.

Chapter 13	# Design of Weldments

In the design of any product the question of whether parts are to be welded, cast, or press formed frequently needs to be resolved. Too often the use of welding is minimized when in reality it may have decided advantages over other fabricating techniques. For example, it is often more expedient to build a welded structure by using rolled plates and structural shapes instead of resorting to a casting because greater savings in weight and cost can be achieved. Similarly, heavy framed units for large machine components will often be easier to construct by welding. The paramount issue in any design function is to achieve maximum savings in cost but at the same time develop a product that is pleasing in appearance, possesses sufficient strength, and effectively fulfills its intended function safely.

Advantages of Welded Construction

It is inconceivable to expect that welding provides the optimum solution to all design problems. Nevertheless, welding frequently has definite advantages in many applications. Specifically these advantages can be summarized as follows:

1. Welding is a great deal more flexible and has numerous basic cost savings over casting and press forming operations. However, this is not necessarily true in the case where high quantity production runs are required and parts are small.

2. Welding is often three or four times stronger than other fabricating processes and consequently can absorb higher loads without increasing the weight of the material.

3. A weldment is frequently two to three times stiffer than castings yet possesses greater ductility.

4. There is less tendency for a welded product to crack due to porosity and shrinkage stresses which are often encountered in a casting.

5. Capital equipment is generally less for welding than for other manufacturing processes.

6. Welded structures provide a freedom of design which is not readily approached by other methods and still meet functional requirements.

7. Welded parts are much more resistant to shock and impact forces than castings and therefore have higher safety factors.

Welding Design Factors

Correct joint design is necessary if welded assemblies are to perform in accordance with the established service and safety requirements. One of the first requisites of joint design is the function of the part. Knowing the type of service it is to render, consideration must then be given to the kind of loads imposed on the welded section. Thus, an analysis must be made whether the applied stress is one of tension, compression, shear, torsion, or bending. Joint configuration is next determined on the basis of the design that provides the greatest resistance to the applied load. For example some types of joints are best if the stress comes from one direction while other joint designs are more suitable for stresses that are of a more complex nature. The problem of whether a joint is under a static or dynamic load is another factor which affects joint design. Finally the efficiency of a joint, that is, its total strength as compared with the actual base metal, is an important influencing factor.

The principal design issues which are to be covered here deal with such items as joint design, joint location, weld size, and joint strength.

Design Optimization

To reach the goal of producing a part at the lowest overall cost it is necessary to evaluate the design to make certain that optimum use is made of the metallurgical and physical properties of the materials. Factors that need to be considered are:

1. The design should satisfy the strength and rigidity requirements of the product. Designing beyond these requirements means that more material and labor will be used thereby adding to the costs.

2. The safety factor designated for the design should be sufficient to meet all contingencies. Using too large a safety factor again means added material and labor.

3. The appearance of the product should be pleasing but where areas are hidden from view appearance should not be a critical factor. Therefore, it is often more expedient to use less expensive grades of material where appearance is unimportant. For example hot-rolled steel will frequently prove as effective as cold-rolled steel.

4. Since more weight means higher costs the members should be analyzed to see if stiffeners could be employed for rigidity as a means of reducing weight.

5. Wherever possible non-premium steel should be specified—except that a high grade steel is best where the added strength reduces the cost.

6. Standard sizes and shapes should be specified wherever it is practical.

7. When a hard surface is required the use of hard surfacing, heat treating, or plating should be noted.

Joint Design

Joint design is greatly influenced by the cost of preparing the joint, the accessibility of the weld, its adaptability for the product being designed, and the type of loading the weld is required to withstand.

The basic joint configurations which are applicable for shielded metal-arc, gas metal-arc, and submerged-arc welding are broadly classified as grooved and fillet. Each group incorporates several variations to provide for different service requirements.

The *square butt joint* is intended primarily for materials that are ⅜ in. or lighter and require full and complete fusion for optimum strength. The joint is reasonably strong in static tension but is not recommended when tension due to bending is concentrated at the root of the weld. It should never be used when it is subjected to fatigue or impact loads, especially at low temperatures. The preparation of the joint is relatively simple since it requires only matching edges of the plates, consequently the cost is low. See Fig. 13–1A.

The *single vee butt joint* is used on plate ⅜ in. or greater in thickness. Preparation is more costly because a special beveling operation is required and more filler material is necessary. The joint is strong in static loading but as in the square butt joint it is not particularly suitable when tension due to bending is concentrated at the root of the weld. See Fig. 13–1B.

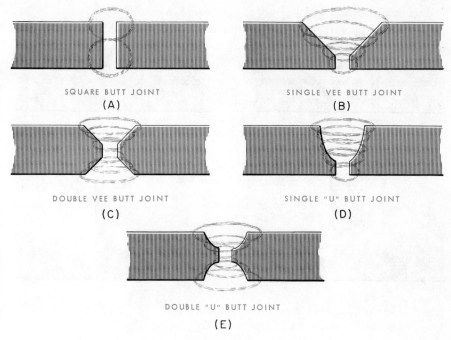

SQUARE BUTT JOINT
(A)

SINGLE VEE BUTT JOINT
(B)

DOUBLE VEE BUTT JOINT
(C)

SINGLE "U" BUTT JOINT
(D)

DOUBLE "U" BUTT JOINT
(E)

Fig. 13-1. Types of butt joint designs.

The *double vee butt joint* is best for all load conditions. It is often specified for stock that is heavier than stock used for a single vee. For maximum strength the penetration must be complete on both sides. The cost of preparing the joint is higher than the single vee but less filler material is required. To keep the joint symmetrical and warpage to a minimum the weld bead must be alternated, welding first on one side and then the other. See Fig. 13–1C.

The *single U butt joint* readily meets all ordinary load conditions and is used for work requiring high quality. It has greatest applications for joining plates ½ to ¾ in. See Fig. 13–1D.

The *double U butt joint* is suitable for plate ¾ in. or heavier where welding can readily be accomplished on both sides. Although the preparation cost is higher than the single U butt joint, less weld metal is needed. The joint meets all regular loading conditions. See Fig. 13–1E.

The *square tee joint* requires a fillet weld which can be made on one

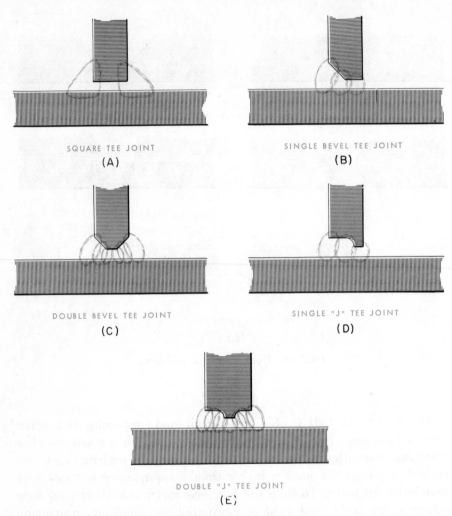

SQUARE TEE JOINT
(A)

SINGLE BEVEL TEE JOINT
(B)

DOUBLE BEVEL TEE JOINT
(C)

SINGLE "J" TEE JOINT
(D)

DOUBLE "J" TEE JOINT
(E)

Fig. 13-2. Types of tee joint designs.

or both sides. It can be used for light to reasonably thick materials where loads subject the weld to longitudinal shear. Since its stress distribution is not uniform, care must be taken in specifying this joint where severe impact or heavy transverse loads are encountered. For maximum strength considerable weld metal is required. See Fig. 13–2A.

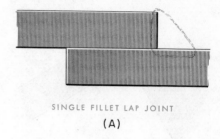

SINGLE FILLET LAP JOINT

(A)

Fig. 13-3. Types of fillet lap joint designs.

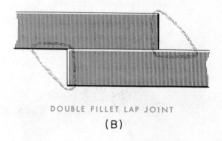

DOUBLE FILLET LAP JOINT

(B)

The *single bevel tee joint* will withstand more severe loadings than the square tee joint due to better distribution of stresses. It is generally confined to plates ½ in. or less in thickness where welding can be done from one side only. See Fig. 13–2B.

The *double bevel tee joint* is intended for use where heavy loads are applied in both longitudinal and transverse shear and where welding can be done on both sides. See Fig. 13–2C.

The *single J tee joint* is used on plates one inch or more in thickness where welding is limited to one side. It is especially suitable where severe loads are encountered. See Fig. 13–2D.

The *double J tee joint* is suitable in heavy plates 1½ in. or more thick where unusually severe loads must be absorbed. Joint location should permit welding on both sides. See Fig. 13–2E.

The *double fillet lap joint* can withstand more severe loads than the single fillet lap joint. It is one of the more widely used joints in welding. See Fig. 13–3A & B.

Design of Weldments 339

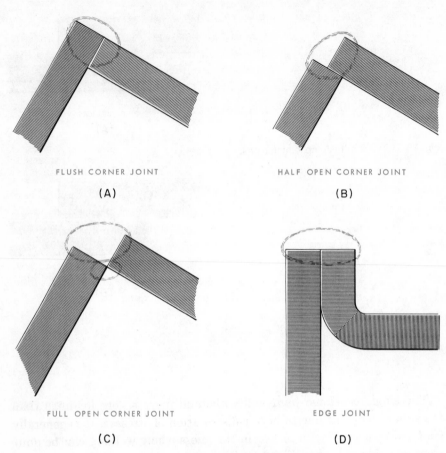

FLUSH CORNER JOINT

(A)

HALF OPEN CORNER JOINT

(B)

FULL OPEN CORNER JOINT

(C)

EDGE JOINT

(D)

Fig. 13-4. Types of corner joint designs.

The *flush corner joint* is designed primarily for welding sheet 12 gage and lighter. It is restricted to lighter materials because deep penetration is sometimes difficult and it supports only moderate loads. See Fig. 13–4A.

The *half open corner joint* is usually more adaptable for materials heavier than 12 gage. Penetration is better than in the flush corner joint. See Fig. 13–4B.

The *full open corner joint* is used where welding can be done from both sides. Plates of all thickness can be welded. This joint is strong and capable of carrying heavy loads. It is also recommended for fatigue and impact applications because of good stress distribution. See Fig. 13-4C.

The *edge joint* is suitable for plate ¼ in. or less in thickness and can sustain only light loads. See Fig. 13-4D.

Selecting the Proper Joint Design

The square groove joint is undoubtedly the most economical but it is often necessary to confine it to products which are to be made of light gage material. Whenever metal is beveled, costs increase. However, there is often no alternative especially if heavy plates are to be welded. Without proper edge chamfering complete penetration is impossible and without adequate penetration the resulting weld joint will be weak. The common practice for the greatest economy is the selection of a root opening and groove angle that requires the least amount of filler material.

The type of groove angle is frequently governed by the accessibility of the weld. For example the position of the members may be such that only a single bevel groove joint is practical. The accessibility of the weld also determines whether a weld is to be made on one side or on both sides.

The required strength of a joint must always be considered a major factor of joint configuration. Only by an actual understanding of the type and magnitude of various loadings is it possible to select the joint that will meet the necessary service requirements.

Fig. 13-5 illustrates the recommended proportions for various joints to produce complete and consistent penetration.

Weld Location

Welding is generally considered a low-cost method of joining parts but unless careful thought is given to joint design and location of welds much of the cost saving can be nullified. Weld location is an integral part of joint design and as such its value must not be minimized. There are many aspects affecting proper weld location. Only a few can be illustrated here.

1. The location of the weld in its relationship to the members being joined has a decided effect on the strength of the joint. Normally welds

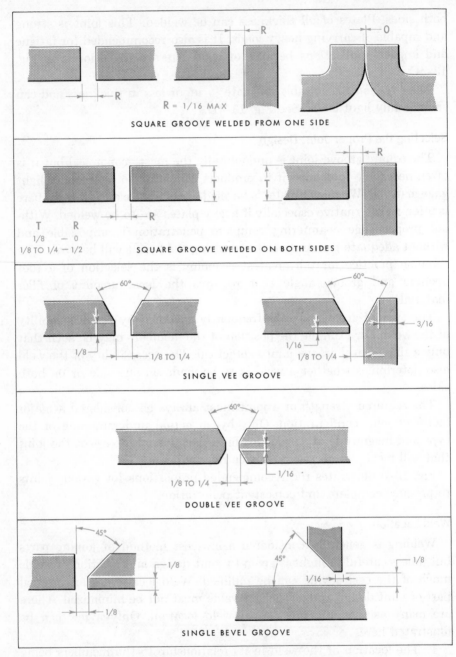

Fig. 13-5. Recommended proportions for weld joints.

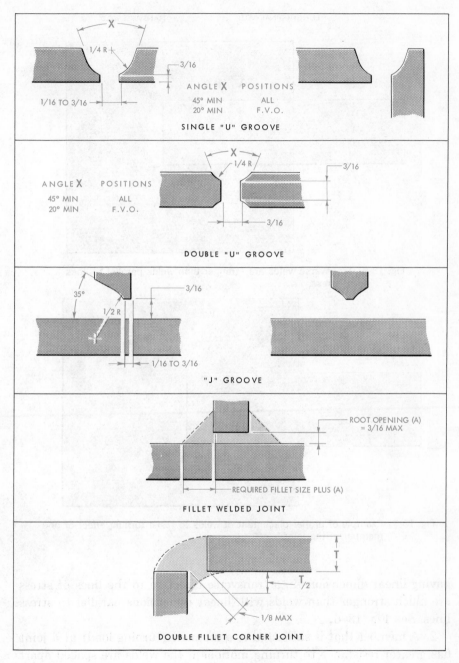

SINGLE "U" GROOVE

X

1/4 R

3/16

1/16 TO 3/16

ANGLE X   POSITIONS
45° MIN   ALL
20° MIN   F.V.O.

DOUBLE "U" GROOVE

ANGLE X   POSITIONS
45° MIN   ALL
20° MIN   F.V.O.

X

1/4 R

3/16

3/16

"J" GROOVE

35°

1/2 R

3/16

1/16 TO 3/16

FILLET WELDED JOINT

ROOT OPENING (A)
= 3/16 MAX

REQUIRED FILLET SIZE PLUS (A)

DOUBLE FILLET CORNER JOINT

T

T/2

1/8 MAX

Fig. 13-5. Continued.

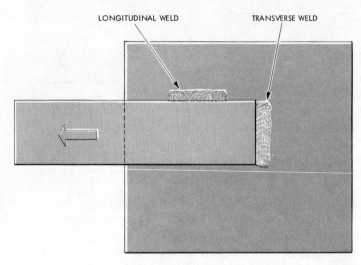

LONGITUDINAL WELD   TRANSVERSE WELD

Fig. 13-6. Transverse welds are stronger than welds parallel to lines
of stress.

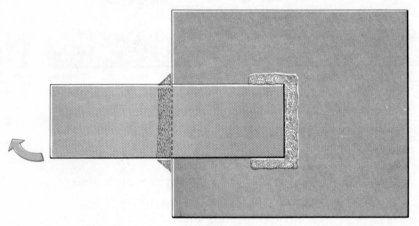

Fig. 13-7. Example of proper placement of welds to resist turning effect of one
member at the joint.

having linear dimensions in a transverse direction to the lines of stress
are much stronger than welds with linear dimensions parallel to stress
lines. See Fig. 13–6.

2. A member that is subjected to a moment (turning load) at a joint
has greater resistance to turning motions if the welds are spaced apart
instead of being close together. See Fig. 13–7.

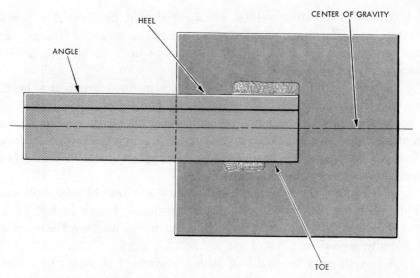

Fig. 13-8. Example of correct lengths of weld for equal load distribution.

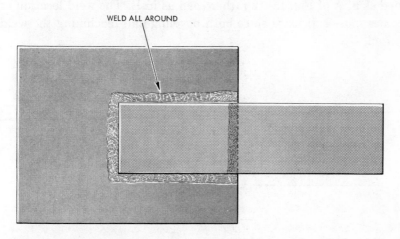

Fig. 13-9. Load distribution in linear welds is better if welded around all sides.

3. When joining special shape sections such as channels, angles, etc., consideration must be given to proper distribution of loads on the welds. Notice in Fig. 13–8 that the length of the weld at the heel of the angle is greater than at the toe. This eliminates the tendency for the angle to turn and generate highly eccentric loads at the joint.

4. Wherever possible welded members should be located to avoid bending or shearing actions. Greater strength is achieved if joints are kept symmetrical. Load distribution is more uniform on symmetrical than non-symmetrical joints.

5. To attain uniform distribution of loads in linear welds it is better to weld all around the joint than simply on two sides. See Fig. 13–9.

6. Abrupt surface changes in a weld joint should be avoided since greater local stresses will be concentrated at the point of surface change. There is a more uniform transfer of stress through a weld if the surface flow is gradual. See Fig. 13–10.

7. Weld joints should be located so they are readily accessible and welding can be carried out with comparative ease. Notice in Fig. 13–11 that the weld at A is almost impossible to make whereas the weld at B is easily accessible.

8. When an angle is used as a stiffener where it is joined to a plate that serves as a locating surface it provides greater economy if positioned as in A of Fig. 13–12 rather than as in B. The weld location at B becomes more expensive since both beveling and machining the weld is required.

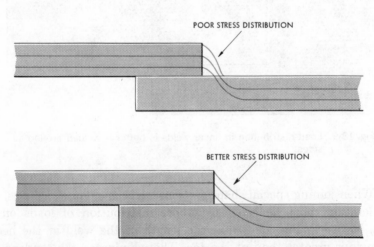

Fig. 13-10. For even stress distribution, avoid abrupt surface changes in a joint.

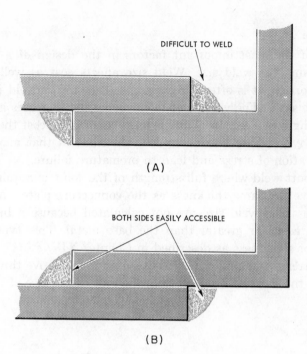

(A)

(B)

Fig. 13-11. Weld joints should be readily accessible.

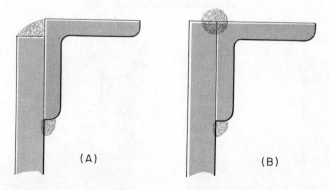

(A)

(B)

Fig. 13-12. Proper location of stiffeners contributes to
greater cost economy.

Design of Weldments 347

Weld Size

One of the most important factors in the design of a weldment is determining the weld size. Weld size affects cost as well as strength and distortion. It is often a common practice to overweld just to be on the "safe side". The volume of metal deposited increases as the square of the length of a leg. In addition to increasing the cost the excess weld metal may generate an uneven load distribution that may result in a concentration of stress and lead to premature failure.

In a butt weld where full strength of the joint is required the weld must have the same thickness as the connecting plates. As a rule the unit stress and weld size are not calculated because a butt weld has strength equal or greater than the base metal. This type of joint is tested for soundness as described in Chapter XIV.

The strength of a fillet weld is based on the effective throat thickness which is the shortest distance from the root to the face of the weld. For

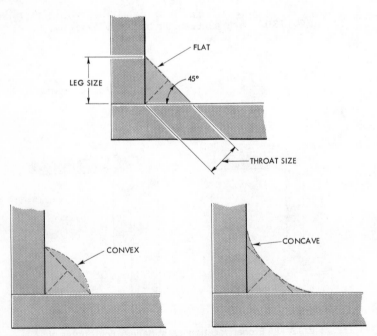

Fig. 13-13. Types of fillet contours.

an equal leg (45°) fillet weld the throat is .707 (sin of 45°) times the leg size of the weld. See Fig. 13–13.

To check the size of fillet welds it is necessary to make gages to fit the required contour of the weld deposit. Weld contours may be concave, flat, or convex with equal or unequal legs. The correct fillet size is determined by finding the leg-length of the largest isosceles triangle that can be inscribed within the cross-section of the required weld metal. See Fig. 13–13.

A flat fillet weld has a flat surface and is considered to be the most ideal weld. A concave fillet weld produces a smooth weld flow from one member to another but is often subject to shrinkage cracks. A convex fillet weld may result in excess weld metal but has less tendency to crack from shrinking forces.

A fillet weld to develop the full strength of the plates being connected must have a *parallel loading* equal to the shear strength of the plate as computed on the basis of a minimum unit shear stress of 15,000 psi. It must also have a *transverse loading* equivalent to the tensile stress of the plate as computed at a unit tensile stress of 20,000 psi. In general this relationship is achieved by maintaining a fillet width equal to ¾ plate thickness. See Fig. 13–14 and Table I.

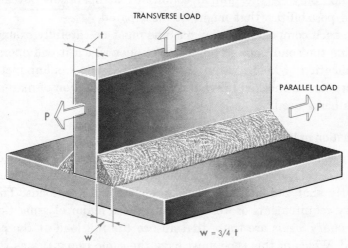

Fig. 13-14. A full strength fillet weld must have a proper parallel and transverse loading relationship.

Table I.   Recommended Fillet Weld Size to Develop Plate Strength.
(Welded both sides.)

(W) FILLET WELD SIZE (INCH)	(T) PLATE THICKNESS (INCH)
5/32	3/16
3/16	1/4
1/4	5/16
5/16	3/8
3/8	1/2
1/2	5/8
5/8	3/4
3/4	1

STRENGTH ANALYZATION

The first step in analyzing the design of a machine is to determine the loads which are applied and generated by the machine. These loads should not only be the normal conditions of operation but also any overload possibilities that might be anticipated.

Next, each component of the machine must be carefully examined to determine the conditions or limitations under which it will operate. By the application of known strength of materials and welding metallurgy each part can be checked to see if the right combination of material and stress is being used.

Classification of Welds

When welds are subjected to loads various types of stresses develop which are either absorbed or transmitted to other members. Welds are frequently analyzed on the basis of their reactions to loads. The serviceability requirements of welds fall into these major classifications:

1. *Primary welds* are those that absorb the full load at the plane of the weld. Welds of this type must have the same properties as the base metal. See Fig. 13–15A.

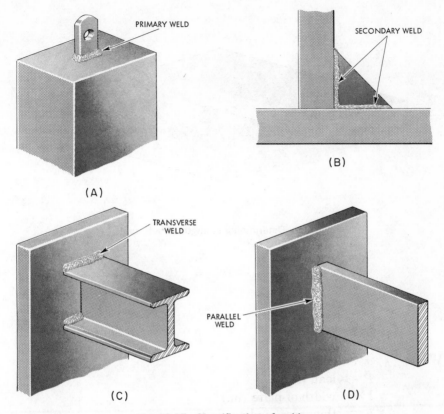

Fig. 13-15. Classification of welds.

2. *Secondary welds* are those that do not carry the full load of connecting members but must absorb forces which are in the member at the point of weld. See Fig. 13–15B.

3. *Transverse welds* are those that transmit tension and compression forces from one member to another. See Fig. 13–15C.

4. *Parallel welds* are those that transmit shear loads from one member to another. See Fig. 13–15C.

Strength of Welded Joints

To properly design welded products the strength of the welded joints must be analyzed. Stress on a butt joint is calculated by the formula:

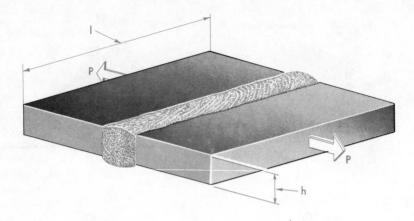

Fig.13-16. Determining strength of butt joints.

$$\sigma = \frac{P}{lh}$$ *Equation* 1

Where σ = Stress (psi)
       P = load (lb)
       l = width of plate (in.)
       h = thickness of plate (in.)

Notice in Fig. 13–16, that in using the above equation the weld material extending above the plate is neglected. Therefore the thickness is taken to be the same as the thickness of the plate. The American Welding Society recommends 20,000 pounds per square inch as the allowable stress for butt joints.

Example:

If two ¼ in. thick plates are to be welded with a butt joint, determine the required width of the plate if the joint is subjected to a tensile load of 10,000 pounds.

Solution:

Using equation 1

$$l = \frac{P}{\sigma h}$$

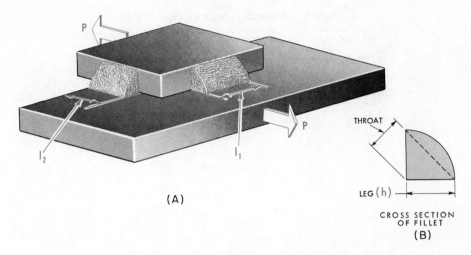

(A)

THROAT

LEG (h)

CROSS SECTION
OF FILLET
(B)

Fig. 13-17. Determining strength of fillet joints.

$$l = \frac{10,000}{20,000 \times .250}$$

$$l = 2 \text{ inches}$$

In considering a typical fillet weld such as is shown in Fig. 13–17, it is noted that the opposing forces tend to make the members slip and cause shearing stresses in the welds. The stress is highest on the plane where the thickness of the weld is the smallest which occurs at the throat. Hence the stress on the weld is determined by equation:

$$\tau = \frac{P}{.707h} \qquad \textit{Equation 2}$$

Where
τ = stress (psi)
P = load (lbs)
h = leg of weld (in.)
l = total length of weld (in.)

The American Welding Society recommends 13,600 pounds per square inch as the allowable shearing stress for 45° fillet welds. This stress is based on the application of loads which are not too severe. Dynamic or

Table II. Allowable Loads for Fillet Welds.

SIZE OF FILLET WELD (LEG) (INCHES)	POUNDS PER LINEAL INCH
1/8	1200
3/16	1800
1/4	2400
5/16	3000
3/8	3600
1/2	4800
5/8	6000
3/4	7200

vibratory loads are considered to be more severe and usually require slightly lower allowable stresses. Table II lists the safe loads for fillet welds. The values are obtained by using equation 2 and the allowable stress of 13,600 psi.

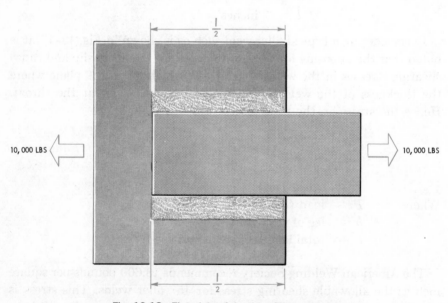

Fig. 13-18. Flat plates joined by fillet welds.

Example.

If two ¼ in. thick plates are to be joined by fillet welds, what is the required length of the weld if a load of 10,000 pounds is applied as shown in Fig. 13–18?

Solution:

Using equation **2**

$$l = \frac{P}{.707h\tau}$$

$$l = \frac{10,000}{.707 \times .250 \times 13,600}$$

$$l = 4.16 \text{ in.}$$

Therefore, in this case the length of the weld on each side of the plate should be half the total length or 2.08 inches.

Additional Strength Formulas. The strength characteristics of several common welded joints were analyzed by means of equations **1** and **2**. Fig. 13–19 provides other basic joint designs and their corresponding stress formulas.

Evaluating Safety Factor

In evaluating the strength of any structure, the factor of safety must never be ignored. The examples which follow indicate the procedure which can be used in considering factors of safety (F.S.).

Example 1.

Determine the F.S. for the joint shown in Fig. 13–20. The material is steel and the fillet weld is ³⁄₁₆ in. (leg size).

Solution. In this case the beam is so short that the bending moment is neglected and the joint is studied for vertical shear only. By using equation **2**:

$$\tau = \frac{P}{.707hl}$$

$$\tau = \frac{2000 \times 12}{.707 (.1875) (2 \times 12)}$$

$$\tau = 7,550 \text{ psi.}$$

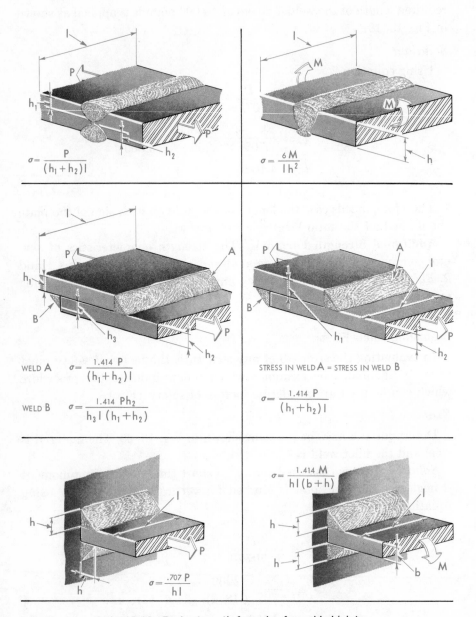

$$\sigma = \frac{P}{(h_1 + h_2)l}$$

$$\sigma = \frac{6M}{lh^2}$$

WELD **A** $\quad \sigma = \dfrac{1.414\,P}{(h_1 + h_2)l}$

WELD **B** $\quad \sigma = \dfrac{1.414\,Ph_2}{h_3\,l\,(h_1 + h_2)}$

STRESS IN WELD **A** = STRESS IN WELD **B**

$$\sigma = \frac{1.414\,P}{(h_1 + h_2)l}$$

$$\sigma = \frac{.707\,P}{hl}$$

$$\sigma = \frac{1.414\,M}{hl(b+h)}$$

Fig. 13-19. Basic strength formulas for welded joints.

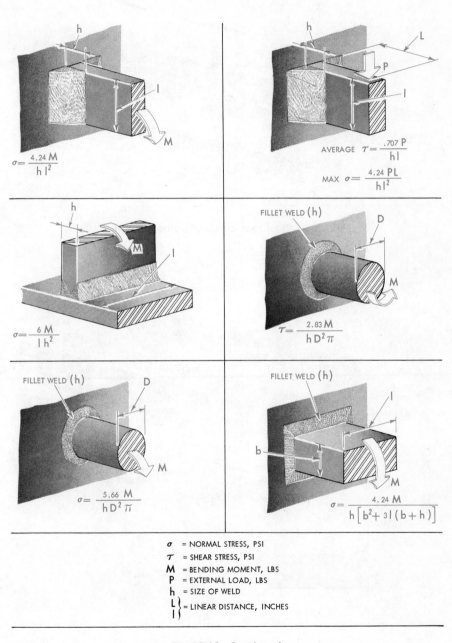

$$\sigma = \frac{4.24\,M}{h\,l^2}$$

$$\text{AVERAGE} \quad \tau = \frac{.707\,P}{h\,l}$$

$$\text{MAX} \quad \sigma = \frac{4.24\,P\,L}{h\,l^2}$$

$$\sigma = \frac{6\,M}{l\,h^2}$$

FILLET WELD (h)

$$\tau = \frac{2.83\,M}{h\,D^2\,\pi}$$

FILLET WELD (h)

$$\sigma = \frac{5.66\,M}{h\,D^2\,\pi}$$

FILLET WELD (h)

$$\sigma = \frac{4.24\,M}{h\left[b^2 + 3l(b+h)\right]}$$

σ = NORMAL STRESS, PSI
τ = SHEAR STRESS, PSI
M = BENDING MOMENT, LBS
P = EXTERNAL LOAD, LBS
h = SIZE OF WELD
L ⎫
  ⎬ = LINEAR DISTANCE, INCHES
I ⎭

Fig. 13-19. Continued.

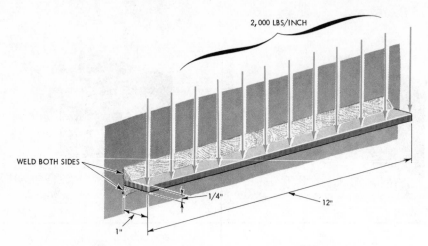

Fig. 13-20. Short cantilever with distributed load.

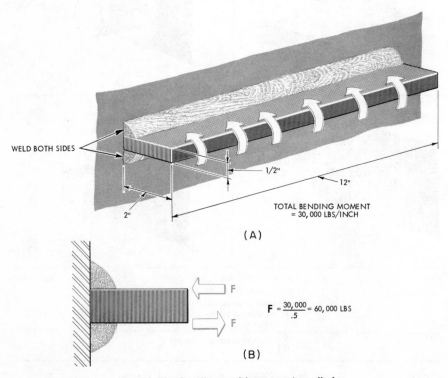

$$F = \frac{30,000}{.5} = 60,000 \text{ LBS}$$

(B)

Fig. 13-21. Cantilever with moment applied.

Since the allowable shear stress for fillet welds is 13,600 psi.,

$$\text{F.S.} = \frac{13,600}{7,550} = 1.8$$

Example 2.

Determine the F.S. for the joint shown in Fig. 13–21A. The material is steel and the fillet weld is ⅜ in. (leg size).

Solution:

The bending moment is first changed into two parallel forces acting at the base of the welds as shown in Fig. 13–21B. By using equation **2**:

$$\tau = \frac{P}{.707hl}$$

$$\tau = \frac{60,000}{.707\ (.3875)\ (2 \times 12)}$$

$$\tau = 9,130 \text{ psi.}$$

Since the allowable shear stress is 13,600 psi, F.S. $= \dfrac{13,600}{9,130} = 1.49$

Example 3:

Determine the F.S. for the joint shown in Fig. 13–22. The material is steel and the fillet weld is ¼ in. (leg size).

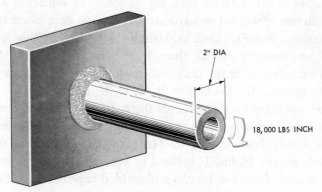

2" DIA

18,000 LBS INCH

Fig. 13-22. Hollow shaft subjected to twisting.

Solution:

In this situation there is a twisting of the shaft which causes a shearing action at the weld. Therefore according to equation **2**:

$$\tau = \frac{P}{.707hl}$$

$$\tau = \frac{18,000 \times 1}{(.707)\ (.250)\ (\pi \times 2)}$$

$$\tau = 16,200 \text{ psi.}$$

Since the allowable stress in shear is 13,600 psi, the welded joint is over stressed. Therefore to determine the minimum weld size (F.S. = 1), it is necessary to solve for h.

$$h = \frac{P}{.707l\tau}$$

$$h = \frac{18,000}{(.707)\ (\pi \times 2)\ (13,600)}$$

$$h = .298 \text{ in.}$$

Accordingly the fillet weld that is required should be ⁵⁄₁₆ in. (to nearest ¹⁄₁₆ in.)

Replacement of Existing Designs

A designer is often faced with the problem of replacing an existing design with one of welded construction. A typical example is the mounting of bearings. See Fig. 13–23A. Usually a housing of this kind can be made of a thinner material than the section where the bearing is mounted by redesigning the unit and employing walls that are just thick enough to support the loads. See B, C, D, E in Fig. 13–23.

Another important use of welds is the reduction of stock removal from parts that are turned on a lathe or screw machine. A good example is the hydraulic piston shown in Fig. 13–24A made from a blank of 6⅛ in. dia. steel which weighs 76 lbs. If instead of turning the piston from a solid blank, it is made from two blanks and welded together as in Fig. 13–24B there is considerable savings not only in metal waste but also machining

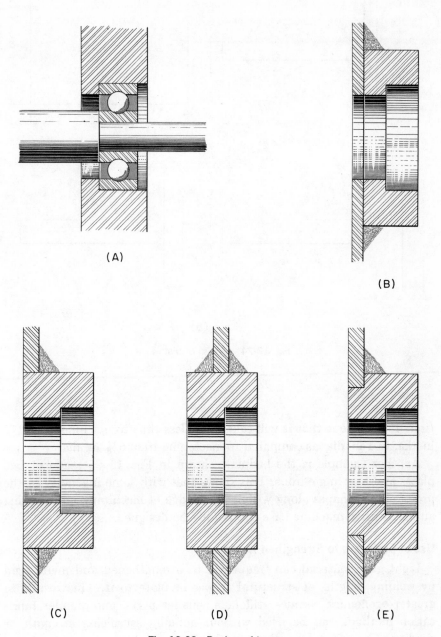

Fig. 13-23. Design of bosses.

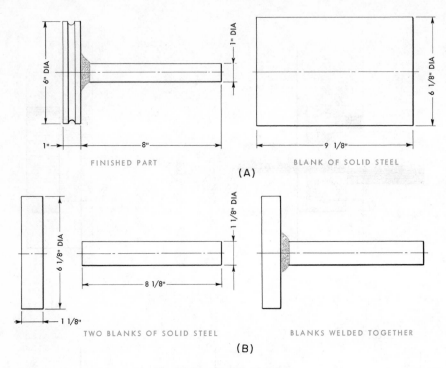

Fig. 13-24. Design of piston.

time. It is obvious that it will take much less time to machine from 1⅛ in. dia. to 1 in. dia. as compared to machining from 6⅛ in. dia.

Another example is the bedplate shown in Fig. 13–25 made of two plates and four angle irons. This shows that with some ingenuity in the use of simple shapes along with a knowledge of machining and welding an inexpensive machine base can readily be designed.

Use of Stiffeners to Strengthen Weldments

Light gage material can frequently be strengthened and made rigid by welding lengths of structural shapes or plate to it. This results in greater economies because stiffeners, smaller parts, and simpler fabrication methods can be used without actually sacrificing strength or rigidity. Thus, if a long thin member is required to support a com-

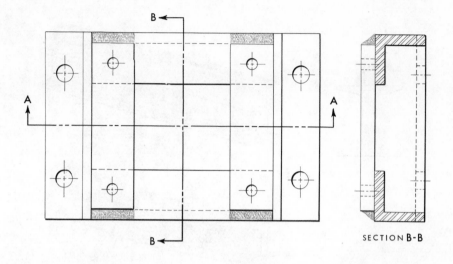

SECTION B-B

SECTION A-A

Fig. 13-25. Design of machine base.

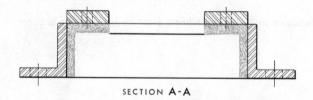

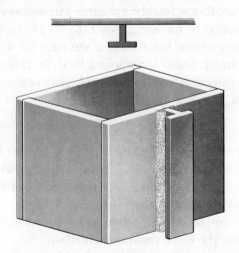

Fig. 13-26. Shape welded to a plate for stiffness.

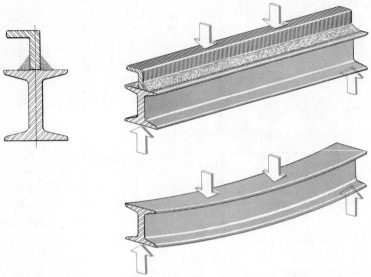

Fig. 13-27. Reducing deflection by adding a structural shape to a beam.

pressive load and there is possible danger of buckling, the member can be stiffened by welding a readily available structural shape or plate to it along the direction of buckling. See Fig. 13–26.

Frequently the material has sufficient strength for a given application but deflects too much under the applied load. In this case, rather than to increase the thickness of the material, stiffeners can be applied to reduce deflection. See Fig. 13–27.

Subassemblies

A subassembly is a group of parts which are fastened together. Welding often simplifies the process. Subassemblies provide a convenient point for evaluating the relationship of parts. Some of the advantages of working with welded subassemblies are:

1. Inspection can be accomplished after subassembly so errors are determined before the work progresses further.

2. Testing can be performed such as checking the adequacy of chambers that are to be free of leaks.

364 *Welding Technology*

3. Operations such as heat-treating and stress relieving can be completed.

4. The parts can be spread out in the shop so it is possible to work on various subassemblies simultaneously, thereby reducing the elapsed time for the entire job.

5. Other operations such as machining, plating, and painting can be completed before assembly into the final product.

6. Simplifies the tooling required.

Assembly

The final grouping of parts that are fastened together is known as an assembly. One important way of joining these parts is by welding. In order to weld an assembly properly good planning is necessary. The following items need to be considered in such planning:

1. Determine if a jig would be of value in holding parts and subassemblies while welding.

2. If preheat is required of certain areas be sure these areas are accessible.

3. Make provision for thoroughly cleaning the parts and particularly the surfaces to be welded.

4. Develop a sequence of welding such as joining the more flexible sections first so they can be straightened prior to welding them to the heavier sections.

5. Select a location for the work so the parts and subassemblies can be easily moved to the area and the completed assembly readily handled after the work is completed.

6. If some portion of the assembly is to have an enclosure, make arrangements for a quality control check prior to the enclosing operation.

REVIEW QUESTIONS

1. A 4 in. wide steel plate is to be welded with a butt joint that must withstand a tensile force of 80,000 pounds. What thickness of plate should be used if the working stress is 20,000 pounds per square inch?

2. A cylinder is made by butt welding parallel to the longitudinal axis. If the cylinder is 3 feet in diameter and made of .250 in. steel,

determine the maximum pressure that can safely be used in the cylinder. (Working stress is 20,000 psi).

3. A rectangular steel bar 2 in. by ¼ in. is to be welded to the frame of a steel machine. Side fillet welds are to be used. If the tensile force in the bar is 7,000 pounds, determine the length of fillet weld for each side. (Working stress is 13,600 psi)

4. A rectangular steel bar 4 in. by ½ in. is to carry a tensile force of 30,000 pounds. The bar is to be welded to a steel plate with an end fillet weld and side fillet welds. Find the length of each side fillet weld. (Working stress is 13,600 psi)

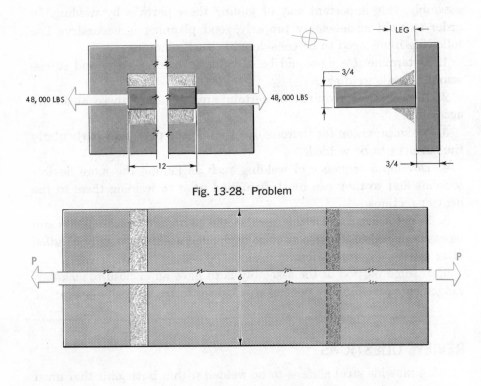

48,000 LBS

48,000 LBS

LEG

3/4

3/4

12

Fig. 13-28. Problem

P

P

6

5/8

Fig. 13-29. Problem

5. Calculate the size of the leg of the weld necessary in Fig. 13–28 in order for the joint to carry a load of 48,000 pounds. (Plates are made of steel.)
6. Determine the allowable load which can be transmitted by the connection in Fig. 13–29.

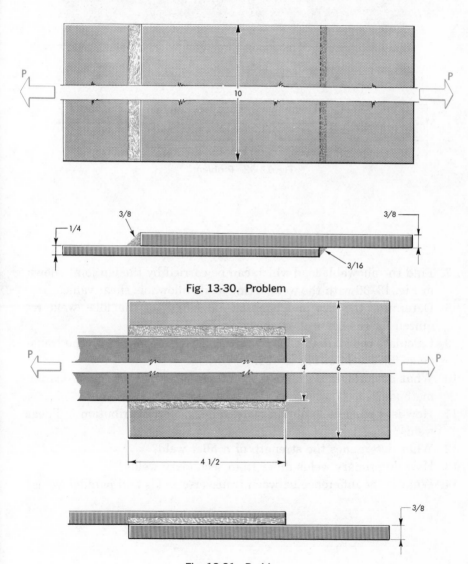

Fig. 13-30. Problem

Fig. 13-31. Problem

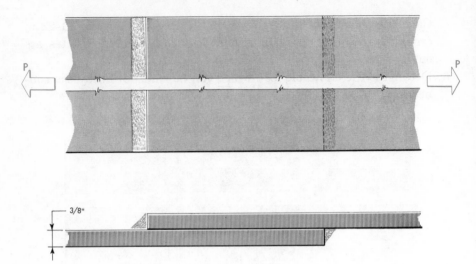

Fig. 13-32. Problem

7. Find the allowable load which can be carried by the lap joint shown in Fig. 13–30 with the weld acting at its allowable shear value.
8. Determine the size and longitudinal length of the fillet weld required for the connection shown in Fig. 13–31.
9. Calculate the allowable load which can be transmitted by the transverse fillet weld in Fig. 13–32.
10. What advantages do welded structures offer over other fabrication methods?
11. How is it possible to obtain more uniform load distribution in linear welds?
12. What determines the strength of a fillet weld?
13. How do primary welds differ from secondary welds?
14. What is the difference between transverse welds and parallel welds?

Chapter 14 | # Testing Welds

The quality and serviceability of a product often depends on the soundness of its welded sections. The effectiveness of a weld can be controlled only by efficient inspection and testing methods. The types of tests used for this purpose are governed to a great extent by the requirements of the finished product.

For some work a visual examination of a weld may be sufficient. In such cases the check consists of looking at the underside of the weld to see if penetration is complete and then noting the size of the weld bead. Weld gages are often used to determine if a weld bead has excessive or insufficient reinforcement. The limitation of this kind of examination is that there is no way of knowing if internal defects exist in the welded area. The outer appearance of the weld may be satisfactory, yet porosity, cracking, lack of fusion, or excessive grain growth may be present which are not externally apparent. Hence, the general practice is to subject welds to one or more mechanical tests. These tests are classified into two broad categories—destructive or nondestructive.

DESTRUCTIVE TESTING

Destructive testing involves subjecting weld specimens to loads until they fail. The common destructive tests include tensile testing, impact testing, fatigue testing, shear strength testing, and ductility testing.

Tensile Testing

To determine the tensile strength of an all weld metal area a specimen is used with dimensions corresponding to those shown in Fig. 14–1.

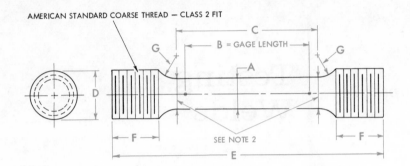

AMERICAN STANDARD COARSE THREAD — CLASS 2 FIT

SPECIMEN	DIMENSIONS OF SPECIMEN						
	A INCHES	B INCHES	C INCHES	D INCHES	E INCHES	F INCHES	G INCHES MIN
C-1	0.500 ± 0.01	2	2 1/4	3/4	4 1/4	3/4	3/8
C-2	0.437 ± 0.01	1 3/4	2	5/8	4	3/4	3/8
C-3	0.357 ± 0.007	1.4	1 3/4	1/2	3 1/2	5/8	3/8
C-4	0.252 ± 0.005	1.0	1 1/4	3/8	2 1/2	1/2	1/4
C-5	0.126 ± 0.003	0.5	3/4	1/4	1 3/4	3/8	1/8

NOTE 1:   DIMENSION A, B AND C SHALL BE AS SHOWN, BUT ALTERNATE SHAPES OF
ENDS MAY BE USED AS ALLOWED BY ASTM SPECIFICATION E-8.

NOTE 2:   IT IS DESIRABLE TO HAVE THE DIAMETER OF THE SPECIMEN WITHIN THE GAGE LENGTH SLIGHTLY SMALLER
AT THE CENTER THAN AT THE ENDS. THE DIFFERENCE SHALL NOT EXCEED 1 PERCENT OF THE DIAMETER.

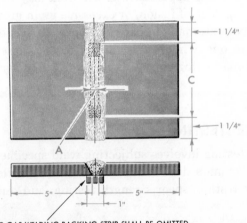

FOR GAS WELDING BACKING STRIP SHALL BE OMITTED.
DOTTED LINES SHOW POSITION FROM WHICH SPECIMEN
SHALL BE MACHINED.

Fig. 14-1. Tensile specimen for all weld metal.

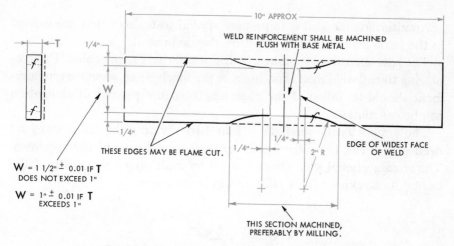

WELD REINFORCEMENT SHALL BE MACHINED
FLUSH WITH BASE METAL

10" APPROX

THESE EDGES MAY BE FLAME CUT.

EDGE OF WIDEST FACE
OF WELD

W = 1 1/2" ± 0.01 IF **T**
DOES NOT EXCEED 1"

W = 1" ± 0.01 IF **T**
EXCEEDS 1"

THIS SECTION MACHINED,
PREFERABLY BY MILLING.

Fig. 14-2. Tensile specimen for flat plate butt weld.

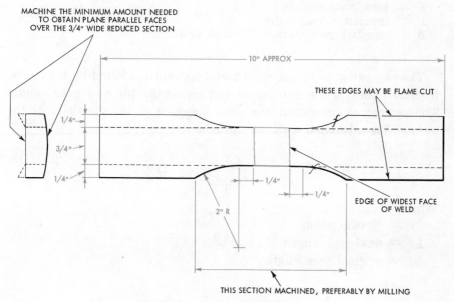

MACHINE THE MINIMUM AMOUNT NEEDED
TO OBTAIN PLANE PARALLEL FACES
OVER THE 3/4" WIDE REDUCED SECTION

10" APPROX

THESE EDGES MAY BE FLAME CUT

EDGE OF WIDEST FACE
OF WELD

THIS SECTION MACHINED, PREFERABLY BY MILLING

Fig. 14-3. Tensile specimen for pipe butt weld.

This specimen should be cut from the welded section so its reduced area, shown as H in Fig. 14–1, contains only weld metal.

Specimens used to determine the tensile strength of a welded butt joint for plate and pipe are shown in Figs. 14–2 and 14–3.

An alternate method is to prepare special test pieces that are welded in the same manner as the structure they represent.

The specimen is now broken in the tensile testing machine. Prior to placing the all weld metal specimen in the machine an accurate measurement should be taken of the gage length so the percent of elongation can be calculated.

The actual tensile strength is found by dividing the maximum load needed to break the piece by the cross-sectional area of the specimen. The cross-sectional area is determined by multiplying the width of the bar by its thickness. This relationship is shown as:

$$ S = \frac{P}{A} $$

Where
S = tensile strength in psi
P = maximum load in lbs.
A = original cross-sectional area in sq in.

The elongation of the all weld metal specimen is found by fitting the fractured ends of the two pieces and measuring the new gage length. The percent of elongation is a good indicator of the plasticity of the weld and is calculated with the formula:

$$ El = \frac{L_f - L_o}{L_o} \times 100 $$

Where
El = % elongation
L_f = final gage length
L_o = original gage length

Shearing Strength

Shearing strength is applicable to either transverse or longitudinal welds. To check the shearing strength of transverse welds, a specimen is prepared as shown in Fig. 14–4. Specimen A is used where only comparative values of linear strength are needed. When absolute values of strength are required specimen B is used.

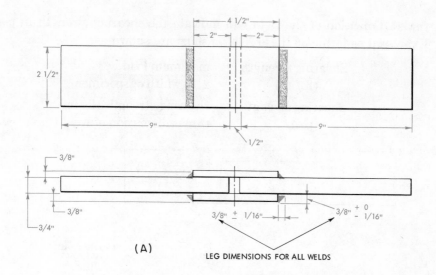

(A)

LEG DIMENSIONS FOR ALL WELDS

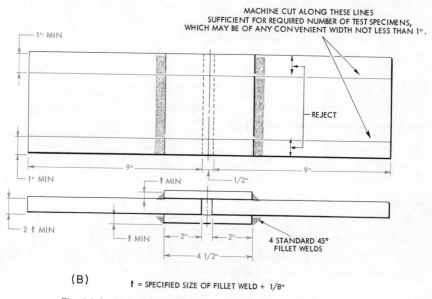

(B)   † = SPECIFIED SIZE OF FILLET WELD + 1/8"

Fig. 14-4. Shearing strength specimens of transverse welds.

The shear strength is found by loading the specimen in tension until it breaks. Dividing the maximum load in pounds by twice the width of the specimen will produce the shearing strength in pounds per linear inch. Dividing the shearing strength in pounds per linear inch by the throat dimension of the weld will indicate the shearing strength in psi. Expressed as formulas these relationships are shown as:

$$\text{Shearing strength (lb/in.)} = \frac{\text{maximum load}}{2 \times \text{width of specimen}}$$

$$\text{Shearing strength (psi)} = \frac{\text{shearing strength (lb/in.)}}{\text{throat dimension of weld}}$$

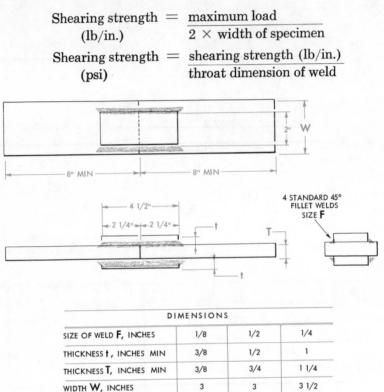

DIMENSIONS

SIZE OF WELD F, INCHES	1/8	1/2	1/4
THICKNESS t, INCHES MIN	3/8	1/2	1
THICKNESS T, INCHES MIN	3/8	3/4	1 1/4
WIDTH W, INCHES	3	3	3 1/2

LONGITUDINAL FILLET-WELD SHEARING SPECIMEN AFTER WELDING

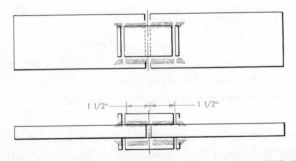

LONGITUDINAL FILLET-WELD SHEARING SPECIMEN AFTER MACHINING

Fig. 14-5. Shearing strength specimen of longitudinal welds.

To determine the shearing strength of longitudinal welds a weld specimen as shown in Fig. 14–5 is required. After the specimen is prepared the length of each weld is measured. The piece is then fractured in a tensile testing machine. The shearing strength in pounds per linear inch is found by dividing the maximum load by the length of the ruptured welds.

Weld Uniformity Test

The nick-break test, the free-bend test, and the guided-bend test are the reliable and convenient methods for determining the soundness or uniformity and ductility of a weld.

Nick-break test. The specimens shown in Fig. 14–6 are used to conduct a nick-break test. The top specimen is designed to test welded butt joints in plates and the center specimen is used to test welded butt joints in pipe.

The prepared specimen is placed on supporting members as illustrated in Fig. 14–6 and a load applied until the piece breaks. The surface of the fracture is then examined for porosity, gas pockets, slag inclusions, overlap, penetration, and grain size. For an accurate check of the soundness of the weld the fractured pieces should be subjected to an etch test. See section on etching test.

Free-bend test. This test is particularly valuable to ascertain the ductility of a weld. A test piece is cut from the plate to include the weld as shown in Fig. 14–7. The top of the weld is ground or machined so it is flush with the base metal surface. The scratches produced by grinding should run across the weld in the direction of the bend as indicated in Fig. 14–8. If the scratches extend along the weld they might cause premature failure and give incorrect results, because they act as stress raisers.

The distance across the weld is layed out as in Fig. 14–8 and lightly marked with a prick punch. The piece is bent by hammering in the vise with the face containing the gage lines on the outside of the bend as in Fig. 14–9A or by imposing a load as in Fig. 14–9B. After the specimen is given a permanent set the final bend is made as shown in Fig. 14–9C. When the bend is completed the distance between the gage lines is measured. Dividing the elongation by the initial gage length and multiplying by 100 will give the percent of elongation.

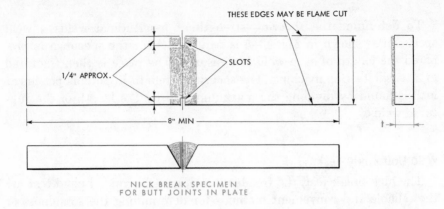

THESE EDGES MAY BE FLAME CUT

SLOTS

1/4" APPROX.

8" MIN

t

NICK BREAK SPECIMEN
FOR BUTT JOINTS IN PLATE

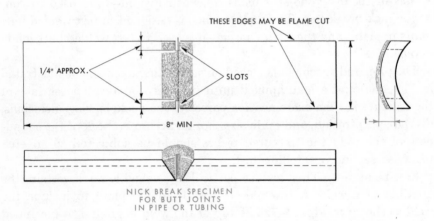

THESE EDGES MAY BE FLAME CUT

1/4" APPROX.

SLOTS

8" MIN

t

NICK BREAK SPECIMEN
FOR BUTT JOINTS
IN PIPE OR TUBING

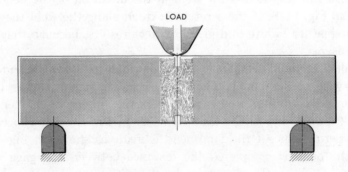

LOAD

METHOD OF RUPTERING NICK BREAK SPECIMENS

Fig. 14-6. Specimens for nick-break test.

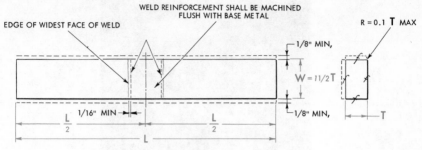

IF FLAME CUT
NOT LESS THAN 1/8"
SHALL BE MACHINED FROM EDGES

WELD REINFORCEMENT SHALL BE MACHINED
FLUSH WITH BASE METAL

EDGE OF WIDEST FACE OF WELD

R = 0.1 T MAX

1/8" MIN,

$W = 1 1/2 T$

1/16" MIN

1/8" MIN,

T

T, INCHES	1/4	3/8	1/2	5/8	3/4	1	1 1/4	1 1/2	2	2 1/2
W, INCHES	3/8	9/16	3/4	15/16	1 1/8	1 1/2	1 7/8	2 1/4	3	3 3/4
L MIN, INCHES	6	8	9	10	11	12	13 1/2	15	18	21
B*MIN, INCHES	1 1/4	1 1/4	1 1/4	2	2	2	2	2	2	3

* SEE FIG 14-9

NOTE — THE LENGTH L IS SUGGESTIVE ONLY, NOT MANDATORY

Fig. 14-7. Specimen for a free-bend test.

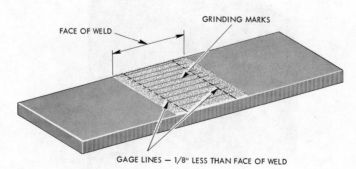

FACE OF WELD

GRINDING MARKS

GAGE LINES — 1/8" LESS THAN FACE OF WELD

Fig. 14-8. How gage lines are located on the weld face of a free-bend specimen.

(A)

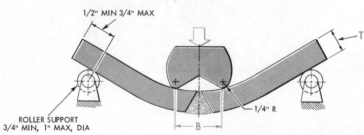

1/2" MIN 3/4" MAX

T

ROLLER SUPPORT
3/4" MIN, 1" MAX, DIA

1/4" R

FOR DIMENSION B SEE FIG. 14 – 7

B

HARDENED AND GREASED SHOULDER OF SAME SHAPE
MAY BE SUBSTITUTED FOR ROLLER SUPPORT

(B)

(C)

Fig. 14-9. Performing a free-bend test.

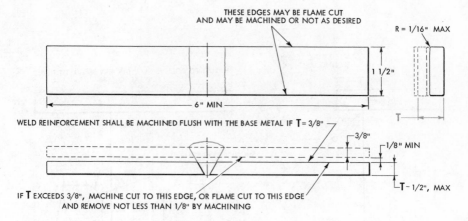

THESE EDGES MAY BE FLAME CUT
AND MAY BE MACHINED OR NOT AS DESIRED

R = 1/16" MAX

1 1/2"

6" MIN

T

WELD REINFORCEMENT SHALL BE MACHINED FLUSH WITH THE BASE METAL IF T = 3/8"

3/8"

1/8" MIN

T - 1/2", MAX

IF T EXCEEDS 3/8", MACHINE CUT TO THIS EDGE, OR FLAME CUT TO THIS EDGE
AND REMOVE NOT LESS THAN 1/8" BY MACHINING

ROOT-BEND SPECIMEN F-1

NOTE
THE FACE BEND AND ROOT BEND TEST ARE NOT APPLICABLE IF THE PLATE THICKNESS T IS LESS THAN 3/8"

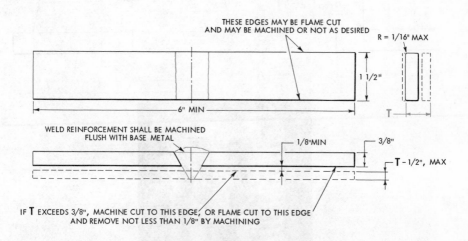

THESE EDGES MAY BE FLAME CUT
AND MAY BE MACHINED OR NOT AS DESIRED

R = 1/16" MAX

1 1/2"

6" MIN

T

WELD REINFORCEMENT SHALL BE MACHINED
FLUSH WITH BASE METAL

1/8"MIN

3/8"

T - 1/2", MAX

IF T EXCEEDS 3/8", MACHINE CUT TO THIS EDGE, OR FLAME CUT TO THIS EDGE
AND REMOVE NOT LESS THAN 1/8" BY MACHINING

Fig. 14-10. Specimen for a guided-bend test.

Guided-bend Test. For this test two specimens are required as illustrated in Fig. 14–10. One piece referred to as the *face-bend specimen* is used to check the quality of fusion; that is, whether the weld is free of defects such as porosity, inclusions, etc. The second piece referred to as the *root-bend specimen* is used to check the degree of weld penetration.

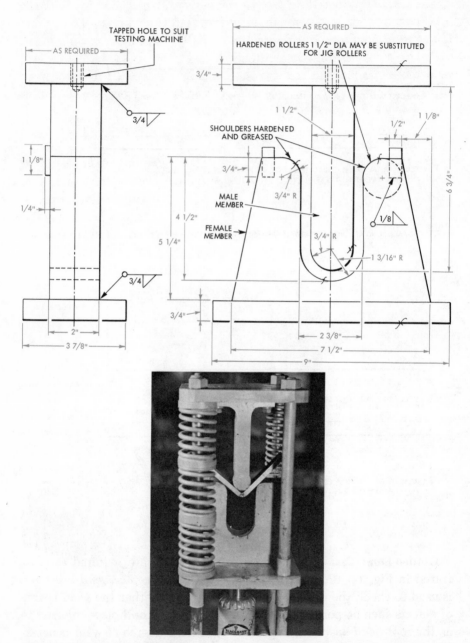

Fig. 14-11. The above view shows a detail working drawing of a jig for a guided-bend test. The view at the bottom shows the jig being used to perform the test.

The test is conducted by placing the face-bend specimen in a guided-bend jig face down and bending until it forms a U-shape. See Fig. 14–11. If upon examination, cracks greater than ⅛" appear in any direction the weld is considered to have failed.

In the root-bend test the specimen is placed in the jig with the root down or in just the reverse position of the face-bend piece. The results must also show no cracks to be acceptable.

Fillet Weld Test

To determine the soundness of a fillet weld a test specimen should be prepared as in Fig. 14–12A. Then a load is applied on the point, Fig. 14–12B, until a rupture occurs. The force may be applied by a press, tensile testing machine, or hammer blows. In addition to checking the fracture weld for soundness this specimen should be subjected to an etching test so the transverse section of the weld can be checked.

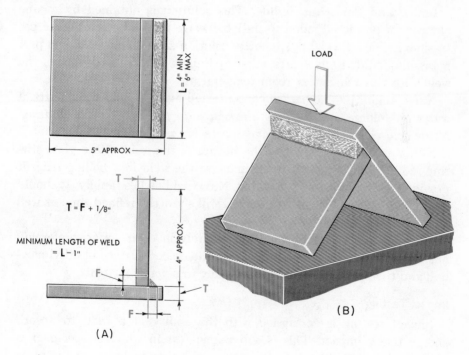

Fig. 14-12. Testing the strength of fillet welds.

Etching Test

The etching test is used to determine the soundness of a weld and to make visible the boundary between the weld metal and the base metal.

To make a test a specimen is cut from the welded joint so it displays a complete cross section of the weld. The piece may be cut either by sawing or flame cutting. The face of the cut is filed and polished with grade 00 abrasive cloth. Then the specimen is placed in one of the following etching solutions:

Hydrochloric Acid. This solution should contain equal parts by volume of concentrated hydrochloric (muriatic) acid and water. Immerse the weld in this reagent at or near the boiling point. Hydrochloric acid will etch satisfactorily on unpolished surfaces. It will usually enlarge gas pockets and dissolve slag inclusions, enlarging the resulting cavities.

Ammonium Persulphate. Mix one part of ammonium persulphate (solid) to nine parts of water by weight. Vigorously rub the surface of the weld with cotton saturated with this reagent at room temperaure.

Iodine and Potassium Iodide. This solution is obtained by mixing one part of powdered iodine (solid) to twelve parts of a solution of potassium iodide by weight. The latter solution should consist of one part potassium iodide to five parts water by weight. Brush the surface of the weld with this reagent at room temperature.

Nitric Acid. Mix one part of concentrated nitric acid to three parts of water by volume. *Always pour the acid into the water when diluting. Nitric acid causes bad stains and severe burns.*

Either apply this reagent to the surface of the weld with a glass stirring rod at room temperature or immerse the weld in a boiling reagent provided the room is well ventilated. Nitric acid etches rapidly. It should be used on polished surfaces only and will show the refined zone as well as the metal zone.

After etching wash the weld immediately in clear water, preferably hot; remove the excess water; immerse the etched surface in ethyl alcohol; and then remove and dry, preferably in a warm air blast.

Impact Testing

Impact testing is concerned with the ability of a weld to absorb energy under impact. This is a dynamic test in which a specimen is broken by a single blow and the energy absorbed in breaking the piece

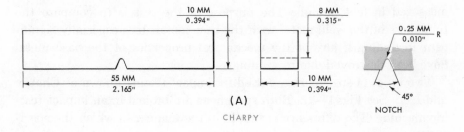

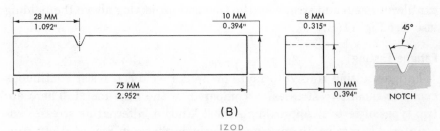

Fig. 14-13. Specimens for impact testing.

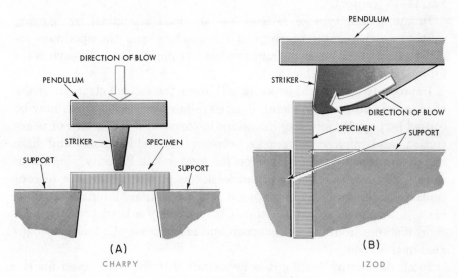

Fig. 14-14. Performing impact tests.

measured in foot pounds. The purpose of the test is to compare the toughness of the weld metal with the base metal. It is especially significant in finding if any of the mechanical properties of the base metal have been destroyed due to welding.

Two types of specimens are used for impact testing known as Charpy and Izod. See Fig. 14–13. Both specimens are broken in an impact testing machine. The difference is simply in the manner in which the specimens are anchored. The Charpy piece is supported horizontally between two anvils and the pendulum allowed to strike opposite the notch as shown in Fig. 14–14(A). The Izod specimen is supported as a vertical cantilever beam and struck on the free end projecting above the holding vise. See Fig. 14(B).

Fatigue Testing

Fatigue testing is used to find how well a weld can resist continuous cyclic or repetitive stresses as compared to the base metal. There are two types of tests to approximate most kinds of alternating stresses encountered in service. In one type a specimen shown in Fig. 14–15 (top) is bent back and forth; that is, subjected to alternating tension and compression in a regular fatigue testing machine. Force is applied by a rotating action which moves the specimen by a vertical vibratory platen. See Fig. 14–15 (center).

In the second type of fatigue test the load is applied by hanging weights on the center bearings of the machine and the specimen rotated so the stress changes from tension to compression with each revolution. See Fig. 14–15 (bottom).

Improperly made weld deposits will lower the fatigue strength of the weld metal and the base metal. Fatigue failures in welded joints may be caused by porosity and slag inclusions in the weld deposit, lack of penetration, and microscopic cracks between the weld deposit and base metal, and the effects of heat upon the base metal. Porosity, slag inclusions, lack of penetration, and cracks act as points for stresses to concentrate and failure may be initiated at these points. Materials susceptible to hardening by heat treatment may develop a hard brittle zone a short distance from the weld deposit and fail because of a lack of toughness in this zone.

In all destructive testing it is important that the welds used for the tests be as representative as possible.

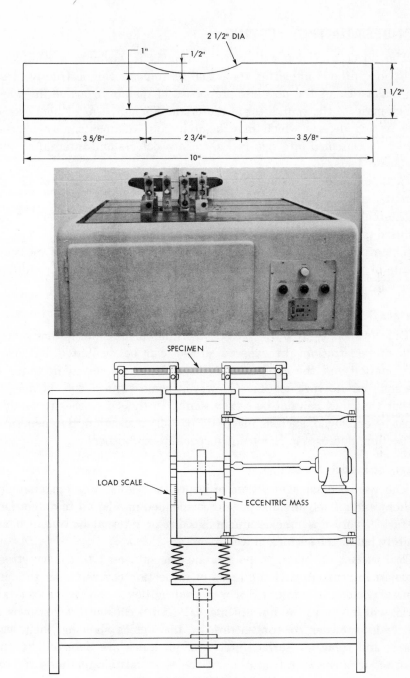

Fig. 14-15. Fatigue testing.

NONDESTRUCTIVE TESTING

The purpose of nondestructive testing is to evaluate a completed structure without impairing its actual usefulness. Nondestructive testing makes use of the physical properties of metals to arrive at values which indicate the soundness of materials. For purposes of nondestructive testing defects which may be found in weldments may be conveniently classified into two types, surface defects and internal or subsurface defects.

Common types of surface defects are, lack of penetration, cracks, undercuts, and arc burn.

Internal or subsurface defects include, porosity, slag inclusions, and unfused metal in the interior of the weld.

These defects are located and determined on the basis of measurements of physical properties of the metal such as: surface condition, magnetic permeability, electrical properties, and density.

Surface Defects

Defects such as arc burns, large cracks, undercuts, and in many cases lack of penetration (in exposed areas) can be evaluated by visual examination and the condition of the weldment assessed in terms of the appropriate specifications. However, some surface defects may be so fine that they can not be seen visually. In this case other techniques must be employed, such as magnetic particle inspection, dye penetrant inspection, fluorescent penetrant inspection, and others.

Magnetic Particle Inspection

The magnetic particle method of inspection uses a strong magnetizing current and a finely divided powder suspended in a liquid to detect lack of fusion, very fine cracks, and inclusions or internal flaws which are slightly below the surface in weldments.

In this test the piece to be examined is subjected to a very strong magnetizing current and the areas of inspection covered with the suspended powder. Any impurities or discontinuities in the magnetized material will interrupt the lines of magnetic force causing the particles of suspended powder to concentrate at the defect showing their size, shape, and location. Surface cracks of all kinds are detected by this method. It is the oldest and one of the most reliable methods of nondestructive testing. See Fig. 14–16.

Fig. 14-16. Magnetic particle inspection.

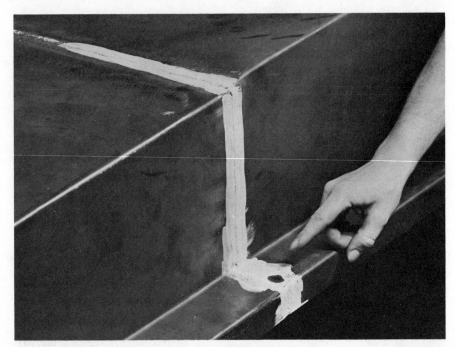

Fig. 14-17. Dye penetrant inspection.

Dye Penetrant Inspection

In dye penetrant inspection surface defects are found by the use of proprietary dyes suspended in liquids which have high fluidity. These liquids have good wetability and are readily drawn into all surface defects by capillary action. Application of a suitable developer brings out the dye and outlining the defect.

In this test the surface of the weldment, which must be clean and dry, is coated with a thin film of the penetrant. After allowing a small amount of time for the penetrant to flow into the defects the part is wiped clean. Only the penetrant in the defects remains. An absorbent material, called a developer, is put on the weldment and allowed to remain until the liquid from the imperfection flows into the developer. The dye now clearly outlines the defects.

Some of the penetrants used contain a fluorescent dye. The method of applying and developing are the same as for the previously mentioned dye penetrants, however the fluorescent penetrant must be viewed under ultraviolet light, commonly referred to as "black light." This light causes the penetrants to fluoresce to a yellow-green color which is more clearly defined than regular dye penetrants. The dye penetrant methods are particularly useful for bringing out defects in nonferrous materials such as aluminum. These materials are nonmagnetic so magnetic particle tests can not be used on them. See Fig. 14–17.

Fig. 14-18. Eddy current testing.

Eddy Current Testing

Eddy current testing uses electromagnetic energy to detect discontinuities in weld deposits and is effective in testing both ferrous and nonferrous materials for porosity, slag inclusions, internal cracks, external cracks, and lack of fusion.

Whenever a coil carrying a high frequency alternating current is brought close to a metal it produces a current in the metal by induction. The induced current is called an eddy current.

The part to be tested is subjected to electromagnetic energy by being placed in or near high frequency alternating current coils. Discontinuities in the weld deposit change the impedance of the coil. The change in impedance is indicated on electronic measuring instruments, and the size of the defect is shown by the amount of this change. See Fig. 14–18.

Radiographic Inspection

Radiographic inspection is a method of determining the soundness of a weldment by means of rays which are capable of penetrating through the entire weldment. X-Rays and Gamma Rays are electromagnetic waves used to penetrate opaque materials and a permanent record of the internal structure is obtained by placing a sensitized film in direct contact with the back of the weldment. When these rays pass through a weldment of uniform thickness and structure they impinge upon the sensitized film and produce a negative of uniform density. If the weldment contains gas pockets, slag inclusions, cracks, or lacks penetration, more rays will pass through the less dense areas and will register on the film as dark areas, clearly outlining the defects, and showing their size, shape and location.

X-Rays are produced by electrons traveling at high speed which are suddenly stopped by impact with a tungsten anode. Gamma Rays are given off when radium or its salts decompose. Gamma Rays are of shorter wave length than X-Rays. See Fig. 14–19.

Ultrasonic Testing

In ultrasonic testing high frequency vibrations or waves are used to locate and measure defects in both ferrous and nonferrous materials. This method is very sensitive, and is capable of locating very fine surface and subsurface cracks, as well as other internal defects. All types of

Fig. 14-19. Radiographic inspection.

joints can be evaluated and the exact size and location of defects measured.

Ultrasonic testing utilizes high frequency vibratory impulses to ascertain the soundness of a weld. If a high frequency vibration is sent through a sound piece of metal a signal will travel through it to the other side of the metal and be reflected back and shown on a calibrated screen of an oscilloscope. Discontinuities interrupt the signal and reflect it back sooner than the signal of the sound material. This interruption is shown

Fig. 14-20. Ultrasonic testing.

as a shorter line on the oscilloscope screen and indicates the depth of the defect. Only one side of the weldment needs to be exposed for testing purposes. See Fig. 14–20.

Hardness Testing

Hardness testing is often used in preference to the more expensive tensile testing since comparable results are obtained.

Hardness tests are effective in determining the relative hardness of the weld area as compared with the base metal. This hardness is indicated by values obtained from various hardness testing machines. Hardness numbers represent the resistance offered by the metal to the penetration of an indentor. The standard hardness machines are known as Brinell and Rockwell. See 14–21.

In the Brinell test a 10 mm diameter ball is forced into the surface of a metal by a load of 3000 kg. The load must remain on the specimen 15 seconds for ferrous materials and 30 seconds for nonferrous materials. Sufficient time is required for adequate plastic flow of the material being tested otherwise the readings will be in error. Brinell hardness numbers are calculated by dividing the applied load by the area of the

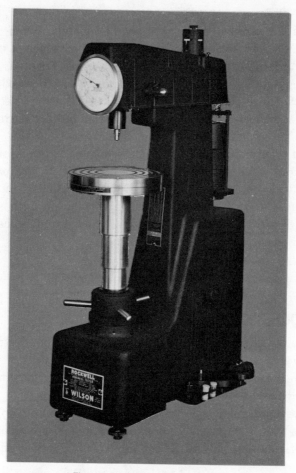

Fig. 14-21. Hardness testing.

surface indentation. The diameter of the indentation is read from a calibrated microscope and this number is then converted to a Brinell hardness number from a chart.

The Rockwell hardness tester employs a variety of loads and indentors consequently different scales can be used. These scales are designated by letters. For example, R_c 60 represents a Rockwell scale with a diamond penetrator and a 150 kg load. Since hardness numbers give relative or comparative hardness values of materials the scale must always be specified.

Corrosion Testing

Corrosion is a problem associated with all metals. When a metal has to perform in a corrosive environment the rate at which it corrodes determines to a large extent the choice of materials selected.

In welding a metal which has been alloyed and processed to produce corrosion resistance great care must be taken to insure that a weld deposit is equal or better than the base material and the corrosion resistance of the parent metal is maintained.

Corrosion tests simulating service environments are performed in the laboratory and weld metal deposits are compared to the parent metal under identical corrosive conditions. Data accumulated in the laboratory are carefully evaluated to establish correlation with service conditions. Any defect of weld material will show a difference in rate of corrosion when compared to the parent metal.

REVIEW QUESTIONS

1. What is the difference between destructive and nondestructive testing?
2. What is the tensile strength of a weld having a maximum load of 78,000 lbs and a cross-sectional area of .125"?
3. Of what significance is the result obtained from an elongation test?
4. What is the percentage of elongation of a weld specimen having 2" as its original length and 2.785" as its final gage length after fracturing?
5. How is the shear strength of a weld determined?
6. What is the shearing strength in pounds per linear inch of a weld in which the specimen is 2½" wide and subjected to a maximum load of 55,000 lbs?
7. In what ways can the uniformity of a weld be ascertained?
8. How does a free-bend test differ from a guided-bend test?
9. How is the strength of a fillet weld determined?
10. Why are weld specimens often subjected to etching tests?
11. Of what value are impact tests?
12. How are fatigue tests conducted?
13. What is the basic principle of the magnetic particle inspection test?

14. How is a dye penetrant used in determining the soundness of a weld area?
15. Why are magnetic types of tests impractical for checking aluminum welds?
16. What is the principle of eddy current testing?
17. How will radiographic inspection show hidden defects in a weld?
18. What is the principle of ultrasonic testing?
19. Of what value are hardness tests?
20. How does a Brinell tester differ from a Rockwell tester?
21. How are corrosion tests conducted?

Production Economy
Chapter 15 **and Cost Estimating**

There is no single and magical formula which can be used to determine accurate welding costs for all products. Although basic estimating principles have to be incorporated in any cost estimating project, in the final analysis realistic costs frequently depend on the sound judgment and perceptiveness of the estimator. A well conceived estimating pattern may be very effective for one job yet fail to produce actual costs for another. In essence, reliable estimates can only be made by individuals possessing broad experience, technical understanding of fabricating processes, and a reasonable knowledge of design engineering.

In the production of any welded component there are two major factors affecting cost, economy of fabrication, and cost of materials.

PRODUCTION ECONOMY

Production economy covers numerous variables such as design of weldments, positioning of welded parts, welding processes, and skill of workers. Each of these elements has a direct cost relationship and failure to recognize the significance may possibly jeopardize basic economies in final production costs.

Design

Undoubtedly the design of the welded unit is one of the most important factors in achieving production economy. For all intent and purposes the design may be excellent, but if the man in the shop cannot produce it economically, then the full value of the design is reduced. Designs often have to be reworked simply because the design engineer failed to consider actual production operations. When this happens added costs become inevitable because redesign means more man hours of work and man hours always must be evaluated in terms of costs.

To avoid excessive redesigns the first requisite in any manufacturing function is to make certain that a close relationship exists between design and manufacturing. The welding engineer or welding supervisor should work closely with the design engineer during the development stages of a new product. Frequent consultation between the two groups will reduce to a minimum work processes that cannot be performed efficiently or effectively. Because of his knowledge and experience in welding techniques the welding engineer can forestall any design feature which would prove impractical in the mass production of the product. Very often a design engineer lacks extensive manufacturing experience and what to him may be entirely feasible on paper may turn out to be unacceptable from the standpoint of manufacturing.

It is also conceivable that the viewpoints of the welding engineer may not always be compatible with the design function of the unit. In other words, the recommendations of the welding engineer could alter the configuration of the design to such an extent that the service function of the product is seriously impaired. Good design and planning require compromises and changes achieved through compromise usually result in a better product.

Positioning of Weldments

A fundamental practice which should be consistently followed in all production welding is the positioning of the work so all welding is performed in a flat or nearly flat position. This may not always be possible but if out-of-position welding can be kept to a minimum production rate generally is increased. Deposition rate in a vertical, horizontal, or overhead position is frequently slower and is considerably more fatiguing to the operator than welding in a downhand or flat position. With greater fatigue there is a decrease in operator efficiency in the quantity of work produced and in the soundness of completed weldments. Invariably, more weld defects will result when welding is done in other than flat positions. A further limitation of out-of-positioning welding is the need for greater operator skill. Unless the available welders are fully qualified to make horizontal, vertical, or overhead welds, they will have to be retrained if welding is to be done in these positions. In terms of cost and time it may often be more advantageous to invest in elaborate positioning fixtures to make possible flat position welding than resorting to a retraining program.

Jigs and Fixtures. The use of jigs and fixtures in production welding is a must requirement. Jigs and fixtures are intended to (1) achieve more effective welding positions and (2) align and hold parts in their proper relationship during the welding process. See Figs. 15–1, 15–2, and 15–3. Without holding devices parts must be assembled by hand and tacked in place. Such procedures are time consuming and often produce errors.

Jigs and fixtures should be relatively simple in design so the operator can assemble the components quickly and accurately and at the same time remove the welded assembly speedily without undue effort. There is no basic jig or fixture that can be used for all welding purposes. The great variation in welded products necessitates holding devices designed to meet specific situations. However this does not rule out the usage of standard parts in the construction of a jig or fixture. Units such as

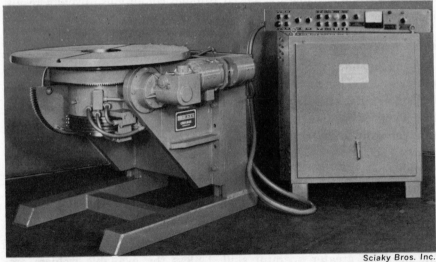

Sciaky Bros. Inc.

Fig. 15-1. A typical fixture which permits the rotation of a weld in a variety of positions.

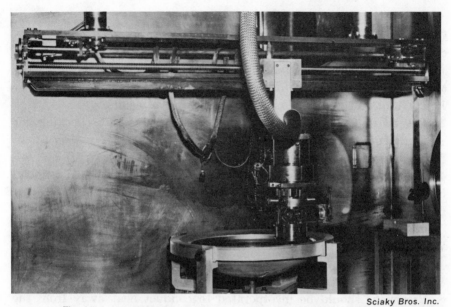

Fig. 15-2. A simple fixture on a turntable holds parts for welding.

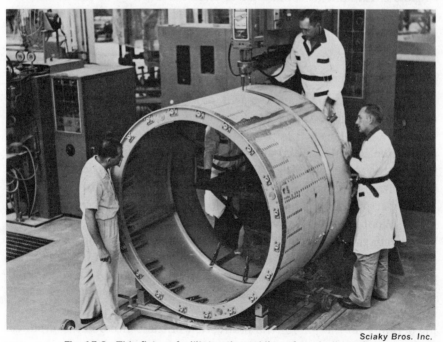

Fig. 15-3. This fixture facilitates the welding of a missile part.

C-clamps, pins, saddles and wedges, jack screws, etc., can often be incorporated in the basic jig or fixture design.

In the planning of any holding device the following requirements should be kept in mind:

1. The unit must be simple and produce positive results.
2. The parts must be held securely without frequent checking for alignment.
3. No measurement of part relationship should be necessary.
4. The clamping and releasing of parts must be accomplished quickly.
5. All alignment points should be readily accessible and visible.
6. If the product contains premachined parts these areas should be protected from welding splatter.
7. If welding is to be done on several surfaces the holding fixture should be mounted on a trunnion frame or spinner so it can be rotated to bring all welded joints in a flat position.
8. Provisions should be incorporated to conduct heat away from the welded joint. This is particularly significant for certain types of metals. Rapid heat conduction is often achieved by means of massive structural jig or fixture members, copper blocks, or by using hollow holding units that permit circulation of water.

Welding Process

Somewhere during the stages of product planning the question of what welding process to be used will have to be answered. See Table I. What, for example, will be the most practical and economical for mass production, shielded metal-arc, gas shielded-arc, submerged arc or some form of resistance welding? The values of semi-automatic versus fully mechanized equipment also will require careful consideration.

In most production problems there is never one obvious, simple solution. What may be relatively basic in one type of production will often be entirely unsuitable in another situation. In attempting to arrive at some reasonable solution to the welding process problem several issues must be analyzed. First the available equipment has to be considered. For example, assuming that a plant is equipped with shielded metal-arc power supply units the question then is—will this type of arc welding actually do the job for the production of the contemplated product? If the answer is yes then the next problem is—will it be the most pro-

Table I.   Recommended Processes for Welding Metals and Alloys.

Welding Process	Low carbon, mild steel — types SAE 1010 and 1020	Medium carbon steel — types SAE 1030 and 1050	Low alloy steel — types SAE 2340, 3145, 4130, and 4340	Austenitic stainless steels — types AISI 301, 310, 316 and 347	Ferritic and martensitic stainless steels — types AISI 405, 410 and 430	High strength, high temperature alloys — types 17-14 CuM, 16-25-6 and 19-9 DL	Cast iron and gray iron	Aluminum and aluminum alloys	Magnesium and magnesium alloys	Copper and copper alloys	Nickel and high nickel alloys	Silver	Gold, platinum and iridium	Titanium and titanium alloys
Shielded metal-arc	R	R	R	R	R	R	S	S	NA	NR	R	NR	NR	NA
Submerged-arc	R	R	R	R	S	S	NR	NR	NA	NR	S	NR	NR	NA
Tig welding	S	S	S	R	S	S	S	R	R	R	R	R	R	R
Mig welding	S	S	S	R	S	S	NR	R	S	R	R	S	S	S
Flash welding	R	R	R	R	S	S	NR	S	NR	S	S	S	S	S
Spot welding	R	R	R	R	S	S	NA	R	S	S	R	NR	S	S
Gas welding	R	R	S	S	S	S	R	S	NR	S	S	R	R	NA
Furnace braze	R	R	S	R/S	S	NR	NR	R	NR	S	R	S	S	S
Torch brazing	S	S	NR	S	S	NR	R	R	NR	R	R	R	R	S

R — Recommended   S — Satisfactory   NR — Not recommended   NA — Not applicable

ductive and economical? After a careful examination it may be found that the purchase of new gas shielded-arc equipment will be far more economical in the long run than attempting to use the existing shielded metal-arc equipment.

The mistake is too often made in retaining existing equipment simply because someone lacks full understanding of the economies of other welding techniques or does not have sufficient knowledge of other forms of welding to suggest changes. Many shops still rely on manual stick welding for mass production when semi-automatic or fully automatic Mig welding would increase production and cut cost by ten percent or more.

Another aspect of the welding process problem is the type of metal to be used in the fabrication of the product. Thus, if the metal is to be light gage mild steel then some form of Tig or Mig welding may be the most appropriate. On the other hand if heavy steel plate is required the most suitable welding may be by submerged arc. If the welding is to be on medium or heavy plate and the appearance of the weld bead is not important the most productive process in terms of fast deposition rate could be the buried arc CO_2. Welding light gage aluminum or stainless steel will usually be more satisfactory if the Tig process is used. For some type of work the relative merits of fluxcore wire over solid wire is a very important consideration. Although fluxcore wire is somewhat more expensive, when evaluated in terms of production rates it may prove to be the most economical.

In arriving at decisions concerning the suitability of welding processes consultation with representatives of welding equipment manufacturers or distributors is generally wise and practical. Most representatives are able to supply facts and figures of one process or another and can demonstrate the advantages and limitations of these welding processes as they apply to a specific production requirement.

Once the welding process is determined the final issue for consideration is the degree of mechanization that should be incorporated. For continuous and large production runs fully automatic equipment undoubtedly is the solution. However, the configuration of the weldments may not always permit complete utilization of automatic equipment. The alternative is the semi-automatic equipment where preset condi-

Bernard Welding

Fig. 15-4. Booms holding Mig semi-automatic equipment are used here to simplify production welding.

tions are possible but individual operators are required to manipulate the gun over the weld seam. See Fig. 15–4. Despite the fact that production economy of semi-automatic equipment may be less than fully automatic it is usually much greater than regular hand operated stick welding.

The Work Force

A very significant factor in ascertaining production economy is the skill of the work force involved in performing the welding. Some types of welding require a greater degree of proficiency than others. If the available manpower lacks the required skills then suitable measures must be instituted to train them. The evaluation of this factor should take into consideration the fact that it takes longer to train shielded-metal arc operators than people to use semi-automatic gas shielded-arc equipment. Even less time generally is needed to train operators to run fully mechanized equipment. Since operator proficiency directly influences production cost the problem of training or up-grading the work force cannot be ignored. Aside from the fact that poorly trained welders are unable to meet production schedules their inefficiency too often results in improperly made welds causing innumerable product rejections and a high volume of scrap.

ESTIMATING WELDING COSTS

There is no one all-inclusive formula which can provide accurate costs for all welding jobs. The cost estimating process can be relatively simple for some parts where only a few weld seams are required and welds consist of straight, single, or multiple weld beads. As the number of weld seams, sizes, and parts increase, the more complex becomes the cost estimating process.

In general most cost estimating involves two major items: (1) cost of weld materials, and (2) operating cost.

Cost of Weld Materials

Weld material costs are contingent on the following factors:
1. Weight of weld metal.

Table II. Weight of Weld Metal Based on Joint Design. (Pounds per Foot of Joint)

BUTT WELDS — first "f" Dimension group (columns f = 1/16″, 1/8″, 3/16″, 1/4″, 3/8″, 1/2″); second "f" Dimension group (columns f = 1/16″, 1/8″, 3/16″, 1/4″); plus round‑bar column.

Included Angle columns: 14°, 20°, 60°, 45° (1/2 of 90°), 70°.

FILLET WELDS — Flat, Convex, Concave. *Values below are for leg size 10% oversize, consistent with normal shop practices.* (Leg size increased 10%.)

"d" or "r"	Butt f=1/16	f=1/8	f=3/16	f=1/4	f=3/8	f=1/2	f=1/16	f=1/8	f=3/16	f=1/4	Round	14°	20°	60°	45° (1/2 of 90°)	70°	Fillet Flat	Convex	Concave
1/16″	.027	.053	.080	.106	.159	.212	.027	.053	.080	.106	.021	.0065	.0094	.031	.027	.037	.032	.039	.037
1/8″	.040	.080	.119	.159	.239	.318	.035	.071	.106	.141	.083	.0147	.021	.069	.060	.084	.072	.087	.083
3/16″	.053	.106	.159	.212	.318	.425	.044	.088	.133	.177	.188	.026	.037	.123	.106	.149	.129	.155	.147
1/4″	.066	.133	.199	.265	.398	.531	.053	.106	.159	.212	.334	.041	.059	.192	.166	.232	.201	.242	.230
5/16″	.080	.159	.239	.318	.478	.637	.062	.124	.186	.248	.531	.059	.084	.276	.239	.334	.289	.349	.331
3/8″	.093	.186	.279	.371	.557	.743	.071	.142	.212	.283	.750	.080	.115	.376	.326	.456	.394	.475	.451
7/16″	.106	.212	.318	.425	.637	.849	.080	.159	.239	.318	1.02	.104	.150	.491	.425	.595	.514	.620	.589
1/2″	.119	.239	.358	.478	.716	.955	.089	.177	.266	.354	1.33	.132	.190	.621	.538	.753	.651	.785	.745
9/16″	.133	.265	.398	.531	.796	1.06	.097	.195	.292	.389		.163	.234	.766	.664	.930	.804	.970	.920
5/8″	.146	.292	.438	.584	.876	1.17	.106	.212	.318	.424		.197	.283	.927	.804	1.13	1.16	1.40	1.32
11/16″	.159	.318	.478	.637	.955	1.27	.115	.230	.345	.460		.234	.337	1.11	.956	1.34	1.58	1.90	1.80
3/4″	.172	.345	.517	.690	1.04	1.38	.124	.248	.372	.490		.275	.396	1.30	1.12	1.57	2.06	2.48	2.36
13/16″	.186	.371	.557	.743	1.11	1.49	.133	.266	.398	.530		.319	.459	1.50	1.30	1.82	2.60	3.14	2.98
7/8″	.199	.398	.597	.796	1.19	1.59	.142	.282	.418	.566		.367	.527	1.73	1.50	2.07	3.21	3.88	3.68
15/16″	.212	.425	.637	.849	1.27	1.70	.150	.301	.451	.602		.417	.599	1.96	1.70	2.38	3.89	4.69	4.45
1″	.226	.451	.676	.902	1.35	1.80	.159	.318	.477	.637		.471	.676	2.22	1.92	2.68	4.62	5.58	5.30
1-1/16″	.239	.478	.716	.955	1.43	1.91	.168	.336	.505	.672		.528	.758	2.48	2.15	3.02	5.43	6.55	6.22
1-1/8″	.252	.504	.756	1.01	1.51	2.02	.177	.354	.531	.706		.588	.845	2.77	2.40	3.36	6.29	7.59	7.21
1-3/16″	.265	.531	.796	1.06	1.59	2.12	.186	.372	.557	.743		.651	.936	3.07	2.66	3.72	7.23	8.72	8.28
1-1/4″	.279	.557	.836	1.11	1.67	2.23	.195	.389	.584	.777		.718	1.03	3.38	2.93	4.10	8.23	9.93	9.43
1-5/16″	.292	.584	.876	1.17	1.75	2.34	.203	.407	.610	.814		.789	1.13	3.71	3.21	4.50			
1-3/8″	.305	.610	.915	1.22	1.83	2.44	.212	.425	.636	.849		.863	1.24	4.05	3.51	4.91			
1-7/16″	.318	.637	.955	1.27	1.91	2.55	.221	.442	.664	.884		.938	1.35	4.42	3.82	5.36			
1-1/2″	.332	.664	.995	1.33	1.99	2.65	.230	.460	.690	.920		1.02	1.46	4.79	4.15	5.81			
1-9/16″	.345	.690	1.04	1.38	2.07	2.76	.239	.477	.716	.956		1.10	1.58	5.18	4.49	6.29			
1-5/8″	.358	.716	1.07	1.43	2.15	2.87	.249	.495	.743	.990		1.19	1.71	5.59	4.84	6.80			
1-11/16″	.371	.743	1.11	1.49	2.23	2.97	.257	.513	.770	1.03		1.28	1.84	6.01	5.20	7.29			
1-3/4″	.385	.769	1.15	1.54	2.31	3.08	.266	.531	.796	1.06		1.37	1.97	6.45	5.58	7.81			
1-13/16″	.398	.796	1.19	1.59	2.39	3.18	.274	.549	.823	1.10		1.47	2.10	6.90	5.97	8.36			
1-7/8″	.411	.822	1.23	1.65	2.47	3.29	.283	.566	.849	1.13		1.56	2.25	7.36	6.38	8.94			
1-15/16″	.425	.849	1.27	1.70	2.55	3.40	.292					1.67	2.40	7.85	6.80	9.52			
2″	.438						.301												

2. Cost of filler metal—stick electrode or wire.
3. Deposition efficiency.
4. Cost of gas and flux if used.

Weight of Filler Metal. To determine the cost of filler metal the area of the weld must be calculated. The cross-sectional area of a weld usually varies as the square of the weld size. Thus the greater the weld size the greater will be the cost. This is one reason why good weld design becomes so important. If a weld is made larger than is actually necessary then the additional material required to fill the joint represents wasted money. Very often a slight increase in weld may add up considerably in the total production cost.

The amount of weld metal required for a particular joint is based on the weight of deposited metal per inch of weld. This is found by computing the cross-sectional area of the joint in square inches and then finding the amount of weld required in pounds per linear feet. Manufacturers of filler weld metal usually have tables where this information is readily available. See Tables II and III. Basically the calculation consists of dividing the area to be filled into standard geometric shapes and then finding the area of these shapes. For example, assume that the joint design to be used is as shown in Fig. 15–5. The joint configuration is divided into shapes A, B, C, and D. By consulting Table II, the weight values of each shape can be determined which when totaled amounts to a net weight of 2.77 lbs of weld metal per foot of weld. If weld shapes are fairly standard the required weight of weld metal per foot is easily determined from a welding manufacturer's chart as shown in Table III. Once the weight value is found then this figure is divided by a figure which represents the *deposition efficiency* of the weld metal. Deposition efficiency represents the loss of metal in the welding process and stub discard and is expressed in percent. (See Table III, "Pounds of electrodes required per foot of weld.") For example, if the deposition efficiency is considered to be 60 percent, which is about average for shielded metal-arc welding, dividing 2.77 lbs by .60 will indicate a total of 4.6 lbs of weld metal needed per foot. In the submerged arc process the deposition efficiency of the electrode is nearly 100 percent. With CO_2 gas shielded-arc welding the spatter may approach 8 percent leaving a deposition efficiency of 92 percent. Most other gas shielded-arc processes will have close to a 100 percent deposition efficiency rate.

Table III.　Electrode Consumption.

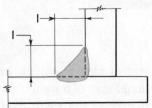

HORIZONTAL FILLET WELD

Size of Fillet l (In Inches)	POUNDS OF ELECTRODES REQUIRED PER LINEAR FOOT OF WELD* (Approx.)		STEEL DEPOSITED PER LINEAR FOOT OF WELD	
	Bare and Thinly Coated	Heavily Coated	Cubic Inches	Pounds
1/8	0.039	0.048	0.094	0.027
5/16	0.090	0.113	0.222	0.063
1/4	0.151	0.189	0.375	0.106
5/16	0.237	0.296	0.585	0.166
3/8	0.341	0.427	0.844	0.239
1/2	0.607	0.760	1.500	0.425
5/8	0.947	1.185	2.340	0.663
3/4	1.365	1.705	3.375	0.955
1	2.420	3.030	6.000	1.698

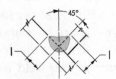

POSITIONED FILLET WELD

Size of Fillet l (In Inches)	POUNDS OF ELECTRODES REQUIRED PER LINEAR FOOT OF WELD* (Approx.)		STEEL DEPOSITED PER LINEAR FOOT OF WELD	
	Bare and Thinly Coated	Heavily Coated	Cubic Inches	Pounds
1/4	....	0.212	0.420	0.119
5/16	....	0.334	0.660	0.187
3/8	....	0.486	0.960	0.272
1/2	....	0.850	1.680	0.475
5/8	....	1.275	2.520	0.713
3/4	....	1.820	3.600	1.020
1	....	3.210	6.350	1.800

Table III. Electrode Consumption. (Continued)

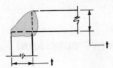

CORNER FILLET WELD

Size of Fillet t (In Inches)	POUNDS OF ELECTRODES REQUIRED PER LINEAR FOOT OF WELD* (Approx.)		STEEL DEPOSITED PER LINEAR FOOT OF WELD	
	Bare and Thinly Coated	Heavily Coated	Cubic Inches	Pounds
⅛ / ³⁄₁₆	0.06 / 0.13	0.07 / 0.16	0.144 / 0.336	0.041 / 0.095
¼ / ⁵⁄₁₆	0.24 / 0.37	0.30 / 0.46	0.588 / 0.923	0.167 / 0.261
⅜ / ½	0.53 / 0.95	0.67 / 1.19	1.335 / 2.350	0.378 / 0.665
⅝ / ¾	1.49 / 2.15	1.86 / 2.68	3.680 / 5.300	1.043 / 1.502
1	3.81	4.77	9.41	2.670

*Includes scrap end and spatter loss

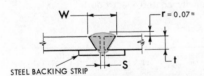

SQUARE GROOVE BUTT JOINT
With Backing Strip

JOINT DIMENSIONS (In Inches)			POUNDS OF ELECTRODES REQUIRED PER LINEAR FOOT OF WELD* (Approx.)				STEEL DEPOSITED PER LINEAR FOOT OF WELD			
			WITHOUT REINFORCEMENT		WITH REINFORCEMENT**		WITHOUT REINFORCEMENT		WITH REINFORCEMENT**	
t	w	s	Bare and Thinly Coated	Heavily Coated	Bare and Thinly Coated	Heavily Coated	Cubic Inches	Pounds	Cubic Inches	Pounds
⅛	¼	0 / ¹⁄₁₆	.... / 0.04	.... / 0.05	0.09 / 0.12	0.11 / 0.15	.... / 0.094	.... / 0.027	0.210 / 0.304	0.060 / 0.086
³⁄₁₆	⅜	¹⁄₁₆ / ³⁄₃₂	0.06 / 0.09	0.07 / 0.11	0.18 / 0.21	0.23 / 0.27	0.140 / 0.211	0.040 / 0.060	0.456 / 0.526	0.129 / 0.149
¼	⁷⁄₁₆	³⁄₃₂ / ⅛	0.12 / 0.15	0.14 / 0.19	0.26 / 0.30	0.33 / 0.38	0.282 / 0.376	0.080 / 0.107	0.649 / 0.742	0.184 / 0.210

*Includes scrap end and spatter loss
**r = Height of reinforcement.

Table III. Electrode Consumption. (Continued)

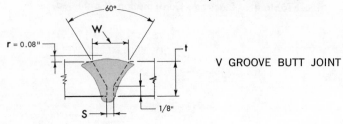

V GROOVE BUTT JOINT

JOINT DIMENSIONS (In Inches)			POUNDS OF ELECTRODES REQUIRED PER LINEAR FOOT OF WELD* (Approx.)				STEEL DEPOSITED PER LINEAR FOOT OF WELD			
			WITHOUT REINFORCEMENT		WITH REINFORCEMENT**		WITHOUT REINFORCEMENT		WITH REINFORCEMENT**	
t	w	s	Bare and Thinly Coated	Heavily Coated	Bare and Thinly Coated	Heavily Coated	Cubic Inches	Pounds	Cubic Inches	Pounds
¼	0.207	¹⁄₁₆	0.12	0.15	0.20	0.25	0.300	0.085	0.504	0.143
⁵⁄₁₆	0.311	³⁄₃₂	0.25	0.31	0.37	0.46	0.611	0.173	0.911	0.258
⅜	0.414	⅛	0.40	0.50	0.56	0.70	0.995	0.282	1.390	0.394
½	0.558	⅛	0.70	0.87	0.91	1.15	1.730	0.489	2.263	0.641
⅝	0.702	⅛	1.08	1.35	1.35	1.68	2.660	0.753	3.330	0.942
¾	0.847	⅛	1.55	1.94	1.88	2.35	3.840	1.088	4.650	1.320
1	1.138	⅛	2.76	3.45	3.20	4.00	6.810	1.930	7.90	2.240

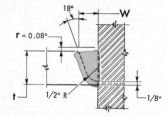

SINGLE J GROOVE

If root of weld is chipped or flame gouged and welded, add 0.19 lb to steel deposited (equal to 0.27 lb of thinly coated or 0.27 lb heavily coated electrode.)

JOINT DIMENSIONS (In Inches)		POUNDS OF ELECTRODES REQUIRED PER LINEAR FOOT OF WELD* (Approx.)				STEEL DEPOSITED PER LINEAR FOOT OF WELD			
		WITHOUT REINFORCEMENT		WITH REINFORCEMENT**		WITHOUT REINFORCEMENT		WITH REINFORCEMENT**	
t	w	Bare and Thinly Coated	Heavily Coated	Bare and Thinly Coated	Heavily Coated	Cubic Inches	Pounds	Cubic Inches	Pounds
1	0.625	....	2.55	....	2.85	5.03	1.43	5.64	1.60
1¼	0.719	....	3.64	....	4.00	7.20	2.04	7.91	2.24
1½	0.781	....	4.80	....	5.15	9.46	2.69	10.20	2.89
1¾	0.875	....	6.12	....	6.55	12.12	3.43	12.95	3.67
2	0.969	....	7.40	....	7.87	14.63	4.15	15.60	4.41
2¼	1.031	....	9.00	....	9.42	17.75	5.03	18.35	5.19
2½	1.094	....	10.60	....	11.10	20.90	5.92	21.95	6.21
2¾	1.188	....	12.30	....	12.92	24.35	6.90	25.55	7.23
3	1.281	....	14.20	....	14.80	28.10	7.95	29.30	8.29
3½	1.438	....	18.40	....	19.10	36.30	10.30	37.80	10.70
4	1.594	....	23.00	....	23.70	45.40	12.90	47.00	13.30

Table III.   Electrode Consumption. (Continued)

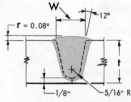

SINGLE U GROOVE BUTT JOINT

If root of weld is chipped or flame gouged and welded, add 0.19 lb to steel deposited (equal to 0.27 lb of thinly coated or 0.34 lb of heavily coated electrode.)

JOINT DIMENSIONS (In Inches)		POUNDS OF ELECTRODES REQUIRED PER LINEAR FOOT OF WELD* (Approx.)				STEEL DEPOSITED PER LINEAR FOOT OF WELD			
		WITHOUT REINFORCEMENT		WITH REINFORCEMENT**		WITHOUT REINFORCEMENT		WITH REINFORCEMENT**	
t	w	Bare and Thinly Coated	Heavily Coated	Bare and Thinly Coated	Heavily Coated	Cubic Inches	Pounds	Cubic Inches	Pounds
½	0.652	....	1.18	....	1.49	2.325	0.659	2.95	0.835
⅝	0.705	....	1.70	....	2.04	3.345	0.947	4.02	1.140
¾	0.758	....	2.24	....	2.61	4.435	1.255	5.17	1.465
1	0.865	....	3.47	....	3.89	6.870	1.945	7.70	2.180
1¼	0.971	....	4.86	....	5.35	9.62	2.72	10.60	3.00
1½	1.077	....	6.41	....	6.95	12.66	3.59	13.72	3.89
1¾	1.173	....	8.08	....	8.65	16.00	4.53	17.10	4.84
2	1.292	....	10.00	....	10.65	19.75	5.60	21.04	5.96
2¼	1.396	....	12.05	....	12.75	23.80	6.75	25.20	7.12
2½	1.502	....	14.25	....	15.00	28.20	7.98	29.65	8.40
2¾	1.608	....	16.60	....	17.40	32.80	9.29	34.65	9.73
3	1.715	....	19.10	....	20.00	37.80	10.70	39.45	11.19
3½	1.927	....	24.70	....	25.50	48.60	13.80	50.50	14.30
4	2.140	....	30.90	....	31.90	61.00	17.30	63.10	17.90

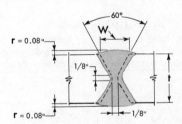

DOUBLE V GROOVE BUTT JOINT

If root of weld is chipped or flame gouged and welded, add 0.10 lb to steel deposited (equal to 0.14 of thinly coated or 0.18 lb of heavily coated electrode.)

JOINT DIMENSIONS (In Inches)		POUNDS OF ELECTRODES REQUIRED PER LINEAR FOOT OF WELD* (Approx.)				STEEL DEPOSITED PER LINEAR FOOT OF WELD			
		WITHOUT REINFORCEMENT		WITH REINFORCEMENT**		WITHOUT REINFORCEMENT		WITH REINFORCEMENT**	
t	w	Bare and Thinly Coated	Heavily Coated	Bare and Thinly Coated	Heavily Coated	Cubic Inches	Pounds	Cubic Inches	Pounds
⅝	0.405	0.72	0.90	1.03	1.29	1.775	0.502	2.56	0.724
¾	0.468	0.98	1.22	1.34	1.68	2.410	0.682	3.31	0.937
1	0.630	1.68	2.10	2.17	2.71	4.150	1.175	5.36	1.520
1¼	0.774	2.53	3.17	3.13	3.92	6.27	1.775	7.75	2.195
1½	0.919	3.56	4.45	4.28	5.35	8.85	2.495	10.59	3.00
1¾	1.063	4.77	5.95	5.58	6.98	11.80	3.335	13.82	3.91
2	1.207	6.13	7.68	7.10	8.88	15.20	4.30	17.58	4.97
2¼	1.352	7.70	9.60	8.75	10.95	19.00	5.38	21.65	6.12
2½	1.496	9.43	11.80	10.60	13.20	23.30	6.60	26.20	7.40
3	1.784	13.36	16.70	14.75	18.50	33.00	9.35	36.50	10.33
3½	2.073	18.10	22.60	19.70	24.60	44.70	12.65	48.70	13.80
4	2.368	23.50	29.40	25.40	31.70	58.15	16.45	62.80	17.80

*Includes scrap end and spatter loss
**r = Height of reinforcement.

Table III. Electrode Consumption. (Continued)

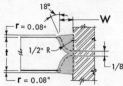

DOUBLE J GROOVE

If root of weld is chipped or flame gouged and welded, add 0.19 lb to steel deposited (equal to 0.27 of thinly coated or 0.34 lb of heavily coated rd.)

JOINT DIMENSIONS (In Inches)		POUNDS OF ELECTRODES REQUIRED PER LINEAR FOOT OF WELD* (Approx.)				STEEL DEPOSITED PER LINEAR FOOT OF WELD			
		WITHOUT REINFORCEMENT		WITH REINFORCEMENT**		WITHOUT REINFORCEMENT		WITH REINFORCEMENT**	
t	w	Bare and Thinly Coated	Heavily Coated	Bare and Thinly Coated	Heavily Coated	Cubic Inches	Pounds	Cubic Inches	Pounds
1	0.500	....	1.87	....	2.37	3.71	1.05	4.67	1.33
1¼	0.563	....	2.48	....	3.03	4.92	1.39	6.00	1.70
1½	0.594	....	3.52	....	4.08	6.95	1.97	8.10	2.29
1¾	0.625	....	4.37	....	5.00	8.635	2.45	9.83	2.79
2	0.656	....	5.47	....	6.11	10.80	3.06	12.06	3.42
2¼	0.688	....	6.55	....	7.21	12.97	3.67	14.29	4.04
2½	0.750	....	7.65	....	8.38	15.12	4.28	16.68	4.69
2¾	0.781	....	8.85	....	9.60	17.52	4.95	19.00	5.38
3	0.813	....	10.10	....	10.85	19.82	5.62	21.45	6.08
3½	0.906	....	12.70	..	13.53	25.05	7.12	26.80	7.58
4	0.969	....	15.70	....	16.60	31.05	8.78	32.80	9.28

*Includes scrap end and spatter loss
**r = Height of reinforcement.

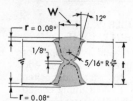

DOUBLE U GROOVE

If root of top weld is chipped or flame gouged and welded, add 0.19 lb to steel deposited (equal to 0.27 lb of thinly coated or 0.34 lb of heavily coated electrode.)

JOINT DIMENSIONS (In Inches)		POUNDS OF ELECTRODES REQUIRED PER LINEAR FOOT OF WELD* (Approx.)				STEEL DEPOSITED PER LINEAR FOOT OF WELD			
		WITHOUT REINFORCEMENT		WITH REINFORCEMENT**		WITHOUT REINFORCEMENT		WITH REINFORCEMENT**	
t	w	Bare and Thinly Coated	Heavily Coated	Bare and Thinly Coated	Heavily Coated	Cubic Inches	Pounds	Cubic Inches	Pounds
1	0.685	....	2.86	....	3.54	5.64	1.60	6.90	1.98
1¼	0.731	....	3.91	....	4.62	7.75	2.19	9.15	2.59
1½	0.784	....	5.05	....	5.83	10.00	2.83	11.55	3.27
1¾	0.838	....	6.30	....	7.12	12.47	3.53	14.10	3.99
2	0.891	....	7.60	....	8.46	15.08	4.26	16.74	4.74
2¼	0.944	....	9.00	....	9.90	17.80	5.04	19.60	5.55
2½	0.997	....	10.45	....	11.45	20.70	5.85	22.60	6.41
2¾	1.050	....	12.00	....	13.05	23.80	6.73	25.80	7.30
3	1.103	....	13.85	....	14.90	27.15	7.75	29.40	8.34
3½	1.211	....	17.20	....	18.40	33.98	9.61	36.30	10.30
4	1.316	....	21.00	....	22.30	41.55	11.75	44.00	12.50

*Includes scrap end and spatter loss
**r = Height of reinforcement.

Table III. Electrode Consumption. (Concluded)

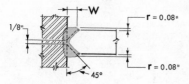

DOUBLE BEVEL GROOVE
If root of top weld is chipped or flame gouged and welded, add 0.19 lb to steel deposited (equal to 0.27 of thinly coated or 0.34 lb of heavily coated electrode.)

JOINT DIMENSIONS (In Inches)		POUNDS OF ELECTRODES REQUIRED PER LINEAR FOOT OF WELD* (Approx.)				STEEL DEPOSITED PER LINEAR FOOT OF WELD			
		WITHOUT REINFORCEMENT		WITH REINFORCEMENT**		WITHOUT REINFORCEMENT		WITH REINFORCEMENT**	
t	w	Bare and Thinly Coated	Heavily Coated	Bare and Thinly Coated	Heavily Coated	Cubic Inches	Pounds	Cubic Inches	Pounds
½	0.188	0.17	0.22	0.32	0.39	0.42	0.120	0.78	0.221
⅝	0.250	0.30	0.38	0.50	0.62	0.756	0.213	1.238	0.350
¾	0.313	0.48	0.59	0.72	0.90	1.175	0.332	1.775	0.503
1	0.438	0.93	1.16	1.27	1.58	2.294	0.648	3.130	0.886
1¼	0.563	1.54	1.92	1.97	2.46	3.790	1.076	4.870	1.38
1½	0.688	2.30	2.87	2.83	3.54	5.670	1.607	7.00	1.98
1¾	0.813	3.21	4.01	3.83	4.78	7.92	2.245	9.47	2.68
2	0.938	4.27	5.33	5.00	6.25	10.53	2.985	12.33	3.50

*Includes scrap end and spatter
**r=Height of reinforcement.

Airco

Without a conversion table the weight of steel deposited can be found by first calculating the volume of deposited metal (area of the groove multiplied by the length). Then this volumetric value is converted to weight by the factor 0.283 pounds per cubic inch for steel.

Stick Electrode Cost. The next step in estimating weld cost is to find the actual electrode cost per foot of weld. This is done by multiplying the cost per pound of electrode by the pounds of weld metal deposited per foot of weld and dividing by the deposition efficiency of the welding process. In formula form the calculation is:

$$\text{Electrode Cost Per Foot} = \frac{(\text{Lbs Deposit/Ft of weld}) \times (\text{Cost/Lb of Electrode})}{(\text{Deposition Efficiency})}$$

Gas Cost. If the welding process requires a shielding gas the cost of gas must be determined. Gas cost is usually based on the cost per foot of weld and is found by multiplying the gas flow in cubic feet per hour

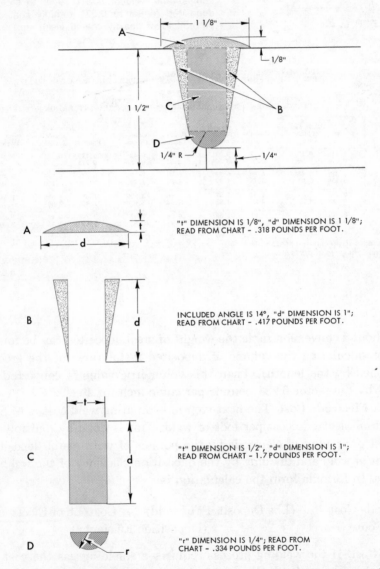

Fig. 15-5. To find the weight of filler metal, the joint design is divided into regular geometric shapes.

by the cost of gas per cubic feet and dividing by the welding speed. Shown as a formula the calculation is:

$$\frac{\text{Gas Cost}}{\text{Per Ft}} = \frac{(\text{Gas flow Cu Ft/Hr}) \times (\text{Cost of Gas/Cu Ft})}{(\text{Travel speed W/Min}) \times \dfrac{(60 \text{ Min/Hr})}{(12 \text{ In/Ft})}}$$

Flux Cost. If brazing or welding operation requires flux the cost of this material can be calculated with the formula:

$$\frac{\text{Flux Cost}}{\text{Per Ft}} = \frac{(\text{Lbs Deposit/Ft of weld}) \times (\text{Flux Ratio}) \times (\text{Cost}}{\text{of Flux/Lb})}$$

Flux ratio is the number of pounds of flux used per pound of electrode consumed. The nominal value of 1.5 lbs of flux per pound of electrode is considered a general average.

Wire Cost. Wire cost for semi-automatic or automatic gas metal-arc or submerged-arc welding is calculated by using the same formula as the one for metallic electrode cost. Sometimes in figuring wire costs the number of inches in a pound of wire is helpful. If the wire feed speed in ipm is known, it is a simple calculation to figure the lbs per weld by using Table IV.

Table IV.   Wire Amount—Inches per Pound.

Decimal	.020	.025	.030	.035	.045	.062	.078	.093	.125
					3/64	1/16	5/64	3/32	1/8
Mg.	50500	34700	22400	16500	9990	5270	3300	2350	1280
Al.	32400	22300	14420	10600	6410	3382	2120	1510	825
Al. Br.	11600	7960	5150	3780	2290	1220	756	538	295
SS.400	11350	7820	5050	3720	2240	1180	742	528	289
Mild St.	11100	7680	4960	3650	2210	1160	730	519	284
SS-300	10950	7550	4880	3590	2170	1140	718	510	279
Si. Br.	10300	7100	4600	3380	2040	1070	675	480	263
Cu-Ni.	9950	6850	4430	3260	1970	1040	650	462	253
Ni.	9900	6820	4400	3240	1960	1030	647	460	252
DO Cu.	9800	6750	4360	3200	1940	1200	640	455	249

Mg. = Magnesium
Al = Aluminum
Al. Br. = Aluminum Bronze
SS-400 = 400 series Stainless Steel
Mild St. = Mild Steel

SS-330 = 300 series Stainless Steel
Si.Br. = Silicon Bronze
Cu.Ni. = Copper-Nickel
Ni. = Nickel
DO Cu. = Deoxidized Copper

Miller Electric Mfg. Co.

Operating Cost

The three essential factors in operating cost are labor, overhead, and power.

Labor. Actual labor cost is governed by the deposition rate of the welding process used. Deposition rate is the amount of weld made per hour. This rate is dependent on (1) positioning of the weld and (2) amperage setting of the welding machine. As indicated previously deposition rates are higher when welds can be made in a flat or nearly flat position. For example, the welding speed for a 1″ fillet weld made in the vertical position with stick electrodes is approximately 6 ipm, whereas if the same weld is made in a flat position the welding speed is 12 ipm or twice as fast. Machine setting is determined by the size of electrode and thickness of the metal to be welded. Thus for shielded metal-arc welding the correct amperage setting is important to achieve maximum deposition rate. By the same token wire feed rate for gas metal-arc welding is also governed by the same welding conditions and must be established for effective speed.

It is significant to note that very often material costs for a particular welding process may be slightly higher than for another welding process. Conceivably Mig welding in some sections of the country may result in higher material costs than shielded metal-arc because of the greater prevailing rates. However, the higher material cost is frequently offset by the faster deposition rate. With increased welding speed there is a corresponding decrease in labor costs. Labor costs can be determined by the formula:

$$\frac{\text{Labor Cost}}{\text{Per Ft}} = \frac{\text{(Hourly Rate)}}{\text{(Travel Speed In/Min)} \times \dfrac{\text{(60 Min/Hr)}}{\text{(12 In/Ft)}} \times \text{(Duty Cycle)}}$$

Travel Speed represents the time it should take to run the weld bead. This time factor depends on several variables such as operator skill, welding process, and weld position. As a rule, travel speed is often established on the basis of actual timing of previous welding jobs. This value then can serve as a basic index for most estimating problems.

Duty Cycle is a percentage indicating the actual arc time as compared with the total hours worked. Duty cycle is affected by the amount of time required for changing electrodes, chipping slag, cleaning welds, setting up equipment, etc.

Arc time is normally established in any production plant by measuring the time a machine is in full welding operation by means of arc time recorders. These instruments read arc time in minutes and tenths of minutes and provide a true picture of the relationship between actual welding time and the other operations. In other words, they tell how many hours per day the equipment is productive.

Overhead. The overhead rate is usually a multiplier factor applied to direct labor or material. It is kept as a separate calculation since this factor frequently distorts the cost estimating process.

Overhead cost is based on the formula:

$$\text{Overhead Cost Per Ft} = \frac{(\text{Overhead Rate/Hour})}{(\text{Travel Speed}) \times (60 \text{ Min/Hr}) \times (\text{Duty Cycle})}{(\text{In/Min})} \quad (12 \text{ In/Ft})}$$

Power Cost. In some cost estimating requirements the cost of electrical power consumed is calculated. For most welding equipment the practice is simply to assume a 50 percent electrical efficiency factor. Inasmuch as the power cost is usually very small this calculation is often eliminated without any significant loss of cost calculation accuracy.

Power can be ascertained by the formula:

$$\text{Power Cost Per Ft} = \frac{(\text{Amps}) \times (\text{Volts}) \times (\text{Power Cost/Kw Hr})}{(\text{Deposition}) \times (\text{Travel Speed In/Min}) \times (\text{Efficiency})}$$
$$\frac{(60 \text{ Min/Hr}) \times 1000}{(12 \text{ In/Ft})}$$

Total Cost Per Foot of Weld. This represents the cost of material and operation multiplied by the total length of weld or:

$$\text{Total Welding Cost} = (\text{Material Cost}) + (\text{Operation Cost}) \times (\text{Length of Weld})$$

Example No. 1 Shielded Metal Arc — $\frac{3}{32}''$ Electrodes

Specifications

Metal	— mild steel
Weld size	— V-Butt joint, $\frac{3}{8}''$ deep, $\frac{7}{16}''$ wide, 20° included angle
Weld metal weight	— .396 lbs per ft
Price of $\frac{3}{32}''$ electrodes	— $.22 per lb
Deposition efficiency	— 60%
Total length of weld	— 24 ft
Hourly labor rate	— $3.00
Overhead rate	— $6.00
Travel speed	— 8″ per min
Duty Cycle	— 30%
Amperage —	80
Volts —	20
Kw hour —	$.03

Material Cost

$$\text{Electrode} = \frac{.396 \times .17}{.60} = \$.112$$

Operation Cost

$$\text{Labor} = \frac{3.00}{8 \times \dfrac{60}{12} \times .30} = \$.25$$

$$\text{Overhead} = \frac{6.00}{8 \times \dfrac{60}{12} \times .30} = \$.50$$

$$\text{Power} = \frac{80 \times 20 \times .03}{.60 \times 8 \times \dfrac{60}{12} \times 1000} = \$.002$$

$$\textit{Total Cost} = .864 \times 24 = \$20.74$$

Example No. 2 Mig — $\frac{3}{64}''$ wire

Specifications

Metal	— Aluminum
Weld size	— Fillet $\frac{1}{4}'' \times \frac{1}{4}''$ convex
Weld metal weight	— .155 lbs per ft
Price of $\frac{3}{64}''$ wire	— $1.80 per lb
Deposition efficiency	— 92%
Total weld length	— 10 feet
Gas	— $.10 per cu ft
Hourly labor rate	— $3.00
Travel speed	— 30'' per min
Duty cycle	— 55%
Overhead rate	— $5.00

Amperage	—	270
Volts	—	26
Kw hour	—	$.03

Material Cost

$$\text{Wire} = \frac{.155 \times 1.80}{.92} = \$.303$$

$$\text{Gas} = \frac{50 \times .10}{\dfrac{30 \times 60}{12}} = \$.03$$

Operation Cost

$$\text{Labor} = \frac{3.00}{\dfrac{30 \times 60 \times .55}{12}} = \$.036$$

$$\text{Overhead} = \frac{5.00}{\dfrac{30 \times 60 \times .55}{12}} = \$.06$$

$$\text{Power} = \frac{270 \times 26 \times .03}{\dfrac{.92 \times 30 \times 60 \times 1000}{12}} = \$.002$$

Total Cost $= .431 \times 10 = \$4.31$

REVIEW PROBLEMS

1. Determine the welding cost of fabricating a welded unit based on the following:

   a. Metal   — Mild steel
   b. Welding Process   — Shielded metal-arc
   c. Joint size   — V-groove butt, $T = \frac{1}{2}''$,
                                   $S = \frac{9}{16}''$ (Table III)
   d. Price of $\frac{1}{8}''$ electrodes — $.16 per lb
   e. Deposition efficiency   — .65%
   f. Total length of weld   — 32 ft
   g. Hourly labor rate   — $3.50
   h. Overhead rate   — $6.25
   i. Travel speed   — 10" per min.
   j. Duty cycle   — 35%
   k. Kw hour   — $.02

(Ascertain from manufacturer's tables amperage and
volts required for $\frac{1}{8}''$ electrode) See Appendix B

2. Find the welding cost for Mig welding based on the following:

   a. Metal   — Stainless steel
   b. Joint size   — Single Fillet Tee Joint $- S = \frac{1}{4}''$
                             (Table III)
   c. Price of .045" wire   — 1.50 per lb
   d. Deposition efficiency   — 95%
   e. Total weld length   — 28 ft
   f. Gas   — $.12 per cu ft
   g. Hourly labor rate   — $3.25
   h. Overhead rate   — $5.00
   i. Travel speed   — 35" per min
   j. Duty cycle   — 65%
   k. Kw hour   — $.03

(Determine amperage, volts and gas flow.) See Appendix B

3. Compare the relative cost of Mig versus shielded metal-arc welding based on the following factors:

a.	Metal	— Mild steel
b.	Joint size	— Partially grooved 60°-single V, Butt Joint $t = \frac{5}{8}''$, $s = \frac{11}{16}''$
c.	Electrode	— $\frac{3}{32}''$ Dia. @ $.16 per lb
d.	Wire	— .045'' Dia. @ $.32 per lb
e.	Total weld length	— 185 ft
f.	Deposition Efficiency	
	Metallic arc	— 60%
	Mig	— 98%
g.	Travel speed	
	Metallic arc	— 8'' per min
	Mig	— 40'' per min
h.	Gas flow	— 25 cu ft. hr
i.	Gas cost	— $.05 per cu ft
j.	Duty cycle	
	Metallic arc	— 30%
	Mig	— 60%
k.	Hourly labor rate	— $3.10
l.	Overhead rate	
	Metallic arc	— $5.50
	Mig	— $5.00
m.	Kw. hour	— $.02

4-7. Estimate the welding costs for the problems shown in Figs. 15–6, 15–7, 15–8 and 15–9.

(See pages 420, 421, 422, and 423)

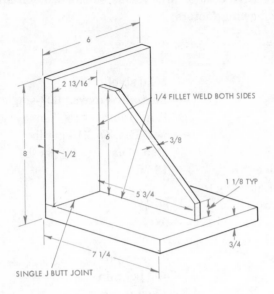

1/4 FILLET WELD BOTH SIDES

Fig. 15-6.

Specifications

a. Metal — Mild steel

b. Welding — Shielded
   Process   metal arc

c. Electrode — $\frac{3}{32}''$
   Dia.

d. Electrode
   Cost   — $.14 per lb

e. Deposition
   efficiency — 65%

f. Hourly
   labor rate — $2.75

g. Overhead
   rate   — $4.75

h. Travel
   speed   — 8″ per min

i. Duty cycle — 35%

j. Kw hour   — $.025

(Ascertain from
manufacturer's charts
amperage and voltage)

Fig. 15-6. Estimating problem.

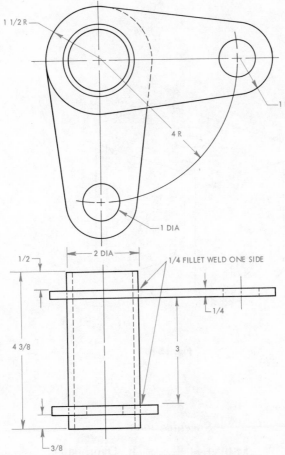

Fig. 15-7.

Specifications

a. Metal —Stainless steel

b. Welding process —Mig

c. Wire Dia. —$\frac{1}{16}''$

d. Wire cost —$1.75 per lb

e. Gas cost —$.12 per cu ft

f. Deposition efficiency —98%

g. Hourly labor rate —$3.10

h. Overhead rate —$5.75

i. Travel speed —35" ipm

j. Duty cycle —60%

k. Kw. hour —$.03

(Determine gas flow, amperage and voltage—See Appendix B)

Fig 15-7. Estimating problem.

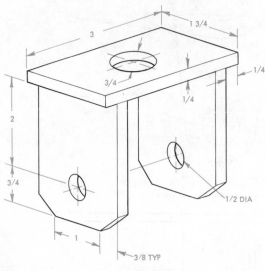

Fig. 15-8.

Specifications

a. Metal —Mild steel
b. Joint —Single
 design J-Butt
 Welded
 one side
c. Welding
 Process —Mig
d. Wire Dia. —.045"
e. Wire Cost —$.32 per lb
f. Gas cost —$.11 per
 cu ft
g. Deposition
 efficiency —95%

h. Overhead
 rate —$6.25
i. Hourly
 labor rate —$3.35
j. Travel
 speed —40" ipm
k. Duty cycle —55%
l. Kw. hour —$.015

(Determine gas flow, amperage
 and voltage—See Appendix B)

Fig. 15-8. Estimating problem.

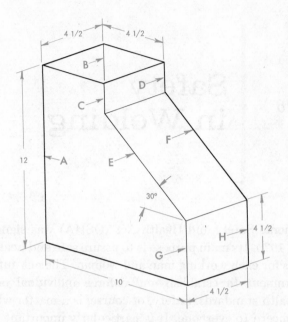

Fig. 15-9.

Specifications

a. Metal —⅛" Aluminum

b. Joint Design —(1) Outside, single-fillet corner joint—(2) Weld edges A, B, C, D, E, F, G, H

c. Welding process —Mig

d. Wire Dia. —¹⁄₁₆"

e. Wire cost —$1.98 per lb

f. Gas cost —$.12 per cu ft

g. Deposition efficiency —96%

h. Hourly labor rate —$3.25

i. Overhead rate —$5.50

j. Travel speed —45" ipm

k. Duty cycle —65%

l. Kw hour —$.02

(Determine gas flow, amperage and voltage—See Appendix B)

Fig. 15-9. Estimating problem.

Chapter 16	# Safety in Welding

The Occupational Safety and Health Act (OSHA) was signed into law in December, 1970. Its main purpose is to assure safe and healthful working conditions for each working man and woman. The act authorizes the federal government to establish and enforce individual occupational safety and health standards. Safety, of course, is a matter which should be of deep concern to everyone. It is particularly important in the field of welding because the operator's own well-being is involved. In industry safety is really significant to the supervisor or foreman because he has to meet a work schedule and, at the same time, bear the responsibility for the welfare of his workers. Injuries inevitably interrupt the work flow. Such disruptions always have serious effects on production.

In a school situation the instructor must be unduly vigilant to ensure that students are in a safe environment. School shop accidents can be prevented providing the instructor initiates and maintains a really functional safety program.

Most welding accidents can be avoided if proper precautions are taken. Generally, accidents happen because someone fails to follow instructions or uses poor judgment in carrying out assigned tasks. Hence, one of the first safety rules which an operator must follow consistently is never to attempt something unless it is fully understood.

Special safety bulletins are available which cover in detail recommended safety practices for welding and cutting. These bulletins can be secured from:

Compressed Gas Association
500 Fifth Avenue
New York, N.Y. 10036

American Welding Society
2501 N.W. 7th Street
Miami, Florida 33125

National Fire Protection Association
60 Batterymarch Street
Boston, Mass. 02110

Union Carbide—Linde Division
270 Park Ave.
New York, N.Y. 10017

General Safety

General safety refers to conditions as they apply to all phases of welding and cutting. Specifically it deals with ventilation, clothing, and equipment.

Ventilation. All welding should be done in well ventilated areas. There must be sufficient movement of air to prevent accumulation of toxic fumes or possibly oxygen deficiency. Adequate ventilation becomes extremely critical in confined spaces where excessive fumes, smoke, and dust are likely to collect.

Where considerable welding is to be done an exhaust system is necessary to keep toxic gases within the prescribed health limits. The general recommendation for adequate ventilation is a minimum of 2000 cubic feet of air flow per minute per welder. If individual movable exhaust hoods can be placed near the work the rate of air flow in the direction of the hood should be approximately 100 linear feet per minute in the welding zone. See Fig. 16–1 and Table I. Hobart Welders has developed

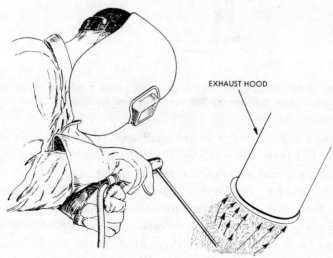

EXHAUST HOOD

Fig. 16-1. An adequate ventilation system should be provided for all welding operations.

Table I. Exhaust Hoods.

WELDING ZONE	MINIMUM AIR FLOW CU FT/MIN	DUCT DIAMETER IN.
4" TO 6" FROM ARC OR TORCH	150	3
6" TO 8" " " " "	275	3 1/2
8" TO 10" " " " "	425	4 1/2
10" TO 12" " " " "	600	5 1/2

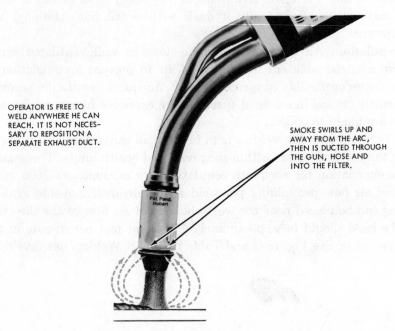

OPERATOR IS FREE TO WELD ANYWHERE HE CAN REACH. IT IS NOT NECESSARY TO REPOSITION A SEPARATE EXHAUST DUCT.

SMOKE SWIRLS UP AND AWAY FROM THE ARC, THEN IS DUCTED THROUGH THE GUN, HOSE AND INTO THE FILTER.

Hobart Brothers Co.

Fig. 16-2. This smoke exhaust system removes the smoke produced at the weld zone and increases operator visibility and thus weld quality.

a new exhaust system which is designed specifically for gas-metal arc welding and flux cored arc welding. The exhaust duct is actually a part of the welding gun and removes smoke from the arc area without disturbing the arc shielding gas. See Fig. 16–2. When toxic fumes from lead, zinc, brass, bronze, cadmium, beryllium-bearing materials, or other substances are present in harmful concentration, a respirator should be used.

Protective Clothing. Fire-resistant gauntlet gloves should be used for most welding operations. See Fig. 16–3. Woolen clothing is preferable to

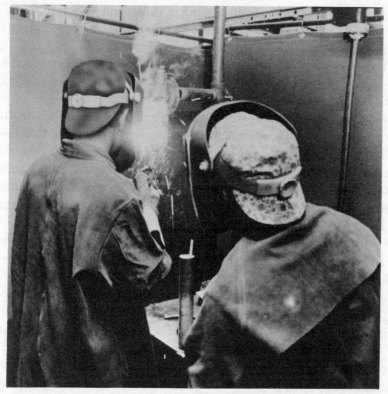

Singer Safety Products, Chicago, III.

Fig. 16-3. OSHA demands an operator be properly clothed and protected when welding. Note use of fire retardant tunics, gloves, face shields, and the specially treated canvas curtains above.

cotton because it does not ignite as readily and protects the skin better from temperature changes. For certain types of work such as heavy cutting or lancing a fire-resisting apron or leggings may be necessary. Rolled-up sleeves and cuffs in overalls or trousers and pockets in front of clothing should be avoided because sparks may become lodged in them. Low shoes with unprotected tops are not suitable for work where sparks and slag may get inside.

Eye Protection. Eye protection with suitable filter lenses is required for all welding and cutting. In some instances, such as in oxyacetylene welding and cutting regular welding goggles are adequate. In any form of arc welding a shield which protects the face as well as the eyes is necessary for protection from the ultra-violet and infrared rays. The danger of these rays is such that an arc should never be looked at with the naked eyes within a distance of 50 feet.

The use of clear lens goggles or head shield should be required when grinding weld metal, or chipping slag off a weld seam. A small flying particle may cause irreparable damage to an eye.

Ear Protection. Federal law (Occupational Safety and Health Act of 1970) now requires that employees be exposed to no more than ninety decibels of sound during their eight hour working day, and exposure to extreme noise be rigidly limited. If noise exceeds these levels, then ear protection must be worn. Ear protection (such as specially manufactured muffs) also prevent dust, sparks, chips and fumes from entering the inner ear, thus reducing the possibility of infection or damage.

Welding. During a welding operation a work piece should never be placed on the concrete floor. Heat may explode a piece of concrete with sufficient force to injure the welder.

Equipment Familiarization. Operators should never be permitted to use any welding equipment until they have received thorough and explicit instructions. Manufacturer's specifications and recommendations are very important and should be complied with at all times. There should also be a definite ruling that in the event of a malfunction in the equipment no attempt should be made by the operator to solve the problem without first consulting with supervising personnel.

Safety in Oxyacetylene Welding

Cylinders.

1. Never move a cylinder by dragging, sliding, or rolling it on its side. Avoid striking it against any object that might create a spark. There may be just enough gas escaping to cause an explosion. To move a cylinder roll it on its bottom edge. See Fig. 16-4.
2. Never permit grease or oil to come in contact with cylinder valves. Although oxygen is in itself non-inflammable if it is allowed to come in contact with any flammable material it will quickly aid combustion.
3. Cylinders should not be exposed to furnace heat, radiators, open fire, or sparks from a torch.
4. Cylinders should not be allowed to lie in a horizontal position nor should the valve-protector cap be used for lifting cylinders.

Fig. 16-4. Correct way to move a cylinder.

5. No attempt should ever be made to repair cylinder valves. If the valves do not function properly, or if they leak, the supplier should be notified.
6. Oxygen should never be used as a substitute for compressed air to operate pneumatic tools, blowing out pipe lines or dusting clothing because a serious accident may result.
7. Cylinders should be chained to a truck or some other steadying structure to prevent them from being knocked over while in use.
8. When opening cylinder valves always stand to one side of and away from the regulator. See Fig. 16–5. A defect in the regulator may cause the gas to blow through shattering the glass.

Fig. 16-5. Always stand to one side when opening cylinder valves.

Piping

All piping and fittings used to convey gases from a central supply system to work stations must withstand a minimum pressure of 150 psi. Oxygen piping can be of black steel, wrought iron, brass, or copper. Only oil free compounds should be used on oxygen threaded connections. Piping for acetylene must be of wrought-iron. Under no circumstances should acetylene gas come in contact with unalloyed copper except in a torch. Any contact of acetylene with high alloyed copper piping will generate copper acetylide which is very reactive and may result in a violent explosion. After assembly all piping must be blown out with air or nitrogen to remove foreign materials.

Testing for Leaks

New welding apparatus needs to be tested for leaks before being operated. Thereafter it is advisable to periodically test in order to be sure that no leakage has developed. Leaky apparatus is very dangerous since a fire may develop. Furthermore, leaks mean wasted gas.

To test for leaks open the oxygen and acetylene cylinder valves and with the needle valves on the torch closed, adjust the regulators to produce about normal working pressure. Apply soapy water with a brush over all connections. Under no circumstances should an open flame be used.

Lighting a Torch

Always use a sparklighter to light a torch never a match. A match brings the fingers too close to the tip and the sudden ignition of the fuel gas may cause a burn. When igniting a torch keep the tip facing downward, Fig. 16-6. Lighting the torch while it is facing outward or upward may result in burning someone nearby as the flame spurts out.

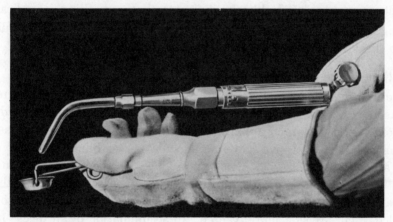

Fig. 16-6. Correct method of lighting a welding torch.

Make no attempt to relight a torch from the hot metal when welding in an enclosed box, tank, drum or other small cavity. There may be just enough unburned gas in this confined space to cause an explosion as the fuel gas from the tip comes in contact with the hot metal. Instead, move the torch into the open, relight it in the usual manner, and make the necessary adjustments.

Backfires and Flashbacks

A *backfire* is a loud noise associated with the momentary extinguishment or re-ignition of the flame at the torch tip. It may be caused by touching the tip against the work, particles entering the tip and obstructing the gas flow, or by over-heating the tip. Sometimes the trouble will clear itself immediately and, if the work is hot enough, the torch will relight automatically. If this does not happen the oxygen and acetylene valves should be closed immediately. The torch is then relighted using standard lighting procedure.

A *flashback* is the burning back of the flame into the tip, torch, or hose. It is often accompanied by a hissing or squealing sound with a smoky or sharp-pointed flame. When this happens the flame should be extinguished immediately by closing first the torch oxygen valve and then the torch acetylene valve. Flashbacks may be caused by failure to purge the lines, improper pressures, distorted or loose tips or mixer seats, kinked hose, clogged tip or torch orifice or over-heated tip or torch.

The occurrence of flashbacks indicates that something is radically wrong with the equipment. Qualified service personnel should be called in to check the equipment before resuming the weld.

Safety in Welding and Cutting Containers

No welding or cutting should be performed on used drums, barrels, tanks, or other containers unless they have been thoroughly cleaned of all combustible substances that may produce flammable vapors or gases. Flammable and explosive materials include (1) gasoline, (2) light oil, (3) acids that react with metal to produce hydrogen, and (4) non-volatile oils or solids that release vapors when exposed to heat. Furthermore, sufficient precaution must be taken to ensure that containers to be welded or cut are sufficiently vented. An accumulation of air or gas in a confined area will expand when heated and the internal pressure may build up to cause an explosion.

Cleaning the Container. Before any cleaning is attempted the content of the container should be ascertained so appropriate cleaning methods can be used. A very small amount of residual flammable gas or liquid may cause a serious explosion.

The following methods are recommended for cleaning containers which are to be welded or cut:

Water Cleaning. This method is satisfactory if the previous substance in the container is readily soluble in water. Examples of water soluble compounds are acetone and alcohol. If the substance is not soluble in water the cleaning must be done by one of the other methods. Water cleaning simply involves filling the container with water and draining several times.

Hot Chemical Cleaning. For chemical cleaning the following procedure is recommended:

1. Flush out any remaining residue of the container with water and drain completely.

2. Dissolve 2 to 4 ounces of trisodium phosphate (strong washing powder) or a commercial caustic cleaning compound into a gallon of boiling water. Pour this solution into the container and fill with water.

3. Attach a steam line to the container and admit steam to maintain the solution at a temperature of 170° to 190° for at least 15 to 20 minutes. During the steaming period add water to allow discharge of volatile liquid, scum, or sludge which may collect by overflowing at the top. At the end of the prescribed period, drain the container.

Steam Cleaning. In regular steam cleaning first flush out the container preferably with hot soda or soda ash (1 lb per gallon of water). After the draining is completed blow steam into the container until it is free from odors and the sides of the container are hot enough to permit steam vapors to flow freely out of the container. Then flush out the container with hot or boiling water.

An alternate cleaning method is to fill the container with an inert gas after it has been thoroughly flushed out. The inert gas will change flammable gases and vapors to a nonflammable and nonexplosive state. The recommended inert gas is either carbon dioxide or nitrogen. To use inert gas first close all openings in the container except the filling connection and vent. Fill the container with as much water as possible and then add the inert gas in the remaining space near the area to be welded or cut.

Welding or Cutting the Container

Before proceeding to weld or cut a container that has been completely cleaned of all volatile substances it is always advisable to take an additional precaution by filling the container with water. The container should be arranged so water can be kept filled to within a few inches of the point where the welding or cutting is to take place. See Fig. 16–7.

Safety in Cutting

Fires often occur in a cutting operation simply because proper precautions were not taken. Too often a worker forgets that sparks and

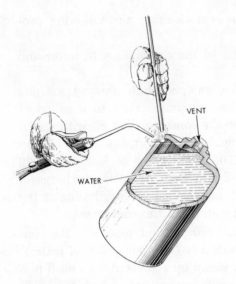

Fig. 16-7. When welding or cutting a
container, fill it with water.

WATER

falling slag can travel as much as 35 feet and can pass through cracks out of sight of the goggled operator. Persons responsible for supervising or performing cutting of any kind should observe the following:

1. Never use a cutting torch where sparks will be a hazard such as near rooms containing flammable materials especially dipping or spraying rooms.

2. If cutting is to be over a wooden floor, sweep the floor clean and wet it down before starting the cutting. Provide a bucket or pan containing water or sand to catch the dripping slag.

3. Keep a fire extinguisher nearby whenever any cutting is done.

4. Whenever possible perform cutting in wide open areas so sparks and slag will not become lodged in confined crevices or cracks.

5. If cutting is to be done near flammable materials, and the flammable materials cannot be moved, suitable fire-resisting guards or partition screens must be used. (See illustration *opposite* p. 1)

6. In plants where a dirty or gassy atmosphere exists extra precaution should be taken to avoid explosions resulting from electric sparks or open fire during a cutting or welding operation.

Safety in Arc Welding

Arc welding includes shielded metal-arc, gas-shielded arc and resistance welding. Only general safety measures can be indicated for these areas because arc welding equipment varies considerably in size and type. Equipment may range from a small portable shielded metal-arc welder to highly mechanized production spot or gas-shielded arc welders. In each instance specific manufacturers recommendations should be followed.

Safety practices which are generally common to all types of arc welding operations are as follows:

1. Welding equipment should be installed according to provisions of the National Electric Code.
2. A welding machine should be equipped with a power disconnect switch which is conveniently located at or near the machine so the power can be shut off quickly.
3. Repairs to welding equipment should never be made unless the power to the machine is shut off. The high voltage used for arc welding machines can inflict severe and fatal injuries.
4. Welding machines must be properly grounded. Stray current can cause severe shock when ungrounded parts are touched.
5. The polarity switch never should be changed when the machine is under a load. Wait until the machine idles and the circuit is open. Otherwise, the contact surface of the switch may be burned and the resulting arcing could cause an injury.
6. Welding cables should not be overloaded or a machine operated with poor connections. Operating with currents beyond the rated cable capacity causes overheating. Poor connections may cause the cable to arc when it touches metal grounded in the welding circuit.
7. Damp areas should be avoided and hands and clothing kept dry at all times. Dampness on the body may cause an electric shock.
8. Do not strike an arc if someone without proper eye protection is nearby. Arc rays are harmful to the eyes and skin. If other persons must work nearby, the welding area must be railed off with a fire retardant canvas curtain to protect them from the arc welding flash.
9. Press type welding machines should be effectively guarded.
10. Suitable spark shields must be used around equipment in flash welding.

11. All operators and attendants of resistance welding should use transparent face shields or goggles.
12. Keep the uninsulated portion of the electrode holder from touching the welding ground when the current is on. This will cause a flash.
13. Keep welding cables dry and free from oil and grease.
14. Never carry welding cable coiled around the shoulders when they are carrying power.

REVIEW QUESTIONS

1. Why should safety be strictly enforced in any welding or cutting situation?
2. What is the recommended practice governing ventilation?
3. Why should a welding supervisor be concerned with safety?
4. What are some of the precautions which should be observed regarding clothing?
5. In arc welding why are proper eye and face shielding devices particularly important?
6. Why should oxygen cylinder valves never come in contact with oil or grease?
7. Why should an operator stand to one side of a regulator when opening cylinder valves?
8. What is the correct procedure for testing leaks in an oxyacetylene welding system?
9. How does a backfire differ from a flashback?
10. What should be done in the event a backfire or flashback does occur?
11. Why should used containers be thoroughly cleaned before any welding or cutting is done on them?
12. What are some of the methods used in cleaning containers for welding and cutting?
13. Why do fires often occur during a cutting operation?
14. Why should all electrical welding equipment be properly grounded?
15. What is the significance of having each welding machine equipped with a power disconnect switch?
16. Why should a welding machine never be overloaded?
17. What precautions should be taken when arc welding in damp areas?

Chapter 17

Welding Symbols

A series of symbols have been standardized by the American Welding Society. These symbols are normally included on the drawing or print which the man in the shop uses in fabricating the required product and indicate to the welder the kind of weld that is to be made, the location of the weld, and the size of the weld.

Base of Weld Symbol

The main foundation of the weld symbol is a reference line with an arrow at one end. See Fig. 17-1. The other data reflecting various characteristics of the weld are shown by abbreviations, figures, and other line arrangements placed around the reference line.

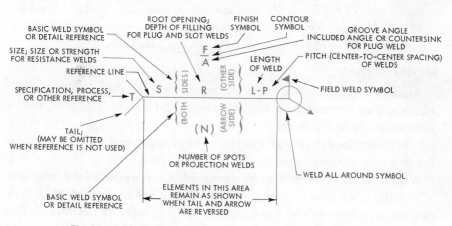

Fig. 17-1. Standard Location of Elements of Welding Symbols

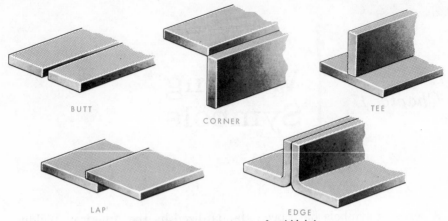

BUTT

CORNER

TEE

LAP

EDGE

Fig. 17-2. Common types of weld joints.

Designating Types of Welds

The first important factor in understanding weld symbols is recognizing the representations used for different types of welds. These welds are used on five basic kinds of joints: butt, corner, lap, tee, and edge. See Fig. 17–2.

Welds are classified as fillet, plug or slot, spot, seam, and groove. Groove welds are further divided according to the particular shape of the grooved joint.

Each type of weld has its own specific symbol. For example, a fillet weld is designated by a right triangle, and a plug weld by a rectangle. All of the basic weld symbols are included in Fig. 17–3.

Location of Welds

Another requirement in understanding weld symbols is the method which is used to specify on what side of a joint a weld is to be made. A weld is said to be either on the arrow or other side of a joint. The arrow side is the surface that is in direct line of vision, while the other side is the opposite surface of the joint. See Fig. 17–4.

Weld location is designated by running the arrowhead of the reference line to the joint. The direction of the arrow is not important, that is, it can run on either side of a joint and extend upward or downward. See Fig. 17–5. If the weld is to be made on the arrow side, the appropriate

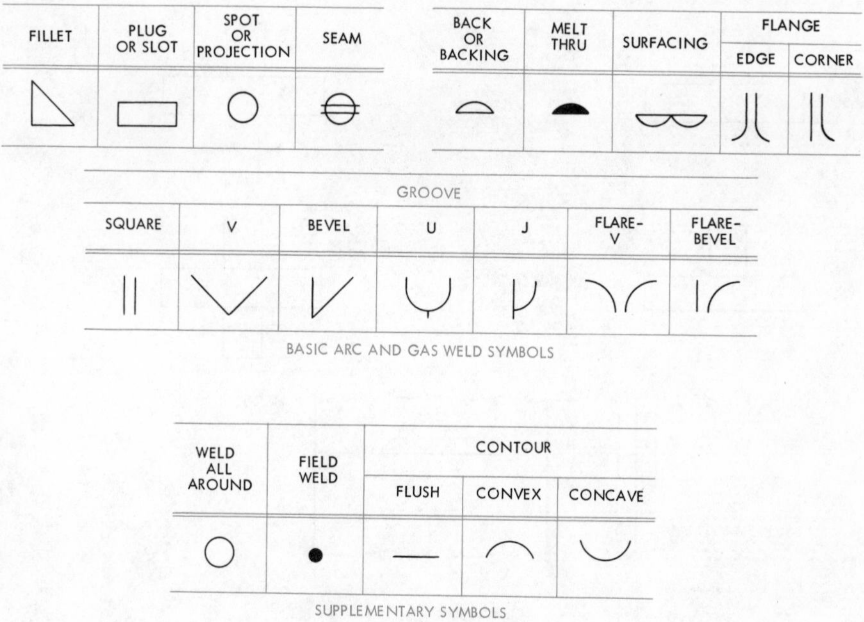

FILLET	PLUG OR SLOT	SPOT OR PROJECTION	SEAM		BACK OR BACKING	MELT THRU	SURFACING	FLANGE	
								EDGE	CORNER
◺	▭	◯	⊖		⌒	⬗	⌣	⎰⎱	⎰⎱

GROOVE							
SQUARE	V	BEVEL	U	J	FLARE-V	FLARE-BEVEL	
‖	V	⌁	∪	∪	⋁	⋁	

BASIC ARC AND GAS WELD SYMBOLS

WELD ALL AROUND	FIELD WELD	CONTOUR		
		FLUSH	CONVEX	CONCAVE
◯	●	—	⌒	⌣

SUPPLEMENTARY SYMBOLS

Fig. 17-3. Weld symbols. (AWS)

Fig. 17-4. Sides of a joint.

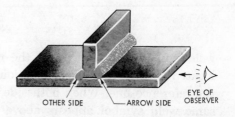

OTHER SIDE   ARROW SIDE   EYE OF OBSERVER

Fig. 17-5. The arrow may run in any direction.

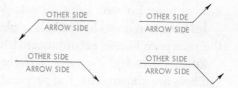

OTHER SIDE
ARROW SIDE

OTHER SIDE
ARROW SIDE

OTHER SIDE
ARROW SIDE

OTHER SIDE
ARROW SIDE

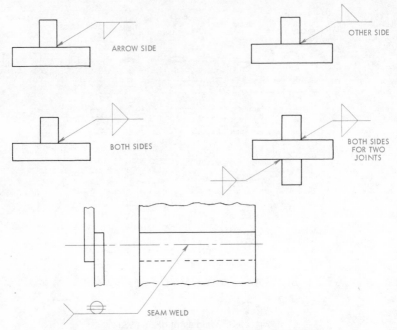

Fig. 17-6. How weld locations are designated. (AWS)

weld symbol is placed below the reference line. If the weld is to be located on the other side of the joint, the weld symbol is placed above the reference line. When both sides of the joint are to be welded, the same weld symbol appears above and below the reference line. See Fig. 17–6.

The only exception to this practice of indicating weld location is in seam and spot welding. With seam or spot welds, the arrowhead is simply run to the centerline of the weld seam and the appropriate weld symbol placed above, below or astride the reference line. See Fig. 17–6.

Information on weld symbols is placed to read from left to right along the reference line in accordance with the usual conventions of drafting.

Fillet, bevel and J-groove, flare-bevel groove, and corner-flange weld symbols are shown with the *perpendicular leg* always to the left.

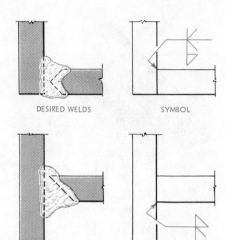

DESIRED WELDS   SYMBOL

Fig. 17-7. Combined weld symbols.

DESIRED WELDS   SYMBOL

Combined Weld Symbols

In the fabrication of a product, there are occasions when more than one type of weld is to be made on a joint. Thus a joint may require both a fillet and double-bevel groove weld. When this happens, a symbol is shown for each weld. See Fig. 17–7.

Size of Welds

Fillet welds. The width of a fillet weld is shown to the left of the weld symbol and is expressed in fractions, decimals or millimeters. When both sides of a filet are to be welded and both welds have the same dimensions, both are dimensioned. If the welds differ in dimensions, both are dimensioned. Where a note appears on a drawing that governs the size of a fillet weld, no dimensions are usually shown on the symbol. The length of the weld is shown to the right of the weld symbol by numerical values representing the actual required length.

When a fillet weld with unequal legs is required, the size of the legs is placed in parentheses as shown in Fig. 17–8.

Intermittent fillet welds. The length and pitch increments of intermittent welds are shown to the right of the weld symbol. The first figure

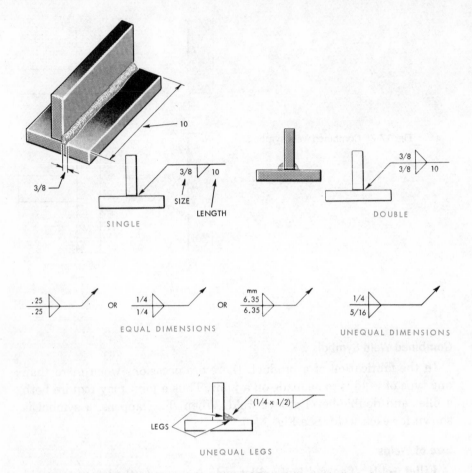

Fig. 17-8. How size and length of fillet welds are indicated.

represents the length of the weld section and the second figure the length of spacing between welds. See Fig. 17–9.

Plug welds. The size of plug welds is shown in fractions, decimals or millimeters to the left of the weld symbol, the depth when less than full on the inside of the weld symbol, the center-to-center spacing (pitch) to the right of the symbol, and the included angle of countersink below the symbol. See Fig. 17–10.

Slot welds. Length, width, spacing, angle of countersink, and location of slot welds are not shown on the weld symbol because it is too cumbersome. This data is included by showing a special detail on the print.

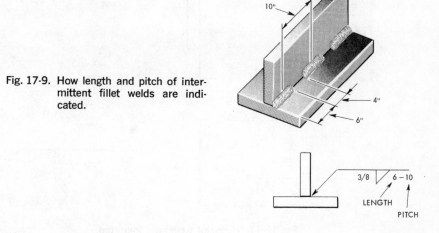

Fig. 17-9. How length and pitch of intermittent fillet welds are indicated.

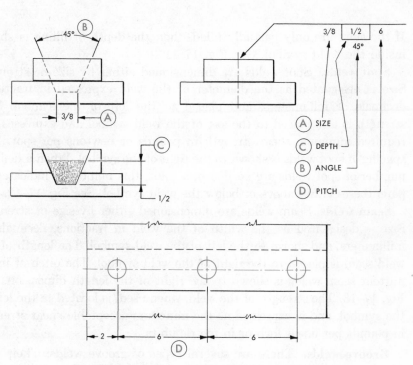

Fig. 17-10. How dimensions apply to plug welds.

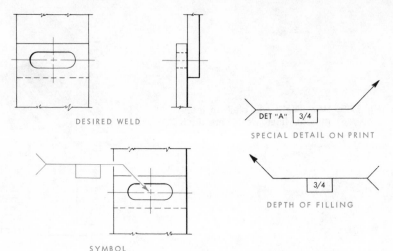

DESIRED WELD

SPECIAL DETAIL ON PRINT

DEPTH OF FILLING

SYMBOL

Fig. 17-11. How slot welds are indicated.

If the slots are only partially filled, then the depth of filling is shown inside the weld symbol. See Fig. 17–11.

Spot welds. Spot welds are dimensioned either by size or strength. Size is designated as the diameter of the weld expressed in fractions, decimals or millimeters, and placed to the left of the symbol. The strength is also placed to the left of the weld symbol and expresses the required minimum shear strength in pounds or newtons per spot. The spacing of spot welds is shown to the right of the symbol. When a definite number of spot welds are needed in a joint, this number is indicated in parentheses either above or below the weld symbol. See Fig. 17–12.

Seam welds. Seam welds are dimensioned either by size or strength. Size is designated as the width of the weld in fractions, decimals or millimeters, and shown to the left of the weld symbol. The length of the weld seam is placed to the right of the weld symbol. The pitch of intermittent seam welds is shown to the right of the length dimension. See Fig. 17–13. The strength of the weld, when used, is located to the left of the symbol, and is expressed as the minimum acceptable shear strength in pounds per linear inch or in metric units.

Groove welds. There are several types of groove welds. Their sizes are shown as follows:

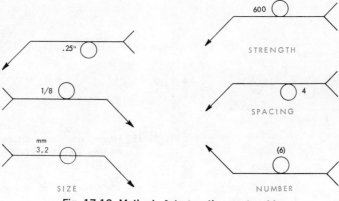

Fig. 17-12. Method of designating spot welds.

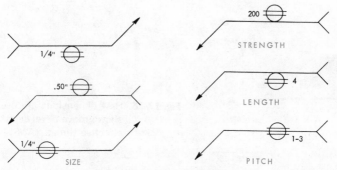

Fig. 17-13. Method of designating seam welds.

1. For single-groove and symmetrical double-groove welds which extend completely through the members being joined, no size is included on the weld symbol. See Fig. 17–14.

Fig. 17-14. Size is not shown for single and symmetrical double-groove welds with complete penetration. (AWS)

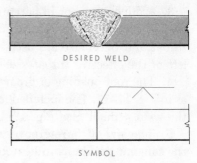

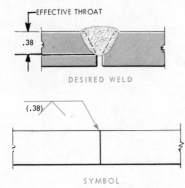

DESIRED WELD

SYMBOL

Fig. 17-15. How size is shown on grooved welds with partial penetration. (AWS)

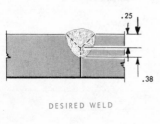

DESIRED WELD

Fig. 17-16. How dimensions are used to show groove bevel depth and effective throat. (AWS)

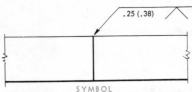

SYMBOL

2. For groove welds which extend only partly through the members being joined or non-symmetrical double groove joints, the weld size (effective throat) is shown in parentheses on the left of the weld symbol.

3. A dimension not in parentheses when placed to the left of the weld symbol indicates the depth of the bevel only. When both the effective throat and bevel size are used, the groove bevel depth is located to the left of the effective throat size as shown in Fig. 17–16.

4. The root opening of grooved joints is shown inside the weld symbol as in Fig. 17–17. The included angle of the bevel is placed above or below the weld symbol. See Fig. 17–17.

5. The size of flare-groove welds is considered as extending only to the tangent points as indicated by dimensional lines. See Fig. 17–18.

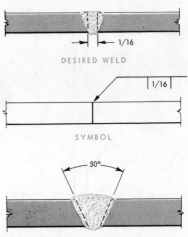

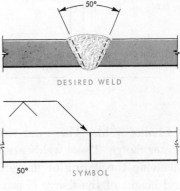

Fig. 17-17. How root opening and included angle are shown for groove welds. (AWS)

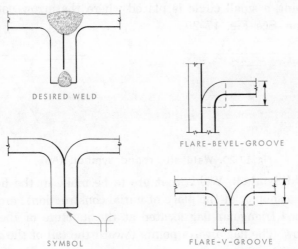

Fig. 17-18. How size of flare-grooved welds are indicated. (AWS)

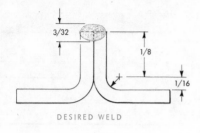

DESIRED WELD

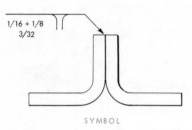

SYMBOL

Fig. 17-19. How dimensions apply to flange welds. (AWS)

Flange welds. There are two types of flange welds—edge flange and corner flange. These welds are used primarily for light gage metal joints involving the flaring or flanging of the edges to be joined. The radius and height of the flange is separated by a plus mark and placed to the left of the weld symbol. The size of the weld is shown by a dimension located outward of the flange dimensions. See Fig. 17–19.

Weld-All-Around Symbol. When a weld is to extend completely around a joint, a small circle is placed where the arrow connects the reference line. See Fig. 17–20.

DESIRED WELD

SYMBOL

Fig. 17-20. Weld-all-around symbol. (AWS)

Field Weld Symbol. Welds that are to be made in the field (welds not made in a shop or at the place of initial construction), are indicated by a darkened triangular flag located at the juncture of the reference line and arrow. The flag always points toward the tail of the arrow. See Fig. 17–21.

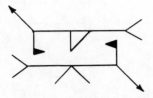

Fig. 17-21. Field weld symbol. (AWS)

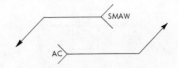

Fig. 17-22. The tail is used to indicate some specific detail or weld process. (AWS)

TABLE I: DESIGNATION OF WELDING PROCESSES*

	Welding Process	Letter Designation
BRAZING	INFRARED BRAZING	IRB
	TORCH BRAZING	TB
	FURNACE BRAZING	FB
	INDUCTION BRAZING	IB
	RESISTANCE BRAZING	RB
	DIP BRAZING	DB
GAS WELDING	OXYACETYLENE WELDING	OAW
	OXYHYDROGEN WELDING	OHW
	PRESSURE GAS WELDING	PGW
RESISTANCE	RESISTANCE-SPOT WELDING	RSW
WELDING	RESISTANCE-SEAM WELDING	RSEW
	PROJECTION WELDING	RPW
	FLASH WELDING	FW
	UPSET WELDING	UW
	PERCUSSION WELDING	PEW
ARC WELDING	STUD WELDING	SW
	PLASMA-ARC WELDING	PAW
	SUBMERGED ARC WELDING	SAW
	GAS TUNGSTEN-ARC WELDING	GTAW
	GAS METAL-ARC WELDING	GMAW
	FLUX CORED ARC WELDING	FCAW
	SHIELDED METAL-ARC WELDING	SMAW
	CARBON-ARC WELDING	CAW
OTHER PROCESSES	THERMIT WELDING	TW
	LASER BEAM WELDING	LBW
	INDUCTION WELDING	IW
	ELECTROSLAG WELDING	EW
	ELECTRON BEAM WELDING	EBW
SOLID STATE	ULTRASONIC WELDING	USW
WELDING	FRICTION WELDING	FRW
	FORGE WELDING	FOW
	EXPLOSION WELDING	EXW
	DIFFUSION WELDING	DFW
	COLD WELDING	CW
CUTTING	ARC CUTTING	AC
PROCESSES	AIR-CARBON-ARC CUTTING	AAC
	CARBON-ARC CUTTING	CAC
	METAL-ARC CUTTING	MAC
	OXYGEN CUTTING	OC
	CHEMICAL FLUX CUTTING	FOC
	METAL POWDER CUTTING	POC
	OXYGEN-ARC CUTTING	AOC

*The following suffixes may be used if desired to indicate the method of applying the above processes:

Automatic Welding	−AU	Manual Welding	−MA
Machine Welding	−ME	Semi-Automatic Welding	−SA (AWS)

Welding Symbols   449

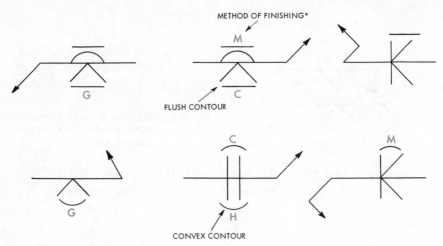

Fig. 17-23. Method of showing surface contour of welds. (AWS)

Reference Tail. The tail is included only when some definite welding specification, procedure, reference, weld or cutting process needs to be called out, otherwise it is omitted. This data is often in the form of symbols and is inserted in the tail. See Fig. 17–22 and Table I. Abbreviations in the tail may also call out some welding specifications which are included in more precise details on some other part of the print.

Surface Contour of Welds. When bead contour is important, a special flush, concave or convex contour symbol is added to the weld symbol. Welds that are to be mechanically finished also carry a finish symbol along with the contour symbols. See Fig. 17–23.

Back or Backing Welds. Back or backing welds refer to the weld made on the opposite side of the regular weld. Back welds are occasionally specified to insure adequate penetration and provide additional strength to a joint. This particular requirement is included opposite the weld symbol. No dimensions of back or backing welds except height of reinforcement is shown on the weld symbol. See Fig. 17–24.

Melt-Thru Welds. When complete joint penetration of the weld through the material is required in welds made from one side only, a special melt-thru weld symbol is placed opposite the regular weld symbol. No dimension of melt-thru, except height of reinforcement, is shown on the weld symbol. See Fig. 17–25.

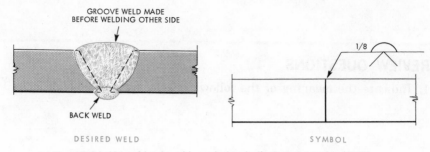

Fig. 17-24. Use of back weld symbol to indicate back weld. (AWS)

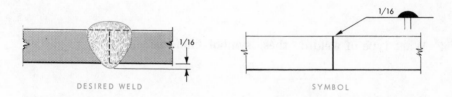

Fig. 17-25. Application of melt-thru symbol.

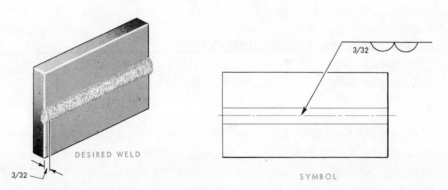

Fig. 17-26. Application of surfacing symbol to indicate surfaces built up by welding.

Surfacing Welds. Welds whose surfaces must be built up by single or multiple pass welding are provided with a surfacing weld symbol. The height of the built-up surface is indicated by a dimension placed to the left of the surfacing symbol. See Fig. 17–26.

REVIEW QUESTIONS

1. Indicate the meaning of the following weld symbols.

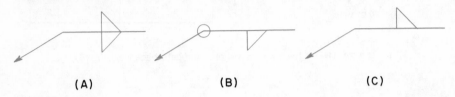

(A) (B) (C)

2. What type of weld do these symbols indicate?

(A) (B) (C) (D)

3. How would you interpret these symbols?

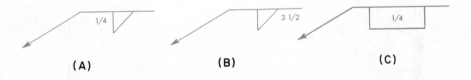

(A) (B) (C)

4. These symbols represent what weld specifications?

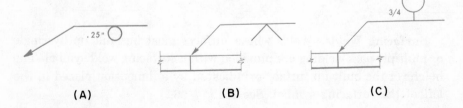

(A) (B) (C)

5. What do welds designated with the following symbols represent?

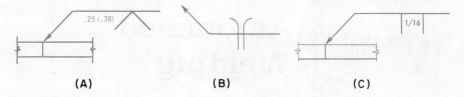

(A) (B) (C)

6. What is the meaning of each of these symbols?

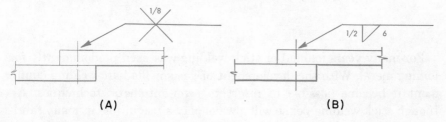

(A) (B)

7. What do these symbols represent?

(A) (B)

8. What do these symbols mean?

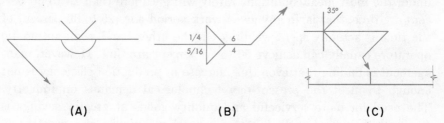

(A) (B) (C)

Chapter 18

Automated Welding

For many years manual or stick welding was used predominantly for joining metal. With the development of gas shielded-arc welding equipment it became possible to resort to semi-automatic techniques. Although stick welding per se will always play a useful role in many fabricating processes, particularly in small job shops and maintenance work, industries in general have to a large extent adopted semi-automatic equipment. Now many metal fabricating manufacturers are beginning to realize the necessity of moving even beyond the semi-automatic processes and going to fully automatic welding systems, even to the extent of using computers to control the automated equipment. See Fig. 18–1.

Advantages of Automated Welding

Studies have shown that the best stick welding operators working under the most ideal conditions rarely will get more than 25 to 30 percent arc time, that is, in any given work period only 25 to 30 percent of the time is actually spent welding. On the other hand, semi-automatic operators usually can achieve 50 to 75 percent arc time. However, management is finding that even this increase in production efficiency is not enough to meet the present-day technological demands on industry. The need for more serviceful and quality goods at greater savings is forcing industries to adopt better control techniques and more functional manufacturing procedures. Specifically, industry is facing the task of boosting production, turning out products of better quality, at lower labor cost and greater safety.

By resorting to fully automated welding, poor and ineffectual welds, operator inefficiency, and high manufacturing cost are reduced. Automated welding not only ensures faster deposition rates but repeatedly

Airco Welding Products

Fig. 18-1. This photograph shows the world's largest flame-cutting machine. It works from punched tape and flame cuts a variety of parts, from circles to complex contoured shapes, in plate between ⅛" and a full 12" in thickness.

produces welds of consistent quality. Although the initial cost of converting to automated welding may be relatively high at the outset, the consensus of welding engineers is that "we cannot afford not to". The eventual savings in manufacturing cost, by and large, will more than make up for any expense incurred in setting up a modern automated system.

Automated Welding Systems

In general, there are two basic systems for automated welding. With one, the welding equipment is stationary and the work flows around the welders. With the alternate plan, the work remains stationary and the welding equipment moves on a track to the designated position where welds are to be made. Either system is designed to give precise control over every welding factor, such as presurge time, hot start level, pulsation delay, initial current, upslope, weld taper delay, weld current, pulse level, high and low pulse time, final taper current, final current, postflow time and post heat. See Fig. 18–2.

There are virtually no limits as to the kind of welding that can be automated. Submerged arc, resistance, and Tig and Mig processes can

Fig. 18-2. This unit is designed for stationary mounting on a manipulator or other permanent fixture. Welding is accomplished by moving the work piece at the desired speed using a positioner or turning rolls. It can be used with gas-metal arc or submerged arc processes.

readily be automated. Mig welding is often preferred because of its greater versatility. This includes spray arc, short arc, cored-wire and CO_2.

For the most part, standard welding power units are used. The number of welding torches in the system is relative: the same results are obtained whether two or a dozen torches are operated. The system is usually tied into a power module which controls the variables in a welding sequence. See Fig. 18–3.

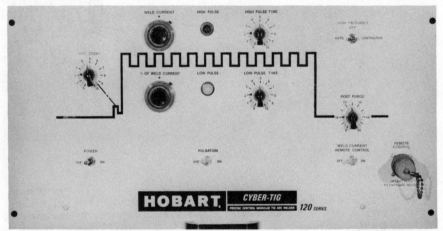

Hobart Brothers Co.

Fig. 18-3. An example of a power control module.

Types of Automated Welding Systems

Generally speaking, there are no conventional or standard types of automated welding systems. Outside of the regular welding units, most systems are specifically designed to function for an existing work flow configuration. Therefore, the type of system used depends on the product being manufactured, the manner in which production is processed, and the existing type of plant facilities. Accordingly, automation is strictly an engineering design problem of individual plants. Several examples to illustrate this point follow.

In one automotive industry, increased production is achieved by means of a traversing robot which moves on floor-mounted tracks or overhead rails and performs welding operations on car bodies or parts on an assembly line. See Fig. 18–4. Affixed to the robot, a spot welding gun is carried down the assembly line and automatically performs the desired welds.

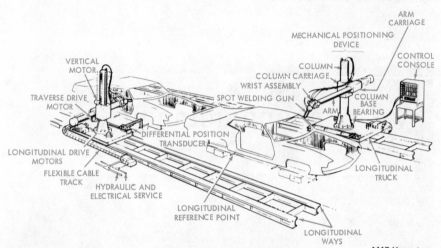

AMF Versatran

Fig. 18-4. Traversing welding robots move on floor-mounted tracks or overhead rails and perform welding operations automatically on car bodies or parts on an assembly line.

Another example is the manufacturer of motor stators which developed an automated system that (1) picks up and orients the correct number of laminations and hands them to the operator, (2) joins the laminations with four simultaneous Tig welds at 90-degree intervals around the circular stator and, (3) presses a cap ring over the stator and gas metal-arc spot welds it in place. Connected to the power sources are four stationary Tig welding torches. Welding is accomplished under the torches on a turntable. A ram smoothly lifts the stator unit past the torches. See Fig. 18–5.

A further example is that of a manufacturer of hydraulic equipment. Here an automated submerged arc system was designed to weld cylinders. This welding system moves on a floor track 50 feet long. The ram which carries a welding head has an eight-foot stroke. Welding current is supplied by a constant voltage power source which is mounted on the

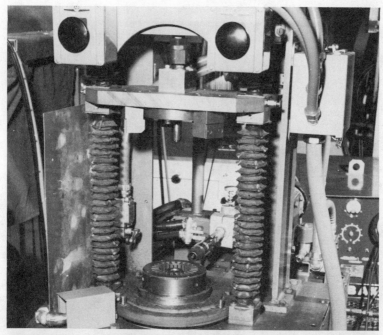

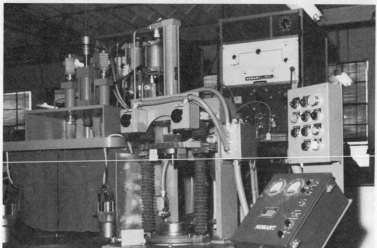

Fig. 18-5. The automatic stator welding system consists of an expanding mandrel which picks up the proper number of laminations, a Tig welding system, and a Mig spot welder. The Tig welder simultaneously makes four welds at equal intervals around the stator. In the Mig spot unit, a cap ring is forced around the stator, and the mandrel turns to bring each plug weld hole in the ring under the torch.

Automated Welding 459

Hobart Brothers Co.
Fig. 18-6. A large automatic welding system moves on a floor track 50 feet long.

frame of the headstock to allow it to move with the system. Air brakes hold the entire unit at any point on the track. See Fig. 18-6.

The fact that one man can often do 80 percent of the work of two men led a manufacturer of dual wheels for all types of farm equipment to develop an automated welding system. In this case, the automatic welders are used to join strips of metal into cylindrical rim blanks. These blanks are first run through an expander to form the proper configuration. A fixture clamps the rim blanks in place and holds the joint sides

460   *Welding Technology*

together during welding. A welding head which is mounted on a carriage moves over the rim and welds the joint. See Fig. 18–7.

Programmed Welding

The ultimate in any automated welding system is programmed welding where numerically controlled (N.C.) equipment is utilized. Just as N.C. is currently used in many machining operations, industries are finding that further economies are achieved by resorting to numerically controlled units for production welding.

Fig. 18-7. This automatic welding system joins strips of metal into cylindrical rim blanks.

Superior Electric Co.
Fig. 18-8. Heat exchangers are welded ten times faster with numerical control.

With programmed welding, all of the welding operations are preset and the entire welding process is initiated with the press of a button. The sequence of operations can be pre-programmed on punched tape to weld numerous different configurations automatically.

A typical example is the manufacturer of heat exchangers who, when converting to automatic numerical control, found that welds could be executed ten times faster. The N.C. unit automatically aligns the fixture containing the heat exchanger, inserts a shot pin, clamps the fixture, positions the assembly for welding, initiates the welding torch guidance system, and returns the exchanger to the start position. The only manual operation required is turning the fixture 180 degrees for welding the opposite end of the exchanger. While one unit is being welded, another is being loaded in a second fixture. This new system requires only 10 percent of the 70 to 80 minutes needed to do the work manually. The old method required putting parts together manually, placing them on a hand-turned spindle, then Mig-welding manually one side of the spiral at a time. With the new system, the parts are rotated past two stationary Mig welding torches feeding $\frac{1}{16}''$ flux-core wire. Torches positioned on both sides of the helix operate simultaneously, producing a stronger weld than before because of better heat penetration and the flux-core shield. See Fig. 18–8. Fig. 18–9 is a numerical control series program for cutting the part shown in Fig. 18–10.

G03	Counterclockwise arc
G04	Dwell .001 to 99.999 sec.
	Ex: G04 x 01 = one sec. dwell
G91	Incremental
G90	Absolute
G92	Preset of position register. If no selection is made. G91 is assumed.
G40	Cancel cutter comp.
G41	Left cutter comp.
G42	Right cutter comp.
M00	Program stop
M01	Optional stop
M02	End of program-machine stops-clears the register
M03	Initiate cutting
M04	Lower torches
M05	Oxygen off
M06	Torches and punch mark up
M07	Punch mark down
M08	Punch mark on
M09	Punch mark off
M14	Low speed range
M15	High speed range
M30	Rewind tape

N/C 900 Series Director

This director system employs integrated circuit modules and offers the following features:

1. Linear interpolation — with .01 to 327" block length limits.
2. Circular interpolation — full 360° with .01 to 327" radii limits.
3. Photo-electric reader with tumble box for 300 characters per second.
4. Incremental program command with reversible reader.
5. Dialed in feed rate at the control panel — no limitation on up or down requirements.
6. Program reverse within block of information.
7. Corner slow down — automatic with rapid traverse and at cutting speeds by tape command or by manual selection as required.
8. 8-channel punched tape to E1A RS 244 with tab sequential Essi format.
9. Kerf correction by dial control for up to 5/16" kerf widths (optional).

A sample N/C 800 Series program for cutting the part shown above follows:

Sequence Number	Preparatory Command	Torch Position Commands				Speed	Auxiliary Functions	Description of Operation
								Place torch above corner of plate
N001	G01	X010				F100		Displace torch w/marker
N002		X00725	Y00625					Depart to 1st position
N003							M07	Lower punch
N004							M08	Marker on
N005							M09	Marker off
N006		X022						Depart to next position
N007							M08	Marker on
N008							M09	Marker off
N009							M06	Raise punch
N010		X-010						Return torch to position
N011	G42							Right kerf on
N012	G01	X-011	Y016					Depart to first cut
N013							M04	Lower torches, initiate cutting
N014			Y-001			F020		Torch cutting, lead in
N015	G92	X0	Y0			F100		Set absolute zero posit.
N016	G90							Go to absolute
N017	G02	X-004	Y004		J004	F020		Cut clockwise circle
N018		X0	Y008	I004				Cut clockwise circle
N019		X004	Y004		J004			Cut clockwise circle
N020		Y0	Y0	I004				Cut clockwise circle
N021							M05	02 off
N022	G01	X-018	Y-007			F100		Depart to next cut
N023							M03	02 on
N024			Y-021			F020		Cutting
N025		X-018	Y-021			F100		*
N026		X018				F020		Cutting
N027		Y-008						Cutting
N028		X018	Y-008			F100		*
N029		X00424	Y00824			F020		Cutting
N030	G03	X0	Y01	I00424	J00424			Cutting
N031		X-00424	Y00824		J00424			Cutting
N032	G01	X-018	Y-008					Cutting
N033							M05	02 off
N034							M02	End of program

*Return blocks

National Cylinder Gas Div. of Chemetron Corp.

Fig. 18-9. A numerical control series program used for cutting the part shown in Fig. 18-10.

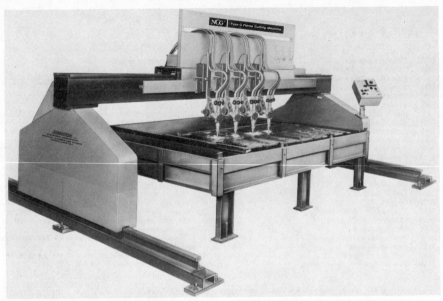

Fig. 18-10. Dual Drive Gantry flame cutting machine with photo-electric scanner or numerical control.

REVIEW QUESTIONS

1. What development has brought about a gradual replacement of "stick" welding for industrial production purposes?
2. Shielded metal-arc welding will always continue to play an important role in what type of work?
3. Why is semi-automatic or fully automatic welding more efficient than "stick" welding?
4. What are the two basic systems used for automated welding?
5. In automated welding, what are some of the variables that are precisely controlled?
6. Automated welding is adaptable for what kinds of welding?
7. Why is Mig welding often preferred for automated systems?
8. What determines the number of welding torches used in an automated system?

9. Why are there really no conventional or standard types of automated welding systems?
10. What determines the particular type of automated welding system used in a plant?
11. What is "programmed" welding?
12. How is programmed welding achieved?

Appendix A

METRIC CONVERSION

Rapid expansion of trade and industry on an international basis in the past two decades has increased the need for understanding of both the *metric* or CGS (Centimeter-Gram-Second) system used by nearly all countries of the world and the *English* or FPS (Foot-Pound-Second) system used by the United States and some other English-speaking countries.

In view of the increasing need for a universal system to measure lengths, areas, volumes, weights, temperatures, etc., it now seems likely that the CGS system will ultimately replace the FPS system. Actually, metric measurement should result in more accuracy because it eliminates multiplying and converting of fractions.

To convert a quantity from *English* to *metric* units:

1. If the English measurement is expressed in fractional form, change this to an equivalent decimal form.
2. Multiply this quantity by the factor shown on the Metric Conversion Table.
3. Round off the result to the precision required.

Relatively small measurements, such as 17.3 cm, are generally expressed in equivalent millimeter form. In this example the measurement would read as 173 mm.

CONVERSION OF ENGLISH TO METRIC UNITS

LENGTHS:		WEIGHTS:	
1 INCH	= 2.540 CENTIMETERS OR 25.40 MILLIMETERS	1 OUNCE (AVDP)	= 28.35 GRAMS
1 FOOT	= 30.48 CENTIMETERS OR 304.8 MILLIMETERS	1 POUND	= 453.6 GRAMS OR 0.4536 KILOGRAM
1 YARD	= 91.44 CENTIMETERS OR 0.9144 METERS	1 (SHORT) TON	= 907.2 KILOGRAMS
1 MILE	= 1.609 KILOMETERS	LIQUID MEASUREMENTS	
AREAS:		1 (FLUID) OUNCE	= 0.02957 LITER OR 28.35 GRAMS
1 SQ IN.	= 6.452 SQ CENTIMETERS OR 645.2 SQ MILLIMETERS	1 PINT	= 473.2 CU CENTIMETERS
1 SQ FT	= 929.0 SQ CENTIMETERS OR 0.0929 SQ METER	1 QUART	= 0.9463 LITER
1 SQ YD	= 0.8361 SQ METER	1 (U.S.) GALLON	= 3785 CU CENTIMETERS OR 3.785 LITERS
VOLUMES:		POWER MEASUREMENTS	
1 CU IN.	= 16.39 CU CENTIMETERS	1 HORSEPOWER	= 0.7457 KILOWATT
1 CU FT	= 0.02832 CU METER	TEMPERATURE MEASUREMENTS	
1 CU YD	= 0.7646 CU METER	TO CONVERT DEGREES FAHRENHEIT TO DEGREES CENTIGRADE, USE THE FOLLOWING FORMULA: DEG C = 5/9 (DEG F –32)	

Appendix B

WELDING TABLES

Table 1. Current Values for Standard Coated Electrodes.

E6010

STANDARD SIZE	AMPERAGE	VOLTAGE
3/32" x 12"	30-80	22-26
1/8" x 14"	80-120	24-28
5/32" x 14"	120-160	24-28
3/16" x 14"	140-220	26-30
7/32" x 18"	170-250	26-30
1/4" x 18"	200-300	28-32

E6011

STANDARD SIZE	AMPERAGE	VOLTAGE
3/32" x 12"	30-80	24-28
1/8" x 14"	80-120	24-28
5/32" x 14"	120-160	26-30
3/16" x 14"	140-220	26-30
7/32" x 18"	170-250	28-32
1/4" x 18"	225-325	28-32

E6012

STANDARD SIZE	AMPERAGE	VOLTAGE
3/32" x 12"	30-90	17-21
1/8" x 14"	80-120	18-22
5/32" x 14"	120-190	18-22
3/16" x 14"	140-240	20-24
7/32" x 18"	170-325	20-24
1/4" x 18"	250-400	20-24
5/16" x 18"	350-500	22-26

Table 1. Current Values for Standard Coated Electrodes. (Continued)

E6013

STANDARD SIZE	AMPERAGE	VOLTAGE
1/16" x 9"	20-40	16-20
5/64" x 12"	25-60	17-20
3/32" x 12"	30-80	18-20
1/8" x 14"	80-120	18-22
5/32" x 14"	120-190	18-22
3/16" x 14"	140-240	20-24
7/32" x 18"	225-300	22-26
1/4" x 18"	250-350	22-26

E6020

STANDARD SIZE	AMPERAGE	VOLTAGE
5/32" x 14"	130-190	30-32
3/16" x 18"	175-250	32-34
7/32" x 18"	225-325	32-34
1/4" x 18"	250-350	32-34
5/16" x 18"	325-450	34-36
3/8" x 18"	450-600	36-38

E6027

STANDARD SIZE	AMPERAGE	VOLTAGE
3/16" x 18"	250-325	28-32
7/32" x 18"	275-350	28-32
1/4" x 18"	375-450	28-32

E7014

STANDARD SIZE	AMPERAGE	VOLTAGE
3/32" x 12"	80-120	18-22
1/8" x 14"	100-140	18-22
5/32" x 14"	140-190	20-24
3/16" x 14"	180-260	22-26
7/32" x 18"	250-325	24-28
1/4" x 18"	300-400	26-30
5/16" x 18"	450-550	26-30

Table 1. Current Values for Standard Coated Electrodes. (Continued)

E7016

STANDARD SIZE	AMPERAGE	VOLTAGE
3/32" x 12"	60-100	20-24
1/8" x 14"	80-120	22-26
5/32" x 14"	140-190	22-26
3/16" x 14"	170-250	22-26
7/32" x 18"	240-325	24-28
1/4" x 18"	300-400	24-28

E7018

STANDARD SIZE	AMPERAGE	VOLTAGE
3/32" x 12"	70-120	18-22
1/8" x 14"	100-150	22-24
5/32" x 14"	120-200	22-24
3/16" x 14"	200-275	22-24
7/32" x 18"	275-350	24-26
1/4" x 18"	300-400	24-26

E7028

STANDARD SIZE	AMPERAGE	VOLTAGE
3/16" x 18"	250-325	24-30
7/32" x 18"	325-400	24-30
1/4" x 18"	400-500	26-32

E8016-C1

STANDARD SIZE	AMPERAGE	VOLTAGE
1/8" x 14"	80-120	22-24
5/32" x 14"	150-185	22-24
3/16" x 14"	200-250	24-26
1/4" x 18"	300-425	24-26

E8018-C3

STANDARD SIZE	AMPERAGE	VOLTAGE
3/32" x 12"	70-110	18-22
1/8" x 14"	100-150	18-22
5/32" x 14"	120-180	18-22
3/16" x 14"	180-275	20-24
7/32" x 18"	275-350	22-26
1/4" x 18"	300-400	22-26

Table 1. Current Values for Standard Coated Electrodes. (Continued)

E9018-M

STANDARD SIZE	AMPERAGE	VOLTAGE
3/32" x 12"	70-110	18-22
1/8" x 14"	100-150	18-22
5/32" x 14"	120-180	18-22
3/16" x 14"	180-275	20-24
7/32" x 18"	275-350	22-26
1/4" x 18"	300-400	22-26

E10016-D2

STANDARD SIZE	AMPERAGE	VOLTAGE
3/32" x 12"	60-100	20-24
1/8" x 14"	80-120	22-26
5/32" x 14"	140-190	22-26
3/16" x 14"	180-250	22-26
1/4" x 18"	300-400	24-28

E11018-M

STANDARD SIZE	AMPERAGE	VOLTAGE
3/32" x 12"	70-120	18-22
1/8" x 14"	100-140	18-22
5/32" x 14"	120-200	18-22
3/16" x 14"	180-275	20-24
7/32" x 18"	275-350	22-26
1/4" x 18"	300-400	22-26

Stainless Steel

STANDARD SIZE	AMPERAGE	VOLTAGE
1/16" x 14"	20-40	14-18
3/32" x 14"	30-60	16-20
1/8" x 14"	75-125	18-22
5/32" x 14"	125-150	20-24
3/16" x 14"	140-175	22-28
1/4" x 14"	170-225	24-32

Table 1. Current Values for Standard Coated Electrodes. (Continued)

Aluminum Bronze

STANDARD SIZE	AMPERAGE	VOLTAGE
5/64" x 11"	40-80	22-26
3/32" x 11"	50-90	22-28
1/8" x 14"	90-130	24-30
5/32" x 14"	130-150	24-32
3/16" x 14"	150-210	26-34
1/4" x 18"	210-275	28-36

Aluminum

STANDARD SIZE	AMPERAGE		VOLTAGE	
1/16"	20	40	22	24
5/64"	30	60	23	25
3/32"	60	90	23	25
1/8"	90	120	24	26
5/32"	120	160	24	26
3/16"	150	190	25	27
1/4"	225	300	26	28
5/16"	275	350	26	30

Table 2. Tig Welding—Aluminum.

Stock Thickness (inch)	Type of Joint	Amperes, AC Current			Electrode Diameter (inch)	Argon Flow 20 psi		Filler Rod Diameter (inch)
		Flat	Horizontal & Vertical	Overhead		lpm	cfh	
1/16	Butt	60-80	60-80	60-80	1/16	7	15	1/16
	Lap	70-90	55-75	60-80	1/16	7	15	1/16
	Corner	60-80	60-80	60-80	1/16	7	15	1/16
	Fillet	70-90	70-90	70-90	1/16	7	15	1/16
1/8	Butt	125-145	115-135	120-140	3/32	8	17	1/8
	Lap	140-160	125-145	130-160	3/32	8	17	1/8
	Corner	125-145	115-135	130-150	3/32	8	17	1/8
	Fillet	140-160	115-135	140-160	3/32	8	17	1/8
3/16	Butt	190-220	190-220	180-210	1/8	10	21	5/32
	Lap	210-240	190-220	180-210	1/8	10	21	5/32
	Corner	190-220	180-210	180-210	1/8	10	21	5/32
	Fillet	210-240	190-220	180-210	1/8	10	21	5/32
1/4	Butt	260-300	220-260	210-250	3/16	12	25	3/16
	Lap	290-340	220-260	210-250	3/16	12	25	3/16
	Corner	280-320	220-260	210-250	3/16	12	25	3/16
	Fillet	280-320	220-260	210-250	3/16	12	25	3/16

(Linde Co.)

psi — pounds per square inch
lpm — liters per minute
cfh — cubic feet per hour

Table 3.   Color Code for Covered Electrodes.

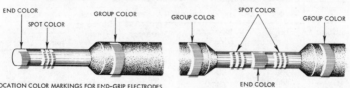

END COLOR · SPOT COLOR · GROUP COLOR · GROUP COLOR · SPOT COLOR · GROUP COLOR

LOCATION COLOR MARKINGS FOR END-GRIP ELECTRODES

END COLOR

LOCATION COLOR MARKINGS FOR CENTER-GRIP ELECTRODES

Mild Steel and Low-Alloy Steel
XX10, XX11, XX14, XX24, XX27, XX28 and all 60XX
GROUP COLOR — NONE

CLASS	END COLOR	SPOT COLOR
E6010	None	None
E6012	None	White
E6013	None	Brown
E6020	None	Green
E6015	None	Red
E6011	None	Blue
E6024	None	Yellow
E6016	None	Orange
E6030	None	Violet
E6027	None	Silver
E7010G	Blue	None
E7010-A1	Blue	White
E7011G	Blue	Blue
E7011-A1	Blue	Yellow
E7014	Black	Brown
E7024	Black*	Yellow
E7028	Black	Black
E8010G	White	None
E8010-B1	White	Brown
E8010-B2	White	Green
E8011G	White	Blue
E8011-B1	White	Black
E9010G	Brown	None
E9011G	Brown	Blue
E8011-B2	Brown	Black
E10010G	Green	None
E10011G	Green	Blue
E6014	Red	Brown
E6028	Red	Black
E6018	Red	Orange
Mil G6010	Silver	None
Mil G6011	Silver	Blue

Low-Hydrogen Low-Alloy Steel
XX15, XX16 and XX18 Except E60XX
GROUP COLOR — GREEN

CLASS	END COLOR	SPOT COLOR
E7015G	None	Red
E7016G	None	Orange
E7018G	None	Blue
E7015	Blue	Red
E7015-A1	Blue	White
E7016	Blue	Orange
E7016-A1	Blue	Yellow
E9015-B3L	Black	White
E8015-B2L	Black	Green
E8015-B4L	Black	Bronze
E7018	Black	Orange
E7018-A1	Black	Yellow
E8018-C3	Black	Black
8018G	Black	Blue
Mil 10018	Black	Violet
E8018-B4	Black	Gray
Mil 12018	Black	Silver
E8015-C3	White	Red
E8015-C2	White	White
E8015-B1	White	Brown
E8015-B2	White	Green
E8015-C1	White	Bronze
E8016-C3	White	Orange
E8016G	White	Yellow
E8016-B1	White	Black
E8016-C1	White	Blue
E8016-C2	White	Violet

CLASS	END COLOR	SPOT COLOR
E8016-B2	White	Gray
E8015G	Gray	Red
Mil 94LC-16	Gray	White
Mil 52LC-16	Gray	Green
Mil 82LC-16	Gray	Bronze
E8018-B1	Gray	Black
E8018-C1	Gray	Blue
E8018-C2	Gray	Violet
E8018-B2	Gray	Gray
E9015G	Brown	Red
E9015-D1	Brown	White
E9015-B3	Brown	Green
E8015-B4	Brown	Bronze
E9016G	Brown	Orange
E9016-D1	Brown	Yellow
E9016-B3	Brown	Blue
E8016-B4	Brown	Violet
Mil 9018	Violet	Orange
E9018-B3	Violet	Black
E9018G	Violet	Blue
E9018-D1	Violet	Violet
E10015G	Green	Red
E10016G	Green	Orange
E10015-D2	Green	Yellow
E10018G	Green	Blue
E10018-D2	Green	Violet
E10016-D2	Green	Gray
Mil 11018	Red	None
E11015G	Red	Red
Mil 11015	Red	White
Mil 9018	Red	Orange
E11016G	Red	Yellow
E11018G	Red	Blue
Mil 260-15	Yellow	Red
Mil 12015	Yellow	White
Mil 260-16	Yellow	Orange
Mil 12016	Yellow	Yellow
E12015G	Orange	Red
E12016G	Orange	Orange
E12018G	Orange	Blue
Mil 230-15	Silver	Red
Mil 230-16	Silver	Orange

Low Alloy Steel
All XX13 and XX20 Except E60XX
GROUP COLOR — SILVER

CLASS	END COLOR	SPOT COLOR
E7020G	Blue	Green
E7020-A1	Blue	Yellow
E8013G	White	Brown
E8013-B1	White	White
E9013G	Brown	Brown
E8013-B2	Brown	White
E10013G	Green	Brown

Chromium and Chromium-Nickel Steel
GROUP COLOR — BLACK FOR DC, YELLOW FOR AC-DC

CLASS	END COLOR	SPOT COLOR
Mil 308-MoL-15	None	None
Mil 308-MoT-15	None	None
Mil 307-L-15	None	Black
Mil 307-T-15	None	Black
Mil 16.8.2-15	Blue	Yellow
E308LC-15	Brown	None

CLASS	END COLOR	SPOT COLOR
Mil 202-LC-15	Brown	Blue
E316LC-15	Brown	White
E330-15	Green	None
E312-15	Green	Red
Type 330HiC	Green	Black
E310-15	Red	None
E310Cb-15	Red	Blue
E310Mo-15	Red	White
20-29 Cu Mo	Red	Brown
E308-15	Yellow	None
E347-15	Yellow	Blue
E316-15	Yellow	White
E317-15	Yellow	Brown
E318-15	Yellow	Green
Mil 308-HC-15	Yellow	Red
Mil 347-HC-15	Yellow	Yellow
Type 349	Yellow	Orange
E309-15	Black	None
E309Cb-15	Black	Blue
E309Mo-15	Black	White
E502-15	Gray	Blue
Type 505	Gray	White
E410-15	Gray	Brown
E430-15	Gray	Green
Type 442	Gray	Red
Type 446	Gray	Yellow

Copper and Copper Alloy
GROUP COLOR — BLUE

CLASS	END COLOR	SPOT COLOR
ECu-Ni	None	Blue
ECu	Green	None
ECu-Si	Red	None
ECu-SnA	Yellow	None
ECu-SnC	Yellow	Blue
ECu-A1-A2	Silver	Blue
ECu-A1-B	Silver	Brown
ECu-A1-C	Silver	Green
ECu-A1-D	Silver	Red
ECu-A1-E	Silver	Yellow

Nickel, Nickel-Alloy and High-Temperature-Alloy
GROUP COLOR — WHITE

CLASS	END COLOR	SPOT COLOR
E3N10	Blue	White
E4N10	Blue	Brown
E3N14	Blue	Red
E3N1B	White	Green
Mil 4N1W	White	Yellow
E3N1C	White	Violet
Ni-Cr-60-13	Green	Blue
Ni-Cr-85-15	Green	White
E4N12	Green	Brown
E3N11	Yellow	White
E4N11	Yellow	Brown
ENi	Orange	Blue
ENiCu	Orange	White
ENiFe	Orange	Brown
ENiCuB	Orange	Green
E3N12	Violet	White
E3N19	Violet	Red
Mil 3N1L	Bronze	White
Mil 3N1N	Bronze	Orange

Table 4.   Tig Welding—Stainless Steel.

Stock Thickness (inch)	Type of Joint	Amperes, DC Current—Straight Polarity			Electrode Diameter (inch)	Argon Flow 20 psi		Filler Rod Diameter (inch)
		Flat	Horizontal & Vertical	Overhead		lpm	cfh	
1/16	Butt	80–100	70–90	70–90	1/16	5	11	1/16
	Lap	100–120	80–100	80–100	1/16	5	11	1/16
	Corner	80–100	70–90	70–90	1/16	5	11	1/16
	Fillet	90–110	80–100	80–100	1/16	5	11	1/16
3/32	Butt	100–120	90–110	90–110	1/16	5	11	1/16
	Lap	110–130	100–120	100–120	1/16	5	11	1/16
	Corner	100–120	90–110	90–110	1/16	5	11	1/16
	Fillet	110–130	100–120	100–120	1/16	5	11	1/16
1/8	Butt	120–140	110–130	105–125	1/16	5	11	3/32
	Lap	130–150	120–140	120–140	1/16	5	11	3/32
	Corner	120–140	110–130	115–135	1/16	5	11	3/32
	Fillet	130–150	115–135	120–140	1/16	5	11	3/32
3/16	Butt	200–250	150–200	150–200	3/32	6	13	1/8
	Lap	225–275	175–225	175–225	3/32	6	13	1/8
	Corner	200–250	150–200	150–200	3/32	6	13	1/8
	Fillet	225–275	175–225	175–225	3/32	6	13	1/8
1/4	Butt	275–350	200–250	200–250	1/8	6	13	3/16
	Lap	300–375	225–275	225–275	1/8	6	13	3/16
	Corner	275–350	200–250	200–250	1/8	6	13	3/16
	Fillet	300–375	225–275	225–275	1/8	6	13	3/16

(Linde Co.)

Table 5.   Tig Welding—Magnesium.

Stock Thickness (inch)	Type of Joint	Amperes AC Current	Welding Rod Diameter (inch)	Argon Flow 15 psi		Remarks
		Flat Position		lpm	cfh	
0.040	Butt	45	3/32, 1/8	6	13	Backup
0.040	Butt	25	3/32, 1/8	6	13	No backing
0.040	Fillet	45	3/32, 1/8	6	13	
0.064	Butt	60	3/32, 1/8	6	13	Backup
0.064	Butt and Corner	35	3/32, 1/8	6	13	No backing
0.064	Fillet	60	3/32, 1/8	6	13	
0.081	Butt	80	1/8	6	13	Backup
0.081	Butt, Corner and Edge	50	1/8	6	13	No backing
0.081	Fillet	80	1/8	6	13	
0.102	Butt	100	1/8	9	19	Backup
0.102	Butt, Corner and Edge	70	1/8	9	19	No backing
0.102	Fillet	100	1/8	9	19	
0.128	Butt	115	1/8, 5/32	9	19	Backup
0.128	Butt, Corner and Edge	85	1/8, 5/32	9	19	No backing
0.128	Fillet	115	1/8, 5/32	9	19	
3/16	Butt	120	1/8, 5/32	9	19	1 pass
3/16	Butt	75	1/8, 5/32	9	19	2 passes
1/4	Butt	130	5/32, 3/16	9	19	1 pass
1/4	Butt	85	5/32	9	19	2 passes

(Linde Co.)

Table 6. Tig Welding—Silicon Bronze.

Stock Thickness (inch)	Type of Joint	Amperes, DC Current—Straight Polarity			Electrode Diameter (inch)	Argon Flow 20 psi		Filler Rod Diameter (inch)
		Flat	Horizontal & Vertical	Overhead		lpm	cfh	
1/16	Butt	80-100	70-90	70-90	1/16	5	11	1/16
	Lap	100-120	80-100	80-100	1/16	5	11	1/16
	Corner	80-100	70-90	70-90	1/16	5	11	1/16
	Fillet	90-110	80-100	80-100	1/16	5	11	1/16
3/32	Butt	100-120	90-110	90-110	1/16	5	11	1/16
	Lap	110-130	100-120	100-120	1/16	5	11	1/16
	Corner	100-120	90-110	90-110	1/16	5	11	1/16
	Fillet	110-130	100-120	100-120	1/16	5	11	1/16
1/8	Butt	120-140	110-130	105-125	1/16	5	11	3/32
	Lap	130-150	120-140	120-140	1/16	5	11	3/32
	Corner	120-140	110-130	115-135	1/16	5	11	3/32
	Fillet	130-150	115-135	120-140	1/16	5	11	3/32
3/16	Butt	200-250	150-200	150-200	3/32	6	13	1/8
	Lap	225-275	175-225	175-225	3/32	6	13	1/8
	Corner	200-250	150-200	150-200	3/32	6	13	1/8
	Fillet	225-275	175-225	175-225	3/32	6	13	1/8
1/4	Butt	275-350	200-250	200-250	1/8	6	13	3/16
	Lap	300-375	225-275	225-275	1/8	6	13	3/16
	Corner	275-350	200-250	200-250	1/8	6	13	3/16
	Fillet	300-375	225-275	225-275	1/8	6	13	3/16

(Linde Co.)

Table 7. Tig Welding—Copper.

Stock Thickness (inch)	Type of Joint	Amperes AC Current	Welding Rod Diameter (inch)	Argon Flow 15 psi		Remarks
		Flat Position		lpm	cfh	
0.040	Butt	45	3/32, 1/8	6	13	Backup
0.040	Butt	25	3/32, 1/8	6	13	No backing
0.040	Fillet	45	3/32, 1/8	6	13	
0.064	Butt	60	3/32, 1/8	6	13	Backup
0.064	Butt and Corner	35	3/32, 1/8	6	13	No backing
0.064	Fillet	60	3/32, 1/8	6	13	
0.081	Butt	80	1/8	6	13	Backup
0.081	Butt, Corner and Edge	50	1/8	6	13	No backing
0.081	Fillet	80	1/8	6	13	
0.102	Butt	100	1/8	9	19	Backup
0.102	Butt, Corner and Edge	70	1/8	9	19	No backing
0.102	Fillet	100	1/8	9	19	
0.128	Butt	115	1/8, 5/32	9	19	Backup
0.128	Butt, Corner and Edge	85	1/8, 5/32	9	19	No backing
0.128	Fillet	115	1/8, 5/32	9	19	
3/16	Butt	120	1/8, 5/32	9	19	1 pass
3/16	Butt	75	1/8, 5/32	9	19	2 passes
1/4	Butt	130	5/32, 3/16	9	19	1 pass
1/4	Butt	85	5/32	9	19	2 passes

(Linde Co.)

Table 8. Tig Welding—Carbon and Low Alloy Steels.

Stock Thickness (inch)	Amperes DC Current Straight Polarity	Filler Rod Diameter (inch)	Argon Flow 20 psi	
			lpm	cfh
0.035	100	1/16	4-5	8-10
0.049	100-125	1/16	4-5	8-10
0.060	125-140	1/16	4-5	8-10
0.089	140-170	1/16	4-5	8-10

(Linde Co.)

Table 9. Tig Welding—Gray Cast Iron.

Stock Thickness (inch)	Type of Joint	Position	Welding Current		Filler Rod Diameter (inch)	Argon Flow 20 psi	
			Type	Amps.		lpm	cfh
1/4	Butt	Flat	*ACHF or **DCSP	160	3/16, 1/4	8	16
1/4	Butt	Vertical	ACHF	150	3/16	8	16
1/4	Butt	Overhead	ACHF	150	3/16	8	16
1	Butt	Flat	DCSP	300-350	3/16 - 1/4 for 1st and 2nd passes. Larger rod for remaining passes.	12	24

*AC High Frequency
**DC Straight Polarity

(Linde Co.)

Table 10. Tig Welding—Nickel and Monel.

Metal	Type of Joint	Stock Thickness (inch)	Argon Flow lpm	Welding Current DCSP
Nickel	Butt	1/8	12	200
Monel	Butt	1/8	12	200

(Linde Co.)

Table 11.   Mig Welding—Aluminum (short arc).

PLATE THICKNESS (INCHES)	TYPE OF JOINT	WIRE DIAM. (INCHES)	ARGON FLOW (CFH)	AMPERES (DCRP)	VOLTAGE (VOLTS)	APPROXIMATE WIRE FEED SPEED (IPM)
0.040	FILLET OR TIGHT BUTT	0.030	30	40	15	240
0.050	FILLET OR TIGHT BUTT	0.030	15	50	15	290
0.063	FILLET OR TIGHT BUTT	0.030	15	60	15	340
0.093	FILLET OR TIGHT BUTT	0.030	15	90	15	410

Table 12.   Mig Welding—Aluminum (spray arc).

PLATE THICKNESS	PREPARATION	WIRE DIAMETER (IN.)	ARGON FLOW (CFH)	AMPERES (DCRP)	VOLTAGE
.250	SINGLE VEE BUTT (60° INCLUDED ANGLE) SHARP NOSE BACKUP STRIP USED	3/64	35	180	24
	SQUARE BUTT WITH BACKUP STRIP	3/64	40	250	26
	SQUARE BUTT WITH NO BACKUP STRIP	3/64	35	220	24
.375	SINGLE VEE BUTT (60° INCLUDED ANGLE) SHOP NOSE, BACKUP STRIP USED	1/16	40	280	27
	DOUBLE VEE BUTT (75° INCLUDED ANGLE, 1/16-IN. NOSE). NO BACKUP. BACK CHIP AFTER ROOT PASS	1/16	40	260	26
	SQUARE BUTT WITH NO BACKUP STRIP	1/16	50	270	26
.500	SINGLE VEE BUTT (60° INCLUDED ANGLE) SHARP NOSE. BACKUP STRIP USED	1/16	50	310	27
	DOUBLE VEE BUTT (75° INCLUDED ANGLE 1/16-IN. NOSE). NO BACKUP. BACK CHIP AFTER ROOT PASS	1/16	50	300	27

Table 13.   Mig Welding—Carbon Steel.

PLATE THICKNESS (IN.)	JOINT AND EDGE PREPARATION	WIRE DIAMETER (IN.)	GAS FLOW (CFH)	DCRP (AMPS)	VOLTAGE	WIRE FEED SPEED (IPM)
.035				55	16*	117
.047				65	17*	140
.063	NON-POSITIONED FILLET OR LAP	.030	10-15	85	17*	170
.078				105	18*	225
.100				110	18*	225
1/8				130	19*	300
1/8	BUTT (SQUARE EDGE)	1/16		280	--	165
3/16	BUTT (SQUARE EDGE)	1/16		375	--	260
3/16	FILLET OR LAP	1/16		350	--	230
1/4	DOUBLE VEE BUTT (60° INCLUDED ANGLE, NO NOSE)		MIXTURE (75%A + 25% CO_2)	375 (1ST PASS) 430 (2ND PASS)	27	83 (1ST) 95 (2ND)
5/16	DOUBLE VEE BUTT (60° INCLUDED ANGLE, NO NOSE)		40-50	400 (1ST PASS) 420 (2ND PASS)	28	87 (1ST) 92 (2ND)
5/16	NON-POSITIONED FILLET			400		87
1/2	DOUBLE VEE BUTT (60° INCLUDED ANGLE, NO NOSE	3/32	MIXTURE (95% ARGON + 5% O_2)	400 (1ST PASS 450 (2ND PASS)		87 (1ST) 100 (2ND)
1/2	NON-POSITIONED FILLET			450	28	100
3/4	DOUBLE VEE BUTT (90° INCLUDED ANGLE, NO NOSE)			450 (ALL 4 PASSES)	29	100
3/4	POSITIONED FILLET			475	30	110
1	FILLET			450 (ALL 4 PASSES)	28	100

*SHORT ARC

Table 14. Mig Welding—Stainless Steel.

MATERIAL THICKNESS T INCH	ELECTRODE SIZE INCH	WELDING CONDITIONS C.C.R.P.		GAS FLOW C.F.H.	TRAVEL SPEED I.P.M.
		ARC VOLTS	AMPERES		
.025	.030	15 - 17	30 - 50	15 - 20	15 - 20
.031	.030	15 - 17	40 - 60	15 - 20	18 - 22
.037	.035	15 - 17	65 - 85	15 - 20	35 - 40
.050	.035	17 - 19	80 - 100	15 - 20	35 - 40
.062	.035	17 - 19	90 - 110	20 - 25	30 - 35
.078	.035	18 - 20	110 - 130	20 - 25	25 - 30
.125	.035	19 - 21	140 - 160	20 - 25	20 - 25
.125	.045	20 - 23	180 - 200	20 - 25	27 - 32
.187	.035	19 - 21	140 - 160	20 - 25	14 - 19
.187	.045	20 - 23	180 - 200	20 - 25	18 - 22
.250	.035	19 - 21	140 - 160	20 - 25	10 - 15
.250	.045	20 - 23	180 - 200	20 - 25	12 - 18

SHIELDING GAS: CO_2 WELDING GRADE
TIP-TO-WORK DISTANCE (STICK-OUT) - 1/4 TO 3/8 INCH

Table 15. Mig Welding—Micro-wire Welding.

SHEET OR PLATE THICKNESS (IN.)*	FILLER WIRE DIAM. (IN.)	CURRENT AMPS.	WIRE FEED (IN./MIN.)	GAS AND FLOW CFH	WELDING POSITION**
1/16***	0.035	110/140	230/260	He 20/30	F, H, VD
1/8-3/16	0.035	110/140	230/260	He 20/30	F, H, OH, VD
1/4-1	0.035	110/140	230/260	A+ 1% O_2 20/30	V, OH
1/4-1	0.035	170/190	330/360	A+ 1% O_2 20/30	F, H, OH‡, VD
1/2-1	0.045	140/180	160/200	A+ 1% O_2 20/30	V, OH
3/16-3/8	0.045	190/310	210/340	A+ 1% O_2 30/40	F, H
7/16 & UP	0.045	190/310	210/340	A+ 1% O_2 20/40	OH‡
1/4 & UP	1/16"	280/350	240/330	A+ 1% O_2 30/40	F, H

*ALL JOINT TYPES—BUT, LAP & FILLET
**F—FLAT; H—HORIZONTAL; OH—OVERHEAD; VD—VERTICAL DOWN; V—VERTICAL UP
***GOOD FIT-UP-REQUIRED
‡ WEAVE BEAD
‡ STRINGER BEAD

Table 16. Average Cost Data for Manual Oxyacetylene Welding of Iron and Steel.

Thickness of Steel Inches	Joint Preparation No Spacing	Diameter of Rod Inches	Tip Drill Size	Oxygen—Cubic Feet Per Hour	Oxygen—Cubic Feet Per Linear Foot Welded	Acetylene—Cubic Feet Per Hour	Acetylene—Cubic Feet Per Linear Foot Welded	Pounds of Rods Per Hour	Pounds of Rods Per Foot	Speed Foot per Hour
1/64	Square Butt	1/32	75	0.7	0.03	0.7	0.03			26.0–30.0
1/32	Square Butt	1/32	75–60	1.0	0.05–0.04	1.0	0.05–0.04			22.0–25.0
1/16	Square Butt	1/16	60–56	2.4	0.13–0.11	2.3	0.13–0.11	0.23–0.27	0.013	18.0–21.0
3/32	Square Butt	1/16	60–54	5.1	0.36–0.30	4.9	0.36–0.29	0.42–0.51	0.030	14.0–17.0
1/8	Square Butt	3/32	60–53	8.8	0.80–0.68	8.5	0.77–0.65	0.58–0.69	0.053	11.0–13.0
3/16	90° Single V	1/8	56–53	17.7	2.36–2.08	17.0	2.27–2.00	1.13–1.28	0.150	7.5–8.5
1/4	90° Single V	3/16	53–49	27.0	4.50–3.86	26.0	4.33–3.72	1.59–1.86	0.265	6.0–7.0
5/16	90° Single V	1/4	49–44	33.0	7.40–6.05	32.0	7.11–5.82	1.87–2.28	0.414	4.5–5.5
3/8	90° Single V	1/4	44–40	45.7	11.42–9.13	44.0	11.0–8.80	2.39–2.98	0.597	4.0–5.0
1/2	90° Single V	1/4	43–36	58.2	11.65–9.70	56.0	11.2–9.33	2.90–3.48	0.637	5.0–6.0
5/8	60° Single V	5/16	40–36	73.8	21.10–16.42	71.0	20.30–15.79	3.06–3.92	0.872	3.5–4.5
3/4	60° Single V	5/16	36–32	91.5	36.60–26.16	88.0	35.20–25.17	3.27–4.57	1.307	2.5–3.5

Table 17. Approximate Weight of Weld Metal in 60° and 90° Single Vee Joints.

Thickness of Metal Inches	Weld Metal in 1" Length of 60° Vee Cu. In.	Weight of Weld Metal 1-inch Length of 60° Vee Joint in Pounds — Steel	Armco Iron	Stainless Steel	Nickel	Page Bronze	Weld Metal in 1" Length of 90° Vee Cu. In.	Weight of Weld Metal 1-inch Length of 90° Vee Joint in Pounds — Steel	Armco Iron	Stainless Steel	Nickel	Page Bronze
1/4	0.035	0.0098	0.0099	0.0101	0.0112	0.0105	0.062	0.0174	0.0176	0.0179	0.0198	0.0187
3/8	0.080	0.0224	0.0227	0.0232	0.0255	0.0240	0.140	0.0392	0.0397	0.0405	0.0446	0.0421
1/2	0.144	0.0403	0.0408	0.0417	0.0459	0.0432	0.250	0.0700	0.0709	0.0723	0.0796	0.0751
5/8	0.225	0.0630	0.0638	0.0651	0.0716	0.0676	0.390	0.1092	0.1105	0.1128	0.1241	0.1172
3/4	0.324	0.0907	0.0918	0.0937	0.1031	0.0973	0.562	0.1574	0.1593	0.1625	0.1789	0.1689
7/8	0.441	0.1235	0.1250	0.1275	0.1404	0.1325	0.765	0.2142	0.2168	0.2211	0.2435	0.2298
1	0.577	0.1616	0.1635	0.1668	0.1837	0.1734	1.000	0.2800	0.2833	0.2890	0.3182	0.3004
1 1/8	0.729	0.2041	0.2066	0.2107	0.2320	0.2190	1.265	0.3542	0.3584	0.3656	0.4026	0.3801
1 1/4	0.901	0.2523	0.2553	0.2604	0.2867	0.2707	1.562	0.4371	0.4425	0.4515	0.4971	0.4690
1 3/8	1.090	0.3052	0.3088	0.3151	0.3469	0.3275	1.890	0.5292	0.5355	0.5463	0.6014	0.5678
1 1/2	1.298	0.3634	0.3678	0.3752	0.4131	0.3899	2.250	0.6300	0.6375	0.6503	0.7160	0.6760
1 5/8	1.523	0.4265	0.4315	0.4402	0.4847	0.4576	2.640	0.7392	0.7480	0.7630	0.8401	0.7932
1 3/4	1.766	0.4945	0.5003	0.5094	0.5620	0.5306	3.062	0.8574	0.8675	0.8850	0.9744	0.9200
1 7/8	2.028	0.5679	0.5746	0.5861	0.6454	0.6094	3.515	0.9842	0.9958	1.0159	1.1185	1.0560
2	2.308	0.6462	0.6539	0.6673	0.7345	0.6934	4.000	1.1200	1.1332	1.1560	1.2728	1.2018
2 1/4	2.920	0.8176	0.8273	0.8439	0.9292	0.8773	5.062	1.4174	1.4341	1.4630	1.6108	1.5209
2 1/2	3.606	1.0097	1.0216	1.0422	1.1475	1.0834	6.250	1.7500	1.7707	1.8063	1.9888	1.8778
2 3/4	4.363	1.2217	1.2361	1.2610	1.3884	1.3109	7.562	2.1174	2.1423	2.1854	2.4063	2.2720
3	5.196	1.4549	1.4721	1.5017	1.6534	1.5611	9.000	2.5200	2.5497	2.6010	2.8638	2.7040

The data in the above table considers the metal veed from only one side of the joint; if the metal is veed from both sides of the joint, the volume and weight of the weld metal required are one-half of the above figures.

Index

Bold face figure indicates illustration.

A

Abrasion, 263
Advantages, 334
Air-acetylene welding, 13
Aircomatic welding, 70
Allotropic metals, 158
Allowable
  loads, **354**
  stress, **307**
Alloys, 158
  iron-based, 165
  iron carbide, **166**
  mechanical mixture, 161
  solid solution, 159
Alloy steel, 203
  properties, 206
  weldability, 205
Alternating current, 51
Aluminum
  alloys, 216
  atomic structure, **154**
  designation, **213**, 216
  properties, 212
  weldability, **217**, 218
Ammonium persulphate, 382
Annealing, 172
Arc
  characteristics, **86**
  length, 30
  starting, 27, **29**, 58, 95
Arc cutting, 283, 296
  amperage setting, **296**
  arc-air, **299**, 300
  hole piercing, **298**
  plasma arc, 300, **301**
Arc-air cutting, **299**, 300
Arc welding
  buried, 95, 96
  gas tungsten, 45
  pulsed spray, 97-103
  safety, 435
  submerged, 105, **106**, **107**
  tubular wire, 103

vapor-shielded, **104**
Argon, 84
Argon-CO_2, 88
Argon-helium CO_2, 88
Argon plus oxygen, 85
Assembly, 365
Atom, **154**
Automated welding, 454, **455**
  advantages, 454
  systems, 455-461, **455**

B

Backfire, 431
Basic symbols, 437, **438**
Beads, **30**
Beams, 311, **312**
  deflection, 320
  design, 327
  stress in, 317, **318**
Bending moment, **314**, 315, **316**
Beryllium
  properties, **227**, **228**
  weldability, 229
Bevel joint
  double, **27**
  single, **27**
Blowholes, 189
Body-centered lattice, **157**, 158
Braze welding, 244, **245**
Brazing, 236
  filler metal, 239
  fluxes, **238**, 239
  heating methods, 240
    carbon electrode, 246, **247**, **248**
    furnace, **241**
    induction, **242**
    torch, **240**
  joints, types, 236
  properties, 236
Brinell test, 392
Buried arc CO_2, 95, **96**
Butt joint, 26, **28**

closed, 26, **28**
double U, **337**
double vee, **337**
open, 26, **28**
single U, 337, **337**
single vee 336, **337**
square, 336, **337**

C

Cadmium-bismuth, **161-162**
Carbon, 170, 197
  steels, 200
    high, 201, 202
    low, 200, 201
    medium, 201, 202
Carbon dioxide, 85
Carbon tool steel, 168
Carburizing, 171
Case hardening, 171
Cast iron, 230
  properties
    gray, 230
    malleable, 230
    nodular, 230
    white, 230
  weldability, 231, **232**
Cementite, 163
Chemical properties (metals), 177
Charpy test, **383**
Chromium, 203
  -steel, **208**
Clamp, ground, **24,** 25
Classification of welds, 350
Closed joints, **28**
Clothing, protective, 426, **427**
Cobalt, 204
Cold working, 172
Compounds, 153, 163
Comprehensive strength, 180
Constant potential, 76, **78**
Containers, safety in
  cleaning, 432
    hot chemical, 432
    steam, 433
    water, 433
    cutting, 433, **434**
Contours, weld, **53**
Control of heat input, 191, **191**
Copper and alloys, 221

Corner joint, **28**
  flush, 340, **340**
  full open, 340, **340**
  half open, 340, **340**
Corrosion, 177, 263
Corrosion testing, 394
Cost of materials, 403
Costs, estimating, 403
Cracking, 196
Crater formation, 32
Creep, 185, **188**
Cryogenic properties, 185
Crystal, **156**
Current
  alternating, **51, 52**
    for Mig, 71, **90**
    for seams, 120
  automated, 455
    direct, 47
      reversed polarity, 47
      straight polarity, 50
    initial, 455
    final, 455
    setting, 27
Cutting
  arc, 283
  container, **433**
  flame, 283
  fuel gases, 286
  machine, **290**
  manual, 287
  oxygen, 283, 284
  process, 284
  procedure, 288, **289**
Cyanide hardening, 172
Cylinders
  oxyacetylene, 3, **4, 5**
  safe handling, 428, **429**

D

Defects, 188
Deflection of beams, 320
Deformation, 308
Design (in costs), 396
Design
  beams, 327
  joints, 336
  weldments, 334
Design factor, 335

Design optimization, 335
Designating location, 437, **439**
Destructive testing, 369
  bend, 375, **377, 380**
  break, 375, **376**
  etching, 382
  fatigue, 384, **385**
  fillet, 381, **381**
  impact, 382, **383**
  shear, 372, **373, 374**
  tensile, 369, **370, 371**
Direct current
  reverse polarity, 47
    Mig welding, 72, **73**
  straight polarity, 50
    Mig welding, 72, **73**
Double-
  bevel tee joint, **338**
  fillet lap joint, **339**
  J tee joint, **338,** 339
  U butt, **337**
  vee butt, **337**
Ductility, 185
Duty cycle, 414
Dye penetrant inspection, **388**

E

Eddy current testing, **389,** 390
Edge joint, **340,** 341
Edge preparation, 190
Elasticity, 183
  modulus of, 180, 309
Elastic limit, 183
Electrical resistance, 196
Electro-gas welding, 149, **150, 151**
Electrode consumption, **406, 411**
  cost, 411
Electrode holders, **23**
Electrode position
  metal arc, **33, 34**
  projection, **122**
  gas tungsten-arc, 57, **58**
Electrodes
  fast-freeze, **38**
  fill-freeze, **39**
  selecting, 36, **37,** 38, **39**
  weaving, 34, **35**
  types
    characteristics, **39**

identifying, 37, **37**
projection weld, 122
seam welding, 118, **119**
spot welding, **115,** 116
Electron beam, 132
  advantages, 133
  compared to fusion, **133**
  limitations, 133
  major units, 134, **134**
Electron gun, 134, **134-136**
Elements, 153
  in steel, 197
Endurance limit, *see* Fatigue
Equilibrium diagram, 165
  copper-aluminum, **174**
Equipment, *see* Process (arc, gas,
    etc.)
Estimating costs, 403
Etching test, 382
Eutectic compounds, **167,** 168
  iron-carbon, 168
Evaluating safety factors, 355
Expansion, thermal, 196, **310**
Explosive limits, gases, **12,** 13
Eye protection, 427

F

Fabco, 103
Face-centered lattice, **157,** 158
Factor of safety, 308, **308**
  evaluating, 355
Fatigue
  limit, 184
  strength, 183, **183, 184**
  test, 384, **385**
Feather edge joint, **28**
Ferrite, **167**
Ferrite structure, 167
Fillet lap joint
  double, **28**
  single, **28**
Fillet weld
  contour, **348**
  size, **440,** 441
  test, **381**
Final current, 455
Flame cutting, 283
  fuel gases, **287**
  machine, **290, 455**

oxygen, 283, 284
pressure, fuel, **288**
procedure, 288, **289**
Flashback, 432
Flash welding, **124**
alignment of parts, 120
heat balance, **126**
Flush corner joint, **340**
Fluxcor, 103
Fluxes
brazing, **238,** 239
solder, 253
Forces, 306
Free-bend test, 375, **377, 378**
Fuel gases, temperature, **287**
Full open corner joint, **340**

G

Galvanized metal, 233
Gas cost, 411
Gas flow, **55,** 88
Gas metal-arc (Mig)
application, 71
automated, 455
current, 72
equipment, 76-81, **77**
gas regulation, 88
hardfacing, 272
metal transfer, 73-76
power supply, 76
shielding gas, 82-87
stainless steel, 211
Gas regulators, 56, **56**
Gas tungsten-arc (Tig)
application, **45,** 47, **46**
automated, 455
automatic, **62,** 63
current, 49, 50, 51
equipment, 47
hardfacing, 272
power supply, 47
shielding gas, 57
speed, **71**
technique, 57, **58**
unit, **62**
Gas welding, 1
Gases
cutting, 286, **287**
explosive limits, **12**

shielding
Mig, 82, **83**
Tig, 57
types
Argon, 84
Argon-CO_2, 88
Argon-helium, 88
Argon-oxygen, 85
Carbon dioxide, 85
Helium, 86
Generator, 18, **19**
Globular metal transfer, **74**
Gouging, 295, **295**
Grain (metal), **156**
growth, 189
size, 176
Ground clamp, **24,** 25
Guided-bend test, 379, **379, 380**
Gun
electron, 134, **134-136**
Mig, **80, 81**
Tig, **64, 65**

H

Half open corner joint, **340**
Hardenability, 196
Hardening
annealing, 172
carburizing, 171
case-, 171
cold working, 172
cyaniding, 172
nitriding, 172
steel, 169
Hardfacing, 261, **262**
materials, 265
powders, **275**
procedures, 271, 272
Hardness, 184
testing, 392, **393**
Heat input, control, **191**
Heat treatment, **175,** 191
solution, 173
Heliarc, 46
Helium, 86
Argon-CO_2, 88
Heliwelding, 46
Hexagonal lattice, **157,** 158
High frequency welding, **129**

Holder, electrode, 23, **23, 24**
Horsepower relationships, 329
Hose, oxyacetylene, 7
Hot start level, 455
Hydrochloric acid, 382

I

Impact, 263
  strength, 184
  testing, **186,** 382, **383**
Inclusions, 189
Inertia welding, 147, **148**
Initial current, 455
Intermetallic compounds, 163
Iodine test, 382
Iron-based alloy, 165
Iron-carbide alloy, **166**
Iron, cast, *see* Cast iron
Izod test, **383**

J

Jigs and fixtures, **192, 398, 399**
Joint
  backing, 92
  design, 336
  preparation, 92
  Tig, **61**
  types
    bevel, **338,** 339
    butt, 336, **337**
    fillet, 339, **339**
    lap, 339, **339**
    tee, **338,** 339

L

Lancing, **294**
Lap joint,
  double fillet, **339**
  single fillet, **339**
Laser welding, 137, **138**
Lattice, space
  body-centered cubic, **157,** 158
  face-centered cubic, **157,** 158
  hexagonal close-packed, **157,** 158
Lenses, protective, 26
Limit
  elastic, 183
  fatigue, 184
  proportional, 183

Liquifluxer, **249**
Location, 341, **344, 345, 346**
Location, designating, 438, **439**

M

Magnesium, 219
  alloys, **220**
  properties, 219
  weldability, 220
Magnetic particle, inspection, 386, **387**
Manganese, 197, 203
  -steel, 204
Manual cutting, 287
Mapp gas, 11, **12**
Martensite, **168**
Mechanical mixtures, 161
Melting point, 196
Metal transfer (Mig), 73, **74, 76**
  globular, **73**
  spray, **73**
  short circuiting, 75
Metal-arc, *see* Gas metal-arc
Metallizing, 276, **277**
  guns, **279, 280**
  operation, 279
  process, 277
Metallurgy, 153
Metals
  properties, **155,** 177
    chemical, 177
    corrosion, 177
    mechanical, 178
    structure, 156, **156**
Metric conversion, 466
  table, 467
Mg welding, *see* Gas metal-arc
Mixtures, 154, 161
Modulus of elasticity, 180, **309**
Molybdenum, 203
  -steel, 205
Moment, bending, **314,** 315, **316**
Multiple pass welds, 34, **35**

N

Nick-break test, 375, **376**
Nickel, 203
  -steel, 204
Nickel alloys
  properties, 223

weldability, 224
Nitric acid, 382
Nitriding, 172
Nondestructive testing, 386
  dye penetrant, **388**
  eddy current, **389**, 390
  hardness, 392, **393**
  magnetic particle, 386, **387**
  radiographic, 390, **391**
  ultrasonic, 390, **392**
Normalizing, 173
Number of passes, **192**, 193
Numerical Control, 461, **462-464**

O

Open joint, **28**
Operating cost, 414
OSHA, 424, 427
Overhead, 414
Oxidation, 178, 197
Oxyacetylene welding
  cylinders, 3, **4, 5**
  equipment, 2, 3
  hardfacing, 266
  hose, 7
  regulators, **6**
  safety, 428
  stainless steel, 210
  technique, 8, **9, 10, 11**
  torch, **7**
Oxygen cutting, 283, 284
  gouging, **295**
  lancing, **294**
  piercing, 291, **292**
  powder, **295**
  scarfing, **293**, 294
  stack, 292, **293**
  washing, 294
Oxy-hydrogen welding, 13
Oxy-Mapp welding, 11, **12**

P

Parallel
  loading, 349
  welds, **351**
Parts out of position, **193**
Passes, number, **192**, 193
Pearlite, **167**
Peening, 191
Percussion welding, 128

Phosphorus, 197
Piercing holes
  arc, **298**
  flame, 291, **292**
Placement of arrows, 437, **439**
Plasma arc cutting, 300, **301**
  conditions, **304**
  cutting speeds, **301**
  equipment, 302, **302**
  gases, 303
  principle, 301, **302**
  procedure, 304
Plasma welding, 143, **143, 145, 147**
  hardfacing, 273
Porosity, 190
Position, welding, **27**
  flat, **59**
  torch, **58**
  vertical, **60**
Positioning work, 92, **93, 94**
Positive weldments, 397
Postflow time, 455
Postheating, 199, 200
Powder cutting, **295**
Power cost, 415
Power supply, 16, **17**
  for Mig, 76
  for Tig, 47, **48, 49**
  source, pulsed spray, **98, 99,
    100, 101**
Preheating, 191, 199, 200
Pressure time, 455
Pressure vessels, **330**
Primary welds, 350, **351**
Production economy, 396
Programmed welding, 461, **462-464**
Projection welding, 121, **122**
  electrodes, 122
Proper edge preparation, 190
Proper joint design, 341
Proportional limit, 183
Proportions for joints, **342, 343**
Protection,
  ear, 428
  eye, 427
Protective
  clothing, 426, **427**
  shield, **25**
Pulsation delay, 455

Pulsed-spray arc welding, **97**
  features, 102
  power supply, 98, **99, 100, 101**
Pulse level, 455
Pulse time, 455

R

Radiographic inspection, 390, **391**
Recrystallization, 176
Rectifier, 21, **22**
Regulators
  gas tungsten, **56**
  oxyacetylene, **6**
Replacement of designs, 360
Residual stress, 190
Resistance welding, 109
  stainless steel, 211
Robot, traversing, 457
Rockwell test, 392, **393**
Rod,
  filler (Tig), 61

S

Safety, 424
  arc welding, 435
  containers, 432
  cutting, 433
  cylinders, 428
  lighting torch, 431, **431**
  OSHA, 424, 427
  oxyacetylene, 428
  piping, 430
  testing leaks, 430
Safety factor, **308**
  evaluating, 355
Scarfing, **293**, 294
Seam welding, 118, **119**
  electrodes, 118, **119**, 120
  speed, 120
  types of seams, **121**
Secondary welds, **351**
Segregation, 190
Selecting electrodes, 36, 37, 38, **39**
Setting, current, 27
Shades of lenses, 26
Shear, vertical, **314-316**
Shearing strength, 372, **373, 374**
Shield, protective, **25**

Shielded metal-arc welding
  application, 16
  D.C. generator, 18, 19
  electrode holder, 23, **24**
  equipment, **16**
  hardfacing, 266
  power supply, 16, **19**
  rectifier, 21, 22
  stanless steel, 210
  technique, 26
  transformer, 20, **21**
Shielding gases
  Mig, 82, **83**
  Tig, 57
Shoulder edge joint, **27**
Sigma welding, 70
Silicon, 197
Single
  bevel T joint, **338,** 339
  fillet lap joint, **339**
  J tee joint, **338,** 339
  u butt, **337**
  vee butt, 336, **337**
Single pass welds, 34
Sizes of welds, 348, **350,** 441, **442**
  arc-spot, 444, **445**
  arc-seam, 444, **445**
  fillet, **440,** 441, 442, **442, 443**
  flange, 446, 448, **448**
  flash, 446, **449**
  groove, **443-445,** 445-447, **446, 447**
  intermittent fillet, 441, **443**
  plug, 442, **443**
  seam, 444, **445**
  slot, 442, **444**
Slope-control, 79
Solder fluxes, 253
Soldering, 250
  copper, 253, **254**
  flame devices, **255**
  gun, electric, 253, **254**
  heating, 253, **254**
  iron, electric, 253, **254**
  pencil, 253, **254**
  procedure, 256
Solders, 250
  composition, **252**
Solid solution alloys, **159, 160**
  characteristics, 159

interstitial, 159
substitutional, 159
Solution heat treating, 173
Space lattice, **157**, 158
Speed, travel, 31
Spheroidizing, **173**
Spot welding
  resistance, **110**
    electrodes, **115**
    heat application, 112
    heat balance, 114
    machines, **111-114**
    multiple, **117**
  Tig, **63**
    amperage, 66
    backing, 68
    equipment, **64, 65**
    gas, 67
    surface condition, **68**
    time, 67
    ultrasonic, **141**
Square
  butt joint, 336, **337**
  tee joint, 337, **338**
Stack cutting, 292, **293**
Stainless steel
  properties, 207, **209**
  weldability, 209
Starting the arc, 27, **29**
Steel
  carbon tool, 168
  case hardening, 171
  hardening, 167, 171
  structural, 169
  tempering, 171
Steel, classifying, 198, 199
Stick electrode cost, 411
Stiffeners, **347**, 362, **363, 364**
Strain, **179, 308**
Strap joint
  double, **28**
  single, **28**
Strength
  analyzation, 350
  compressive, 180
  impact, 184
  joints, 351, **352, 353, 356, 357**
  materials, 306
  tensile, 180

yield, 180
Strength, testing for
  compressive, 180
  impact, **186**
  tensile, 180
Stress, 178, 179
  allowable, 307, **307**
  beams, 317, **318**
  distribution, **346**
  residual, 190
  simple, 306
Structural steel, 169
Structure
  atomic, **157**
  change, 197
Stud welding, 40, **41,** 42
  process, 41
  time and current, 42, **43**
  sizes, 42
Subassemblies, 364
Submerged arc, 105, **106, 107,** 455
  hardfacing, 273
Sulphur, 197
Surface condition, 196
Surface defects, 386
Surfacing, 261
  alloys, **264**
  processes, **276**
Symbols, 437

T

Taper current, 455
Taper delay, 455
Tee joint
  double bevel, **338,** 339
  double fillet, **27**
  double J, **338,** 339
  single bevel, **338,** 339
  single fillet, **27**
  single J, **338,** 339
  square, 337, **338**
Temperatures of gases, 12
Tempering, 171
Tensile strength, 180
Tensile testing, 369, **370, 371**
Testing welds, 369, also *see,*
    Destructive testing and
    Nondestructive testing
Thermal conductivity, 196

expansion, 196, **310**
Thin walled vessels, **330**
Tig, *see* Gas tungsten-arc
Titanium, 224
    properties, 224, **225**
    weldability, 225
Torch
    cutting, **287**
    oxyacetylene, **7**
    position, **58**
    Tig, 52, **54**
Torque, 327
Torsion, **327**
Total cost per foot, 415
Toughness, 185
Transformer, 20, **21**
Transverse
    loading, 349
    weld, **350**
Travel speed, 31, 414
Traversing robot, 457
Tubular wire, 103
Tungsten, 203
    arc welding, 46

U

Ultrasonic testing, 390, **392**
    welding, 140, **141**
Upset welding, **127**
Upslope, 455

V

Vee joint
    double, **28**
    single, **28**
Vacuum
    chamber, 136

pumping system, 137
Vanadium, 203
Vapor shielded arc, **104**
    description, 105
    equipment, **104**
Vaporization, 197
Ventilation, 425, **425**
    smoke exhaust system, 426, **426**
Vertical shear, **314-316**
Vertical welding, **60**
Vessels, thin walled, **330**

W

Washing, 294
Weaving electrodes, 34, **35**
Weight of filler metal, **404,** 405
Weld contours, **53**
Weld uniformity test, 375
Weldability, 196
Welding,
    multiple pass, 34
    single pass, 34
Welding defects, 188
Wire cost, **413**
Wire feeding, **80**
Wire size, 89
    feed, 91
    stickout, 91, **92**
Work force, 403

Y

Yield point, 180, **182**
Yield strength, 180

Z

Zirconium, 226
    properties, 226
    weldablity, 226

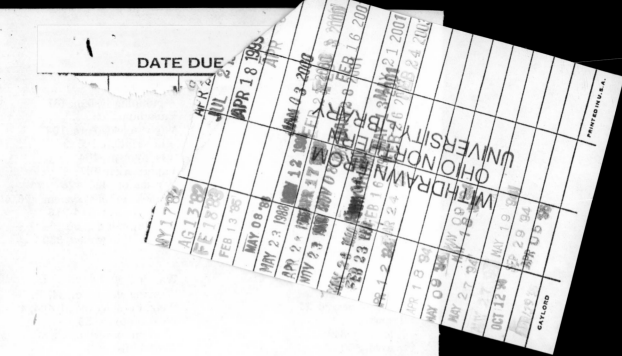